The Protection of Biodiversity and Traditional Knowledge in International Law of Intellectual Property

The relationships between international intellectual property treaties, the United Nations international environmental treaties (first and foremost the convention on Biological Diversity), the relevant customary norms and soft law form a complex network of obligations that sometimes conflict with each other. The first set of treaties creates private rights while the latter affirms the sovereignty rights of States over genetic resources and related knowledge and creates international regimes of exploitation of the same.

Jonathan Curci proposes solutions to the conflicts between treaties through the concept of "mutual supportiveness," including the construction of a national-access and benefit-sharing regime, mandatory contractual provisions in relevant international contracts, a defensive protection when genetic-resource-related traditional knowledge is unjustly patented through the analysis of the concepts of "*ordre public* and morality," "certificate of origin" in the patent application and "novelty-destroying prior art" and positive protection through existing and *sui generis* intellectual property rights and misappropriation regimes.

JONATHAN CURCI is the Legal Counsel of Quantam Business Group, Ltd, Israel and Academic Counsel of Touro International University, Rome.

Cambridge Intellectual Property and Information Law

As its economic potential has rapidly expanded, intellectual property has become a subject of front-rank legal importance. *Cambridge Intellectual Property and Information Law* is a series of monograph studies of major current issues in intellectual property. Each volume contains a mix of international, European, comparative and national law, making this a highly significant series for practitioners, judges and academic researchers in many countries.

A list of books in the series can be found at the end of this volume.

The Protection of Biodiversity and Traditional Knowledge in International Law of Intellectual Property

Jonathan Curci

CAMBRIDGE
UNIVERSITY PRESS

CAMBRIDGE
UNIVERSITY PRESS

University Printing House, Cambridge CB2 8BS, United Kingdom

Cambridge University Press is part of the University of Cambridge.

It furthers the University's mission by disseminating knowledge in the pursuit of education, learning and research at the highest international levels of excellence.

www.cambridge.org
Information on this title: www.cambridge.org/9780521199445

First published 2010

A catalogue record for this publication is available from the British Library

ISBN 978-0-521-19944-5 Hardback

Contents

List of abbreviations

ABS	access and benefit sharing
ACP	African Caribbean Pacific
AIPLA	American Intellectual Property Law Association
AIPPI	Association internationale pour la protection de la propriété intellectuelle
ARIPO	African Regional Industrial Property Organization
ATRIP	International Association for the Advancement of Teaching and Research in Intellectual Property
CAF	Corporación Andina de Fomento
CBD	Convention on Biological Diversity
CBE	Convention sur le brevet européen
CFR	Code of Federal Regulations
CGIAR	Consultative Group on International Agricultural Research
CGRFA	Commission on Genetic Resources for Food and Agriculture
CHM	clearing house mechanism
CIEL	Center for International Environmental Law
CIT	Court of International Trade
CLR	compulsory liability regime
CNRS	Centre nationale de la recherche scientifique
COP	Conference of the Parties
CPGRFA	Commission on Plant Genetic Resources for Food and Agriculture
CRADA	Cooperative Research and Development Agreement
CSD	Commission on Sustainable Development
CSIR	Council for Scientific and Industrial Research
DC/DCs	developing country/developing countries
DDAGTF	Doha Development Agenda Global Trust Fund
DNA	deoxyribonucleic acid
DSB	dispute settlement body
DSU	dispute settlement understanding

EC	European Community
ECHR	European Convention of Human Rights.
ECJ	European Court of Justice
ECOSOC	United Nations Economic and Social Council
EPC	European Patent Convention
EPO	European Patent Office
EU	European Union
FAO	Food and Agriculture Organization
FTA	free trade agreement
FTC	Federal Trade Commission
GAOR	General Assembly Official Records
GATT	General Agreement on Tariffs and Trade
GIs	geographical indications
GMO	genetically modified organism
GNU	general public license
GPA	Global Programme of Action for the Protection of the Marine Environment from Land-based Activities
GR/GRs	genetic resource(s)
GRAIN	Genetic Resources Action International
GRULAC	Group of Countries of Latin America and the Caribbean
HIV	human immunodeficiency virus
IAO	Instituto Agronomico per l'Oltremare
ICANN	Internet Corporation for Assigned Names and Numbers
ICESCR	International Covenant on Economic, Social and Cultural Rights
ICJ	International Court of Justice
ICTSD	International Centre for Trade and Sustainable Development
IGC	intergovernmental committee
ILA	International Law Association
ILC	International Law Commission
ILM	International Legal Materials
ILO	International Labor Organization
INBio	Costa Rican National Biodiversity Institute
INDECOPI	Instituto Nacional de Defensa de la Competencia y de la Protección de la Propiedad Intelectual
IP	intellectual property
IPGRI	International Plant Genetic Resources Institute

IPGRTKF	intellectual property and genetic resources, traditional knowledge and folklore
IPR	intellectual property right
ITPGRFA	International Treaty on Plant Genetic Resources for Food and Agriculture
IUCN	The World Conservation Union
IUPGR	international undertaking on plant genetic resources
LDC	least developed country
MEA	multilateral environment agreement
MFN	most favored nation
MNC	multinational corporation
MTA	material transfer agreement
NAFTA	North American Free Trade Agreement
NGO	non-governmental organization
NIF	national innovation fund
OAU	Organization of African Unity
OEB	Office européen des brevets
OECD	Organization for Economic Cooperation and Development
OSD	open source definition
OSI	open source initiative
PBR	plant breeders' right
PCIJ	Permanent Court of International Justice
PCT	Patent Cooperation Treaty
PGM	plant genetic material
PGR/PGRs	plant genetic resource(s)
PGRFA	plant genetic resources for food and agriculture
PGS	plant genetic systems
PIC	prior informed consent
PLO	Palestine Liberation Organization
PLT	Patent Law Treaty
PVP	plant variety protection
R&D	research and development
RAFI	Rural Advancement Foundation International (now Action Group on Erosion, Technology and Concentration, ETC Group)
RNA	ribonucleic acid
RSA	Recueil des sentences arbitrales
SAA	Round Statement for Administrative Action
SAARC	South Asian Association for Regional Cooperation
SADC	Southern Africa Development Community

SCP	Standing Committee Law of Patents
SINGER	System-wide Information Network for Genetic Resources
SPLT	Substantive Patent Law Treaty
SPS	Sanitary and Phytosanitary Measures
SRISTI	Society for Research and Initiatives for Sustainable Technologies and Institution
TAK	traditional agricultural knowledge
TEK	traditional ecological knowledge
TIPR/TIPRs	traditional intellectual property right(s)
TK	traditional knowledge
TKUP	traditional knowledge of the uses of plants
TMK	traditional medicinal knowledge
TRIPS	Agreement on Trade-Related Aspects of Intellectual Property
UDHR	Universal Declaration on Human Rights
UNCTAD	United Nations Conference on Trade and Development
UNEP	United Nations Environment Program
UNESCO	United Nations Educational, Scientific and Cultural Organization
UNFOF	United Nations Forum on Forests
UNHCR	United Nations High Commissioner for Human Rights
UNTS	United Nations Treaty Series
UPOV	Union internationale pour la protection des obtentions végétales (International Union for the Protection of New Varieties of Plants)
URAA	Uruguay Round Agreements Acts
USPTO	United States Patent and Trademark Office
VCLT	Vienna Convention on the Law of Treaties
WHO	World Health Organization
WIPO	World Intellectual Property Organization
WIPO IGC on IPGRTKF	Intergovernmental Committee on Intellectual Property and Genetic Resources, Traditional Knowledge and Folklore
WSSD	World Summit on Sustainable Development
WTI	World Trade Institute
WTO	World Trade Organization

Acknowledgments

The present work is based on a doctoral dissertation deposited at the Library of the Graduate Institute of International and Development Studies University in Geneva (Switzerland) that is much richer in terms of footnotes and legal reasoning because broader in scope than the present book. I thank the University of Torino (Italy) and the Brigham Young University in Provo, Utah (United States of America) for providing the necessary education, their libraries and work place to draft this work. I owe a special thanks to the Research Fund of the European Patent Office (EPO) for the grant that has allowed me to carry on this work under more comfortable conditions.

I wish to express my gratitude to the supervisor of the underlying doctoral work Marcelo Kohen (Professor of International Law at the Graduate Institute of International Studies in Geneva, Switzerland), Thomas Cottier (Director of the World Trade Institute – WTI – at the University in Berne, Switzerland) and Marco Ricolfi (Professor of Intellectual Property at the Law School of the University of Turin, Italy). Their extensive expertise in the technical aspects of the new and cutting-edge topic of international trade law and the opportunities each of them were able to offer me as a researcher or consultant on projects under their direction provided me with invaluable experience, as well as much-needed motivation.

I also wholeheartedly thank Dr. Anthony Taubman, Acting Director and Head of the WIPO Global IP Issues Division, and Dr. Shakeel Bhatti, former Head of the WIPO Genetic Resource Traditional Knowledge and Biotechnology Section. A special thanks to my friends Daniel Stewart (Attorney-at-Law), David Newell and Peter Bradford (Patent Attorney) for having thought, discussed and researched with me related topics in our professional and research projects.

A very special thanks for the linguistic insights and corrections by Alison Craig and William Marconi.

Acknowledgments

Part I

The main problems

1 Introduction to legal issues related to genetic resources and traditional knowledge in the international intellectual property system

In the new millennium, biotechnology is enabling genetic engineering to yield very important breakthroughs, with immense possibilities for novel organisms to be developed. The myriad biotechnological applications released into the environment for pharmaceutical, agricultural, and medicinal purposes generate transnational concerns that pose an enormous challenge to national and international communities. The means of protection sought for these types of inventions is the patent. Although opinions about how much patent systems contribute to long-term economic growth vary, there can be no dispute that patents are vital to the business models of many companies and are playing an increasing role in society. As human technological prowess has expanded throughout the natural and human worlds, the patent has followed, not far behind. Questions about the proper place of patents in society, some old and some new, have found increasing urgency and importance, especially as patent law extends to societies not accustomed to its peculiarities.

Peoples in developing countries (DCs) denounce the patentability of genes, which reduces the world's genetic resources (GRs) down to mere property rights, resulting in corporate control over access to food, medicinal technology, and other resources essential to mankind's health and welfare. Additionally, potential transnational harm caused by genetic engineering may also arise through the destabilization of regional ecologies via genetic pollution and through an accelerated decline of biological diversity on a global scale. Thus, legal control over biodiversity is an issue of serious international consequence.

The present book focuses particularly on the international legal regime of commercial exploitation and ownership of GRs, on which biotechnological innovation is based. At the core of this study lies the problems of sharing benefits arising from the exercise of intellectual property rights (IPRs) over plant genetic resources (PGRs) and traditional knowledge (TK) under existing treaties and conventions with special attention to the contractual relations between companies from industrialized countries and indigenous communities and genetic resource providing countries.

Although this analysis is conducted through an international law approach, it does not neglect some anthropological and sociological aspects of private ownership of living forms and its interaction with different value systems.

1.1 Defining the problems

This chapter starts with general considerations on the problem of ownership and patents on PGRs; in a second stage, it observes the interaction among the international public domain, the States' sovereign rights over PGRs and private IPRs over the same, and ultimately introduces the new problem of traditional knowledge (TK).

It lays out the methodological aspects of the analysis and presents a brief overview of the theories of creation of the sources of international law that are relevant to this subject-matter and that will be used through the development of analysis. Accordingly, the impact of international law, with particular attention to World Trade Organization (WTO) law, shall be taken into account in a comparative approach. Because the European Union (EU) and United States (US) jurisdictions have developed various laws, policies and judicial decisions on the relationship between protection of biodiversity and intellectual property they offer broad examples of implementation of international law that are worthy to be described and discussed when appropriate.

1.1.1 Patents and ownership of genetic resources

The patentability of biotechnology took off after the US Supreme Court's landmark decision in *Diamond v. Chakrabarty.*[1] By acknowledging that statutorily patentable subject-matter included "anything under the sun that is made by man," the Court encompassed both foreseeable and unforeseeable subject-matter. This *Diamond* standard encompassed the inventive work of biotechnology and gene sequences. Consequently, an "imitation effect" rippled from the US to Europe and other jurisdictions, generating a series of legislative measures to patent living forms. In addition, the WTO Agreement on Trade-Related Aspects of Intellectual Property Rights[2]

[1] *Diamond v. Chakrabarty*, 447 United States, 303–09 (1980), reported also in F. Abbott, T. Cottier and F. Gurry (eds.), *The Intellectual Property System: Commentary and Materials* (Kluwer, The Hague, 1999) 25.

[2] Agreement on Trade-Related Aspects of Intellectual Property Rights (April 15, 1994) Marrakesh Agreement Establishing the World Trade Organization, Annex 1C, Legal Instruments – Results of the Uruguay Round 31–33 *International Legal Materials* (*ILM*) 1197 (1994).

(TRIPS) internationalized the patent protection of biotechnological practices.

In industrialized societies, investment and innovative output in the biotechnology industry has been so conspicuous that the benefits of innovation in this field have generally been viewed as outweighing the costs of the monopolistic restrictions created by patents. Now, not only plant varieties but also micro-organisms and genetically modified animals are patentable. Genetically altered animals, such as the infamous *Onco-Mouse* of Harvard University (bred for cancer research), have also been given patents. Thousands of patent claims have been made and granted on human genetic material, including material that has arguably been altered from its natural state.

The patent is the primary IPR that is sought in the field of biotechnology because it is meant to be a right concerning innovations used in new or improved products or processes. Patents enable the holder to exclude imitators from marketing such inventions or processes for a specified time; in exchange, the holder is required to disclose the formula or idea behind the product or process. After a patent is granted, the owner has a monopoly over commercial exploitation of the invention for a limited period. The stated purpose of a patent is to stimulate innovation by offering higher monetary returns than the market otherwise might provide.[3]

There are two problems that patent protection generates. The first concerns the monopolistic feature of the cost analysis of patent protection in this field. The classical IP scholarship has crafted each protection according to the principle of "allocative efficiency" according to which the long-term benefits flowing to society from the protection granted to a particular class of creators or innovators outweigh the (mainly short-term) costs imposed by the monopolistic structure of the patent grant.[4] And the "mainstream legal literature" has applied this standard principle from IP economics to the patenting of biotechnology as well.[5]

The second problem is generated when formal, industrial, patentable knowledge builds upon prior art of informal TK which is in a quasi-commons

[3] Abbott *et al.*, *The Intellectual Property System*, 25.

[4] P. Torremans and J. Holyoak, *Intellectual Property Law* (Oxford University Press, 2006) 16, 20. N. Carvalho, "From the Shaman's Hut to the Patent Office: How Long and Winding is the Road?" (1999) 40 *Revista da ABPI* 3–28. R. H. Coase, *The Firm, the Market and the Law* (University of Chicago Press, revised edition, 1990), see chapters 1 and 2 "The Firm, The Market, and The Law" and "The Nature of the Firm."

[5] Which includes, in the European literature in the bibliography quoted in M. Ricolfi, "Biotechnology, Patents and Epistemic Approaches" (2002) *Journal of Biolaw & Business, Special Supplement* 77–90.

regime. When it comes to the benefit sharing of the profits arising from the exploitation of this knowledge at the international level these problems are amplified.

A vivid example of benefit sharing illustrates the controversy of private property rights in GRs based on TK held by indigenous groups. Imagine a plant that produces a natural sweetener and has been preserved for several millennia in a local farming micro-culture. This sweetener performs its sweetening function without negative dietary or health side effects. A foreign corporation comes along bioprospecting and secures samples of the local sweetening plant, maps its genome, and then proceeds to genetically engineer a plant that yields sweetener with a potency tenfold that of the original. The corporation then patents the modified plant, and the world quickly forgets the original plant as the patented plant is markedly more productive. Consequently, through commercialization, all of the profits flow to the company patent holder without a farthing going to the indigenous farmers who preserved the plant for millennia. Some 6.5 percent of all genetic research undertaken in agriculture focuses on germ plasma derived from wild species and land races (farmer-developed varieties of crop plants that are adapted to local environmental conditions). Thus, the question is posed: is it fair to give the entire pastry to the one who adds the final cherry to the pie?[6]

This tendency has been popularly called *biopiracy* or *biocolonialism*. The origin of the two terms reveals that the context in which they were formed is the one of political science or sociology. These are not legal terms, let alone technical intellectual property terms. The term *biopiracy* was coined by Mooney as part of a counter-attack strategy on behalf of DCs that, as already said, are accused by industrialized countries of supporting *intellectual piracy*, i.e., counterfeiting all types of goods protected in the industrialized countries by IPRs. In turn, DCs feel that they are no more pirates than corporations that acquire resources and TK from their countries, use them in their Research and Development programs, and acquire patents and other IPRs without compensating the provider countries and communities.[7] This anti-*biopiracy* rhetoric adopted by some DC trade

[6] Ricolfi, "Biotechnology, Patents and Epistemic Approaches", 77; T. Cottier, "The Protection of Genetic Resources and Traditional Knowledge: Towards More Specific Rights and Obligations in World Trade Law", in Abbott *et al.*, *The Intellectual Property System* 1820–27; M. Blakeney, Presentation at the World Intellectual Property Organization (WIPO) – Torino Law School Specialization Course in Intellectual Property, International Property Aspects of Traditional Agricultural Knowledge (TAK) 2 (Nov. 22, 2001), unpublished, on file with the author.

[7] R. Mooney, "Why I Call It Biopiracy", in H. Svarstad and Sh. S. Dhillion (eds.), *Responding to Bioprospecting: From Biodiversity in the South to Medicines in the North*

negotiators has not prevented the legalization of this so-called "conquest" through the TRIPS Agreement. This treaty extends to all the developing and least developed members of the WTO the obligation to grant IPRs (patents, trademarks, and trade secrets, etc.), and, to some extent, also to innovations based on GRs, without mandating any compensation to the local communities who have bred and preserved these resources. At the same time, some 90 percent of genetic information and related TK are found in DCs.[8]

Biocolonialism is another term related to *biopiracy* and it often refers to the pattern whereby the industrialized country corporation extracts raw genetic materials from the DC, patents the genetically modified products based on the raw materials without prior informed consent (PIC) and benefit sharing, and then sells the finished product to the provider country at unaffordably high prices. In addition to these perceptions of injustice and misappropriation, the wide scope of the exclusive patent rights granted in industrialized countries stirs animosity on the part of the consumers in DCs, especially when the patent itself is based on a GR or TK preserved by the consumers of the patented product in DCs.

Even part of the legal doctrine has been vociferously arguing that IP regimes may jeopardize the freedom of countries or communities to choose the way in which they want to deal with the use and protection of biodiversity and the related TK. This issue blatantly arises when the genes are not appropriated by the sovereign State that patents them but by a foreign entity that manipulates and sells the genetically modified product. As a consequence of the double expansion of patent law both from inanimate to animate subject-matter (biotechnological inventions) and from a small group of industrialized countries to most of developing and Least Developed Countries (LDCs), several peoples in DCs are reacting against this kind of "piracy" of indigenous and local community knowledge.

These are some of the reasons for which peoples in DCs allege that IPRs in the field of biotechnology could prevent the Convention on Biological Diversity (CBD) from realizing the full and practical meaning of Article 3[9] on national sovereignty over their natural resources and Article 8(j)[10] on

(Spartacus Press, AS., Oslo, 2000) 37; V. Shiva, *Biopiracy: the Plunder of Nature and Knowledge* (South End Press, 1998) 1–5; A. Story, "Biopiracy and the Dangers of Patent Over-protection", (1999) 149 *New Law Journal* 158.

8 Cottier, "The Protection of Genetic Resources", in Abbott *et al.*, *The Intellectual Property System*, 1827.

9 Article 3 of the CBD, Convention on Biological Diversity (June 5, 1992) UNEP/Bio.Div/N7-INC5/4, 31 *ILM* 818 (1992).

10 WIPO, *Intellectual Property Needs and Expectations of Traditional Knowledge Holders: WIPO Fact Finding Missions on Intellectual Property and Traditional Knowledge* (WIPO, Geneva,

the rights of local and indigenous communities. These provisions aim at fairly distributing the benefits resulting from the use of GRs situated in the territories of the Contracting Parties.

Industrialized countries respond by affirming their effort to develop technology enabling the modification, the innovation, and the marketability of raw genetic materials that otherwise would remain unexploited within developing country indigenous communities that do not have such capacities. The debate is acrimonious and solutions are not easily at hand.

1.1.2 International public domain, sovereign rights, and intellectual property rights over genetic resources

This section moves from the general concept of ownership of GRs to the conflict between a State's public law regime of exercise of sovereignty rights upon GRs and then to the private exercise of IPRs upon the same.

The international exercise of patent rights has an impact both on the ownership regime over the GR *per se* and on the knowledge of the uses for and the characteristics of plant and animal GRs. Biotechnology depends on biological diversity as the basis of innovation. The access to biological diversity in a given country has traditionally been free and open. This led to the basic inequity (already sketched in section 1.1 above) consisting of the freedom of appropriation of GR and of TK on the part of the inventor on one side, while on the other the users in the country in question had to purchase the secondary products subjected to proprietary protection. Profits flow into the hands of right-holders in industrialized countries for the exploitation of biodiversity and related knowledge in DCs.

For example, suppose a researcher were to incorporate into his studies TK that had been generated by a particular community over hundreds of years and not attributable to any particular person. As far as the researcher is concerned, the TK used in his research is, for all intents and purposes, public domain knowledge. Suppose further that the researcher subsequently reports this knowledge with or without acknowledging the intellectual contributions of the initial TK holding community. Should that information ever prove useful in the creation of a patentable good, i.e. the creation of a drug through use of TK on a particular medicinal plant, the community would be without recourse to claim ownership or rights in the TK at the heart of the innovation, merely because that TK was

2001) 50. T. Taubman, "Genetic Resources" in S. Von Lewinski, *Indigenous Heritage and Intellectual Property: Genetic Resources, Traditional Knowledge and Folklore* (Kluwer, The Hague, 2nd edn, 2008) 192–216.

within the public domain at the time it was recorded. Meanwhile, the company owning the patent exclusively reaps all the commercial benefits.

In the systems of protection of IP in industrialized countries, TK related to GRs has until recently been considered as international public domain because of the confusion of the public domain with the international legal concept of *res communis humanitatis* (common heritage of mankind).[11] The assimilation of TK into *res communis humanitatis* was necessary to justify the free accessibility of TK to all private users.[12] While the concept of *res communis humanitatis* covers the ocean floor,[13] Antarctica,[14] the moon,[15] and outer space,[16] it is doubtful, in my view, whether biodiversity in general should be placed under the concept of common heritage of mankind, *stricto sensu*.[17] There is no treaty or customary principle[18] that places TK and GRs under the concept of *res communis humanitatis*. On the contrary, starting from the colonial era, colonial states used to transfer GRs to their masters as contributions to their research centers.[19]

The international community finally discussed the position of GRs in international law during negotiation of the CBD adopted in 1992. At the start of the negotiations, the legal status of GRs *in situ* and *ex situ* was very

[11] C. Joyner, "Legal Implications of the Concept of the Common Heritage of Mankind", (1986) 35 *International and Comparative Law Quarterly* 190; R. Wolfrum, "The Principle of the Common Heritage of Mankind" (1983) 43 *Zeitschrift für ausländisches öffentliches Recht und Völkerrecht, Heidelberg Journal of International Law* 312.

[12] *Matter Concerning Intellectual Property and Genetic Resources, Traditional Knowledge and Folklore – an Overview*, WIPO/kritf/IC1/3, 8–9 (March 16, 2001).

[13] J. Van Dyke and C. Yuen, "Common Heritage v. Freedom of the Seas: Which Governs the Seabed?" (1982) 19 *San Diego Law Review* 493.

[14] F. Francioni and T. Scovazzi (eds.), *International Law for Antarctica* (Kluwer, The Hague 1996); F. Francioni, *International Environmental Law for Antarctica* (Giuffrè, Milano, 1992); C. Joyner, "Antarctica and the Law of the Sea: Rethinking the Current Legal Dilemmas" (1981) 18 *San Diego Law Review* 415.

[15] K. Baslar, *The Concept of the Common Heritage of Mankind in International Law* (Martinus Nijhoff, The Hague, 1998) 307–13. C. Christol, "The Common Heritage of Mankind Provision in the 1979 Agreement Governing the Activities of States on the Moon and Other Celestial Bodies", (1980) 14 *International Lawyer* 429.

[16] P. P. C. Hannapel, *The Law and Policy of Air Space and Outer Space: a Comparative Approach* (Kluwer, The Hague, 2003); L. Tennen, "Outer Space: A Preserve for All Humankind", (1979) 1 *Houston Journal of International Law* 145.

[17] I. Mgbeoji, "Rethinking the Role of International Law in Relation to the Appropriation of Traditional Knowledge of the Uses of Plants" 132, 139, 148, 150, 159, 161, 163–70, 179, 252, 253 (a dissertation submitted for the Degree of Doctor in the Science of Law, Dalhousie University Halifax, November 2001. Copy on file with author).

[18] A. D'Amato, "Trashing Customary International Law in Appraisals of the ICJ's Decision: Nicaragua v. United States", (1987) 81 *American Journal of International Law* 74–75; M. S. McDougal, H. D. Lasswell, and M. Reisman, "The World Constitutive Process of Authoritative Decision", (1967) 19 *Journal of Legal Education* 403.

[19] I. Mgbeoji, "Rethinking the Role", 163–70.

unclear: few national laws had been enacted for the commercial exploitation of the GRs *in situ* and no real international status had been created for the gene banks conserving germplasm[20] *ex situu* (see in more detail section 4.2.3 below). *Ex situ* collections of GRs could be acquired freely; no international obligations existed to share the economic benefits to the communities that provided and conserved the resources, and only very few international breeding programs were set up to develop and distribute crop varieties for use in the DCs.[21]

The status of GRs in international law started to be clarified with the adoption of a United Nations (UN) General Assembly Resolution 1830 (XVII) on 4 December 1962. At that time, the international community focused its efforts on the preservation of biological diversity and on its related knowledge under threat of extinction. Meanwhile, the slow process of globalization of IPRs was considered a successful tool in protecting and encouraging the further development of so-called "modern," "formal," or "technological" knowledge applied to GRs (see the relevant distinctions of TK in section 4.2 below). TK holders, especially in DCs, had felt that this knowledge, passed on from generation to generation, had progressively become an "economic resource." The increasing pace of exploitation of this knowledge through modern technological instruments led the international community to shift the focus of its attention from the "preservation" of GRs to their "utilization." Rapidly, various international fora became involved in the regulation of this matter: United Nations Environmental Program (UNEP), World Intellectual Property Organization (WIPO), United Nations Educational, Scientific and Cultural Organization (UNESCO), International Labour Organization (ILO), United Nations Conference on Trade and Development (UNCTAD), etc.[22]

One of the most important highlights in the chronological development of international public policy on this matter occurred in 1989, when the United Nations Food and Agriculture Organization (FAO) enacted the International Undertaking on Plant Genetics Resources (IUPGRs), which originally defined PGRs as the "heritage of mankind which should be available without restriction." In other words it considered the germplasm collected *ex situ* in gene banks as "common heritage of mankind."

[20] Germplasm is genetic material extracted from a plant.

[21] M. Hassemer, "Genetic Resources" in S. Von Lewinski (ed.), *Indigenous Heritage and Intellectual Property: Genetic Resources, Traditional Knowledge and Folklore* (Kluwer, The Hague, 2004) 159–60.

[22] See Table 1 "The Overview of the Regulatory Framework", in T. Taubman and M. Leistner, "Analysis of Different Areas of Indigenous Resources" in Von Lewinski, *Indigenous Heritage and Intellectual Property* (2nd edn, 2008) 200–1.

This concept was maintained in the IUPGRs of the FAO until 2001 (see section 3.3 below) when the international community adopted the FAO International Treaty on Plant Genetic Resources for Food and Agriculture (ITPGRFA) that facilitated access to a database of PGRs held in trust by a Multilateral System for specific purposes of utilization.[23] This treaty thus establishes PGRs in a combination of a regime of State sovereignty and a regime of multilateral cooperation, although nowhere in the ITPGRFA is this concept explicitly stated.

The legal status of the rest of the biodiversity was defined by the CBD adopted by the UNEP in 1992. In its preamble, it is stated that the preservation of biodiversity is a "common concern of humankind," whereas, in its Articles 3 and 15.1, it acknowledges the principle of permanent sovereignty of the States over their natural resources on their territories. This means that access to GRs has to be regulated by a private law contract, a so-called "material transfer agreement" (MTA) involving the provider State and bioprospecting entity (see chapter 5). The international community has moved from this bilateral-contractual solution envisaged by the CBD to a clarification of the concept of "common concern of humankind" as it relates to the conservation and sustainable exploitation of PGRs.

Finally, the WIPO General Assembly, in creating in 2000 the Intergovernmental Committee on Intellectual Property and Genetic Resources, Traditional Knowledge and Folklore (WIPO IGC on IPGRTKF) has started a new era of diplomatic discussions on the interaction among IP, GRs and TK. This IGC, supported by a Secretariat of technical experts in the field, is paving the way for the negotiation and adoption of a treaty which should clarify the relationship between private rights of intellectual property and TK.

1.1.3 Introduction to the tensions between the exercise of intellectual property rights and preservation of genetic resources

Six months after the CBD entered into force, WTO Members adopted the TRIPS Agreement in 1994[24] that marked the commencement of a new era of globalization of IPRs.[25] This treaty mandates minimum standards of private property protection of all types of "formal" or "modern knowledge," including knowledge developed from GRs. Since then IP scholars have intensely studied the ability of TRIPS-mandated IPRs to protect TK related to GRs, taking into account the parallel evolution of non-IP treaties (e.g. CBD and ITPGRFA). Indeed States' obligations under

[23] www.fao.org/Legal/treaties/033t-e.htm, last viewed November 2007. [24] TRIPS.
[25] As of January 2007 there are 150 Member States in WTO.

non-IP international law drag the patent system into an unprecedented public debate.

Traditionally, the international IP system had been developed on the basis of common economic values belonging to a restricted number of industrialized countries. Since IPRs are territorial – i.e. their enforcement can only be effective within the boundaries and according to the laws of a given nation or region – international conventions were aimed at harmonizing the IP aspects of the national legal traditions of contracting parties. One of the globalizing effects of TRIPS Agreement is to extend these rights and obligations to all the DCs' national legal systems that do not have the same level of industrial development to let them benefit from the international exercise of IPRs. DCs are mainly users instead of holders of IPRs.

It has been observed in section 1.1.1 above that the international patent system has been experiencing a double expansion both sectorial and geographic: (i) from the traditional protection of inanimate matter to the more sophisticated and complex protection of biological matter; (ii) from the small number of Northern industrialized countries to the Southern developing and least-developed countries. This double expansion has been generating a three-dimensional hardship: (i) a certain criticism of the classic patent system as conceived by the initial industrialized countries that created it; (ii) an unprecedented disequilibrium between the immediate interests of industrialized countries and DCs; (iii) an institutional fragmentation and overlap among various UN agencies and other international fora addressing IP, biotechnology, biodiversity conservation and utilization, local farmers' development and the like. These tensions between IP and environment exist *mutatis mutandis* between IP and other fields (for instance, health, cyberspace, etc.).

Major industrialized States have realized the potential gains flowing from this new technology for their national economies spurred on by private industries. Consequently, they are promoting stronger IP standards to be integrated in multilateral and bilateral treaties to which most DCs are parties. At the same time, industrialized countries have been accused of watering down the patentability requirements of biotechnology within their own national jurisdictions in order to accommodate corporate interests without precisely and carefully considering the issues involved and the consequences thereof.

In sum, the transnational behavior of subjects of international law in this field has been regulated by at least two major multilateral treaties which are both legally binding: the CBD and TRIPS Agreement. Since these issues are intrinsically complex and multifaceted, various international institutions (such as WIPO, FAO, UNESCO, ILO to name only a few) are becoming eagerly involved in producing guidelines or even new treaties on the subjects concerned.

With respect to biotechnological inventions, State Parties are bound, under Article 27 of the TRIPS Agreement, to accept: patenting of micro-organisms and "microbiological processes" and providing some "effective" form of IPRs on plant varieties, either patents or some *sui generis* (new) version. But, while the TRIPS Agreement requires that States grant exclusive private rights over biological material, the CBD, on the contrary, affirms the sovereign rights of States to biological material. This is the "epitaph" that utters the core of the dichotomy that will be developing throughout my study (see in particular chapter 3).

Broadly speaking, tensions between the sovereign rights of States and the expansion of IPRs to biological subject-matter has to be primarily seen in the framework of resource allocation at the international level – under the pressure of science and technological innovation within national economies. In this context, an analogy can be drawn between the expansion of sovereignty rights over GRs and the law of the sea.[26] The evolution of the law concerning the continental shelf, exclusive economic zones and the phenomenon of the "State's creeping jurisdiction" during the last 50 years has been due to the invention of the combustion engine, other uses of oil, gas, and mineral resources, and the advancement of fishing technology.[27] The codification of the CBD marks the same tendency of defining the expanding sovereign rights of States over their GRs, even including the information contained therein. The expansion of IPRs from inanimate to animate subject-matter has followed the same pattern. However, the first difference consists in the fact that the law of the sea sets forth the rights of the State whereas the IPRs on biotechnological inventions concern essentially private rights. A second difference, and a more complex one, is that the first kind of conventions deal with physical features of natural resources (land, airspace, fish, gas, oil), whereas IPRs deal with appropriation of ubiquitous information: in our case, genetically encoded, exclusively in nature and untouched by the genetic alterations of man in a laboratory.

The expansion of IPRs to living matter has to be interpreted also in light of a basic structural economic problem. If traditionally States have taken responsibility for the development of biodiversity, e.g. plant varieties,

[26] United Nations Convention on the Law of the Sea (December 10, 1982) UN Doc. A/Conf.62/122, UN Sales No. E.83.V.5 1983, 1833 *United Nations Treaty Series* (*UNTS*) 397. Convention on the Continental Shelf (April 29, 1958) 499 *UNTS* 7302, 312–21. Convention on the Fishery and the Conservation of Biological Resources in the High Seas (April 29, 1958) 1958 *UNTS* 1966, 280–96.

[27] Cottier, "The Protection of Genetic Resources", 1823.

governments no longer can take it for granted that this work will be financed by the tax-payer. During the last decade, many States have fostered privatization of many previously State-owned businesses which has transferred some of the decision-making power of governments to profit-driven private companies. The repercussions on the international law-making process within international organizations, such as WTO, have been evident: since the scope of IPR protection granted to specific subject-matters is one of the most important forms of revenue for companies, they strongly encouraged their States to include TRIPS as a WTO Agreement to assure the respect of their IPRs also in all DCs willing to become members of WTO. And IPRs on living matter are no exception to this trend.

While the expansion of IP protection is including living matters, the world is experiencing an unprecedented loss of biodiversity: one hundred species become extinct every day, many more than the creation of new species. This *"biocide"* (neologism for "biological extermination") is accomplished while few transnational corporations will eventually control the world's food supply because of the disparity of means of research. Indeed biotechnology today is substantially driven by private company research which will be critical in achieving future food security. In turn, the incentive for biotechnological inventions is stimulated by a financial investment that flows only if such an invention is protected by an appropriate kind of IPR. The loss of biodiversity associated with the alleged misappropriation of TK related to GRs creates worries in many DCs whose sustainable development largely relies upon these two essential elements.

1.1.4 A brief introduction to the concept of traditional knowledge

The terminologies "biodiversity-related TK," "TK related to GRs," or TK *tout court* signify the same IP protectable subject-matter. Anthropological science generally prefers to use the expression "traditional ecological knowledge."[28]

For the purpose of IP protection, TK is the information on GRs that people in a given community, based on experience and adapted to local culture and environment, have developed over time; TK constantly evolves. This knowledge is used to sustain the community and its culture and to maintain the biological resources necessary for the continued survival of the community. The Canadian government's Royal

[28] T. Taubman and M. Leistner, "Analysis of Different Areas of Indigenous Resources", in Von Lewinski, *Indigenous Heritage and Intellectual Property*, 69–89.

Commission on Aboriginal Peoples views indigenous knowledge "as a cumulative body of knowledge and beliefs, handed down through generations by cultural transmission, about the relationship of living beings (including humans) with one another and their environment."[29]

Hansen and Vanfleet introduce this basic definition in their handbook:

> Traditional knowledge includes mental inventories of local biological resources, animal breeds, local plants, crop and tree species. It may include such information as trees and plants that grow well together and indicator plants such as plants that show the soil salinity or that are known to flower at the beginning of the rains. It includes practices and technologies, such as seed treatment, storage methods and tools used for planting and harvesting. Traditional knowledge also encompasses belief systems that play a fundamental role in a people's livelihood and in maintaining their health and the environment. Traditional Knowledge is dynamic in nature and may include experimentation in the integration of new plant or tree species into existing farming systems or a traditional healer's tests of new plant medicines.
>
> The term "traditional" – used to describe this knowledge – does not imply that this knowledge is old or untechnical in nature, but rather that it is "based on traditions." It is traditional simply because it is created in a manner that reflects the traditions of the communities wherever they may be found. In this sense TK is easily distinguishable from cosmopolitan knowledge, which is drawn from global experience and combines "western" scientific discoveries, economic preferences and philosophies with those of other widespread cultures.[30]

TK does not relate to the nature of the knowledge itself, but to the way in which that knowledge is created, preserved, and disseminated. Knowledge typically refers to "information held in human memories that is accessible, by recall and the practice of learned skills, in a useful way in day-to-day life."[31] TK is more broadly defined as wisdom, which implies a blend of knowledge and experience integrated with a coherent world view and value system. Within the context of TK, the meaning of "traditional" implies that such knowledge is handed down from one generation to another, and that it has been accumulated by societies in the course of long experience in a particular place, landscape or ecosystem. Therefore, TK, in most cases, is usually collective in nature and considered the property of the entire community. As such, it does not belong to any single individual within the community; it is rather transmitted through specific cultural and traditional information exchange

[29] *Report of the Royal Commission of Aboriginal Peoples*, Canada Communications Group 454 (Ottawa, Vol. 4, 1996).

[30] A. Hansen and J. Vanfleet, *Traditional Knowledge and Intellectual Property*, http://shr.aaas.org/tek/handbook/handbook_1.pdf 13.

[31] *Knowledge, Innovations and Practices of Indigenous and Local Communities: Implementation of Article 8(j)* Doc., UNEP/CBD/COP/3/19 8–9.

mechanisms. Because the relationship between GRs and legal subjects that compose TK-based communities is structured under legal schemes that are different from the concepts of ownership and property rights under the common or civil law systems, many in such legal systems are led to conclude that TK is *res nullius*, the property of nobody until its discovery by explorers, corporate scientists, governments and so on. This attitude ignores the fact, however, that national or tribal customary laws[32] recognize forms of ownership separate from those designated by IP law.[33] As will be discussed in further detail later, TK ownership is viewed in traditional communities as a responsibility rather than an exclusive property right. This responsibility is often maintained and transmitted orally by elders or specialists (breeders, healers, etc.) and often to only a select few people within a community. This is why these indigenous communities are so alarmed when they see their precious and confidentially used TK being commercially exploited by private companies without their PIC.

As Dutfield puts it:

> what is traditional about TK is not its antiquity, but the way it is acquired and used. In other words, the social process of learning and sharing knowledge, which is unique to each indigenous culture, lies at the very heart of its 'traditionality.' Much of this knowledge is actually quite new, but it has a social meaning, and legal character, entirely unlike the knowledge indigenous people acquire from settlers and industrialized societies.[34]

It appears from the above description that the IP protection of TK's holistic nature is fraught with various difficulties. Hence, it is warranted to state at this early stage that current *lex lata* of IP will only be marginally able to satisfy the needs and expectations of TK holders. Furthermore, because of the anecdotal nature of TK, it is generally viewed as unreliable by governments attempting to incorporate TK into their various natural resource management processes. Moreover, since it is hard to dissociate TK from so many other aspects of the cultures of traditional communities, it can be difficult to qualify TK as legitimate protectable material, because, after all, TK can be anything that TK holders claim it to be.[35]

[32] Taubman and Leistner, "Analysis of Different Areas of Indigenous Resources", 89–90.
[33] C. Correa, *Traditional Knowledge and Intellectual Property Issues and Options Surrounding the Protection of Traditional Knowledge – A Discussion Paper* (Quaker UN Office, Geneva, 2001) 3.
[34] G. Dutfield, *Valuing Traditional Knowledge. A Review of the Issues*, background paper for a seminar at the Rockfeller Foundation (November 7, 2000).
[35] A. Howard and F. Widdowson, *Traditional Knowledge Advocates Weave a Tangled Web*, Options Politiques (April, 1997) 46–48.

It goes without saying that economic interests might not constitute either the only or the most important priority of such communities of people. The study of the reasons for which many communities oppose the integration of their inherent rights over their knowledge in a commercial system would require the aid of social sciences and anthropology and lies outside the scope of the present research. This book limits itself to the TK aspects related to intellectual property protection.[36]

1.2 Some methodological aspects

1.2.1 Objectives

The present study seeks to achieve three major objectives:

The first is to identify the international rules governing access to GRs and the acquisition of IPRs over GRs. Part II interprets various IP treaties and multilateral environmental agreements (MEAs) on the conservation and sustainable use of biodiversity and related TK.

The second objective is to study the methods of domestic implementation of internationally mandated obligations on States, both the providers and the recipients of GRs. Accordingly, Part III explores the various options that countries or regional intergovernmental organizations can follow in order to shape a well-balanced IP system related to the protection of GRs and knowledge related to GRs (see chapters 6 and 7), including access- and benefit-sharing (ABS) regimes governing their GRs. IPRs may indeed have a remarkable impact on the attainability of the objective of sustainable use of biodiversity at a variety of levels. IPRs may provide either incentives or disincentives for dynamic conservation of biodiversity, *ex situ* or *in situ*, depending on the factual and institutional settings that may prevail inside a country or region and abroad. Similarly, IPRs' design may either promote or discourage benefit sharing between countries and communities which make available genetic material and information as well as TK, on the one side, and countries and entities which hold IP and technology, on the other side.

The third objective entails the possible adaptation of IP laws to accommodate "new" claims by TK stakeholders under IP international laws and system.

In order to achieve these objectives, it is crucial to use the concept of *mutual supportiveness* to reconcile the TRIPS Agreement with other relevant IP treaties, the CBD, the ITPGRFA, and then other relevant multilateral agreements and sources of international law, including soft law and customary law.

36 Taubman and Leistner, "Analysis of Different Areas of Indigenous Resources", 71–77.

In a broader economic and political science perspective, this study seeks to move beyond the recurrent objection to the exercise of IPRs on biotechnology by postulating solutions based on the comparative advantage that is potentially to be implemented between North and South. These solutions are primarily based on the very fact that while industrialized countries are empowered with technology able to yield important biotechnological inventions, DCs are richly endowed with biological diversity which is progressively lacking in industrialized countries. From this perspective, the solutions proposed here mark the transition from an era of *confrontation* to an era of *cooperation* between developing and industrialized countries. The international exchange of PGRs and IP protection represents one of the IP global issues that requires much synergy and strong interdependence between technologically advanced and biodiversity rich countries. An international law approach is at the core of this synergy.

1.2.2 *An international law approach*

The international law approach influences both the structure of this study and the content thereof. The explanation of the structure will unfold the content in this international law perspective of the problems at stake.

The present work consists of three parts. After the explanation of the main problems in Part I, Part II lays out some theoretical elements forming the international law perspective of this work's subject-matter and provides an overview of the impact of IPRs on the preservation and exploitation of GRs and related TK. It limits the range of observation to the new relations between IP law and environmental obligations. The controversial issue of the alleged inconsistencies arising from the TRIPS Agreement and the CBD coexist with questions about the ethics of property rights in living organisms and concerns about biodiversity in domestic and regional IP systems in industrialized countries. In order to understand how to apply the relevant international law into the regional IP systems, it is also important to analyze the approaches adopted by US and European administrative and judicial bodies towards CBD principles in their decisions on the patentability subject-matter of biotechnological inventions.

The main legal challenges posed by the two-pronged expansion of IPRs (i) from inanimate to biological matter and (ii) from industrialized countries to developing ones aggravate the underlying paradox that these legal challenges pose. Through the adoption of Article 27 of the TRIPS Agreement, Northern hemisphere countries have embarked on a rapid and spectacular race to engage Southern hemisphere countries in international obligations on IP protection for biotechnology when they themselves have yet to define clear guidelines within their own IP systems in order to assure protection of biodiversity. These facts should spur an overall message of extreme caution in dealing with the delicate question of monopolized private ownership of the building blocks of life.

Article 27 of the TRIPS Agreement stands at the top of the pyramid as an overarching international provision from which all the other international, regional and domestic laws flow. This book's international law approach considers international IP law as an area of public international law. The US historically developed the patentability of GRs; the European patent system followed this example along with many other industrialized countries. An assessment of the differences between US and European methods of patentability of genes falls outside the scope of the present work.

For the sake of grounding this legal analysis in a real world scenario, it is important to address the economic rationale behind the forms of protection granted under IP laws and policies in order to understand how industrialized nations view the proper application of IP laws to GRs (see chapter 3).

At this juncture, the analysis will move towards the dynamics of protection of GRs and TK by applying or adapting existing IPRs as mainly contained in the TRIPS Agreement. These concerns have led me to primarily focus on the legal relationships between the TRIPS Agreement and the CBD. The alleged incompatibility between these two main treaties also involves legal relations with other treaties like the International Union for the Protection of New Varieties of Plants (UPOV)[37] and the FAO's ITPGRFA. The Patent Cooperation Treaty (PCT) and the Substantive Patent Law Treaty (SPLT), on the one hand, and the relevant soft law,[38] on the other hand, will fall within the scope of this analysis when interpreting these instruments and finding solutions *de*

[37] The text of all UPOV Acts can be found at www.upov.org (accessed June 20, 2005).

[38] For an evaluation of the value of the source of soft law, see G. Abi-Saab, "Eloge du 'droit assourdi': Quelques réflexions sur le rôle de la soft law en droit international contemporain" (1993) *Mélanges Rigaux* 66. M. Virally, "La distinction entre textes internationaux de portée juridique et textes internationaux dépourvus de portée juridique. Rapport provisoire", (1983) 60 (1) *Annuaire de l'institut de droit international* 332–33. C. Chinkin, "The Challenge of Soft Law: Development and Change in International Law" (1989) 38 (4) *International and Comparative Law Quarterly* 851. P. Weil, "Toward Relative Normativity in International Law" (1983) 77 *American Journal of International Law* 436; R. Baxter, "International Law in 'Her Infinite Variety'", (1980) 29 *International and Comparative Law Quarterly* 550; J. Dupuy, "Droit déclaratoire et droit programmatoire: de la coutume sauvage à la soft law", *Société francaise pour le droit international*, Colloque de Toulouse, *L'Elaboration du droit international public* (Pédone, Paris, 1975) 385; Q. Nguyen, *Droit International Public* (6th edn, LGDJ, Paris, 1999) 386; G. F.Handl, *et al.*, "A Hard Look at Soft Law" (1988) 82 *Proceedings of the American Society of International Law* 372.

lege ferenda (how the law should be) to make the international IP and environmental treaties mutually supportive.

The TRIPS Agreement and the CBD are not legally incompatible *stricto sensu* since they do not have the same object and purpose. States have the obligation to stretch their thought and comply with both simultaneously.

The conflict between TRIPS and CBD lies outside international law, i.e. at the crossroads of stark contrasts of opposing perspectives between industrialized and developing countries, or, oversimplifying, between North and South. The debate over IPRs on biological resources is embedded in a broad context with so many interconnections and competing interests. The complexity of this political debate is accompanied by intense emotions, as the debate is often framed in terms of a battle between "haves" and "have-nots."[39] However, the political or social controversies are hereby dealt with as far as they are very relevant to the legal issues at stake.

While concentrating on how IPRs protect innovations derived from the genetic pool of biodiversity and how the IPRs in return impact biodiversity, this book also identifies the international principles of conservation of biodiversity, especially in the context of the interrelations between international IP legal instruments and MEAs.

Having ranged from international IP law (that protects the private property of the innovation in this field) to the major international law instruments regulating the protection, preservation and conservation of GRs, Part III moves into the analysis of IP methods of redistribution of benefits in the international trade context by using the allocative efficiency principles.

Part III identifies the aspects of TK that are protectable in the international patent system. Therefore, section 4.2 introduces some relevant methodological distinctions on which the subsequent analysis builds. It is not my intent to analyze rights to TK in a holistic manner (which would include natural rights, land claims, etc.), rather only to the extent of its possible integration in the existing IP system. In other words, the scope of observation is limited to the IP-related aspects of the exploitation of GRs and TK, even when it is considered along with the human rights protection of TK.

Protection of TK can be sought through (i) national or regional "access legislation," to GRs, (see chapter 5); (ii) defensive protection by adapting internationally mandated IP laws to prevent misappropriation of GRs

39 J. F. Badimboli Atibasay, "The International Legal Regime for Biotechnology Patenting: An Appraisal from the Standpoint of Developing Countries" (2001) 31 *Revue générale de droit* 294.

through the introduction of the disclosure of origin (section 6.1 below), the patentability exceptions of *ordre public* and morality (section 6.3 below), the full assessment of the novelty criterion (section 6.2 below); (iii) positive protection of TK through marketing, transferring, and licensing aspects of existing IPRs (see chapter 7). This analysis *de lege ferenda*, shall also expand on positive protection through new types of IPRs or liability regimes.

In this respect, Part III will explore the proposals that have been put forward by governments at the relevant fora, the international legal doctrine, the recommendations of international organizations, States' statements and non-governmental organization (NGO) sponsored studies in order to render the IP system more supportive of the benefit-sharing treaty provisions analyzed in Part II.

The objectives stated in section 1.2.1 above lead me to postulate ways in which TK holders can seek protection in EU and US law and judicial practice. Of course, the differences between common law and civil law approaches and the underlying policy options will be highlighted by resort to comparative law with a particular emphasis on the European IP legal framework. Through the analysis of this system of reference, the other main industrialized countries' systems are then compared with the aim of finding the best methods of protection adapted to particular circumstances. IP international treaties will serve as an overarching guide, setting forth the fundamental IP principles that are more precisely applied within national or regional jurisdictions, thus building upon the existent scholarship that identifies and interprets rules governing the IP exploitation of GRs and TK.

The relevant international legal instruments have been adopted by the same States but within different fora. Influential States have been successful in reaching their objectives in one forum and not in another, thus creating various treaties with apparently conflicting provisions. In turn, the applicable international obligations stemming from different treaties remain disarticulated, unless they are interpreted in light of the concept of mutual supportiveness.

1.2.3 The concept of mutual supportiveness and the balance of rights

The concept of mutual supportiveness[40] indirectly stems from the Vienna Convention on the Law of Treaties (VCLT) and belongs to the handful of

[40] L. Boisson de Chazournes and M. Mbengue, "A propos du principe du soutien mutuel – Les relations entre le Protocole de Cartagena et les Accords de l'OMC" (2008) 4 *Revue générale de droit international public* 829–63.

rules of interpretation of treaties. If a conflict among two international treaty provisions arises, it should be solved through the classic rules of *lex posterior* and *lex specialis* on conflicting treaty norms as provided by Article 30.2 of the VCLT:

> When a treaty specifies that it is subject to, or that it is not to be considered as incompatible with, an earlier or later treaty, the provisions of that other treaty prevail.

Although very important in other instances, the application of this Article to the relations between the WTO TRIPS Agreement and the provisions of environmental law treaties would create a hierarchy of norms, a disarticulated situation, and much confusion for the contracting parties.

Avoiding conflicts of norms between the international environmental and trade regimes is vital for the unity of the international legal system. Therefore, the method of interpretation adopted in this study will rely more predominantly on Article 31.3(c) of the VCLT,[41] according to which a treaty has to be interpreted in light of all the other rules of international treaty and general law applicable to the parties.[42] The principle of mutual supportiveness can be also inferred from this treaty provision. Finally, this norm avoids placing one set of norms of an international organization above another merely because of some chronological difference in treaty ratification and it helps avoid conflicts to the utmost extent.

International judicial and policy organs increasingly adopt this approach in order to avoid creating self-contained legal systems totally independent from general norms and from each other. The WTO network of treaties on trade law and the UN body of treaties on environmental law can be considered as separate self-contained systems. The mutual supportive concept renders the commercial or environmental character of the norms irrelevant as regards their objective interpretation and simultaneous application.

Accordingly, the concept of *mutual supportiveness* between environmental and trade regimes requires the application of all the relevant norms among the parties with a presumption of absence of conflict among the

[41] "There shall be taken into account, together with the context: (c) any relevant rules of international law applicable in the relations between the parties," *Vienna Convention on the Law of Treaties* (May 23, 1969) 1155 *UNTS* 331.

[42] G. Marceau, "Conflict of Norms and Conflicts of Jurisdiction – The Relationship between the WTO Agreement and MEAs and other Treaties" (2001) 1081 *Journal of World Trade* 1109; G. R. Tarasofsky, "Ensuring the Compatibility Between Multilateral Environmental Agreements and the GATT/WTO" (1996) 7 *Yearbook of International Law* 52; F. Francioni (ed.), *Environment, Human Rights and International Trade* (Hart Publishing, Oxford, 2001) especially 22–24.

norms *ipso jure*. It enables the parties to simultaneously and harmoniously apply the two apparently conflicting bodies of international law: trade law and environmental law.

This concept has already been implemented in the international law-making process (in declarations, treaties and protocols). It has also been used since 1994 by the WTO Committee on Trade and Environment that has been instituted "with the aim of making international trade and environmental policies mutually supportive."[43] More recently, section 31 of the Doha Declaration of the WTO Ministerial Conference has restated its conviction that "the aims of upholding and safeguarding an open and non-discriminatory multilateral trading system, and acting for the protection of the environment and the promotion of sustainable development can and must be *mutually supportive*" (italics added).[44] It goes without saying that the *raison d'être* of continuing negotiations involving international trade and the environment is the enhancement of this mutual supportiveness.

On the side of environmental law, Article 22.1 of the CBD attempted to mark the superiority of that treaty when it stated that:

> the provisions of this Convention shall not affect the rights and obligations of any Contracting Party deriving from any existing international agreement, except where the exercise of those rights and obligations would cause a serious damage or threat to biological diversity.

The somewhat vague wording of such a safeguard clause could leave room for the superiority of environmental law above other bodies of law, thus jeopardizing the international legal order. This peril has led to the necessity of adopting the Biosafety Protocol to the CBD that has been carefully drafted so to ensure its mutual supportiveness with the WTO treaties, thus including the TRIPS Agreement.[45]

More recently, a less burdensome technique has been used to realize mutual supportiveness: rules on the relationship between international legal instruments are encapsulated in the preamble and not in the substantive norms. This practice has the advantage of being rapidly negotiated but presents the disadvantage of not being incorporated in the substantive provision. The preamble sets at least the interpretative tools

[43] Decision on Trade and Environment, adopted by ministers at the meeting of the Uruguay Round Trade Negotiations Committee in Marrakesh on April 14, 1994, www.wto.org/English/tratop_e/envir_e/issu5_e.htm.

[44] WTO, Ministerial Declaration of November 14, 2001, WT/MIN(01)/DEC/1, 41 *ILM* 746, 751 (2002).

[45] Boisson de Chazournes and Mbengue, "A propos du principe du soutien mutuel", 851–59.

necessary for the implementation of the substantive treaty provision. A preamble can create mutual supportiveness with other treaties. A relevant example of this technique is set forth in paragraph 9 of the preamble of the FAO International Treaty on Plant Genetic Resources for Food and Agriculture (ITPGRFA)[46] that states that "this Treaty and other international agreements relevant to this Treaty should be mutually supportive with a view to sustainable agriculture and food security," and "nothing in this Treaty shall be interpreted as implying in any way a change in the rights and obligations of the Contracting Parties under other international agreements" (preamble paragraph 10), and finally, "the above recital is not intended to create a hierarchy between this Treaty and other international agreements." In spite of the non-substantive character of this wording, this preambular language provides some principles of interpretation of its substantive provisions with the rest of trade law treaties.

This concept will be particularly important in reconciling the TRIPS Agreement and the CBD in implementing the concept of PIC and "benefit sharing." The same is valid when in paragraphs 6 and 7 it is demonstrated how patent rights and the Plant Variety Protection (PVP) (mandated by Article 27.3(b) of TRIPS Agreement) in the DCs is compatible with the exigency of protecting "farmers' rights," as articulated in the ITPGRFA.

The interpretation of relevant international treaties in a mutually supportive way can assist in adapting international patent law so as to accommodate certain needs and expectations of local and indigenous communities with regard to the defensive protection of their TK (e.g. section 6.2 below). Indeed, the mutual supportive principle does not only mean that one treaty *de jure* or *de facto* conflicts with another, but that one body of treaties has to be taken into account to interpret the ordinary meanings of a treaty in another area. Therefore the interpreter of the TRIPS Agreement cannot subordinate the norms of the CBD to those of TRIPS. Excluding the CBD from the interpretation of TRIPS Agreement provisions amounts to subordinating the latter to the first.[47] On the very practical side, the concept of mutual supportiveness leads, in this context, to the indirect applicability of the CBD in a potential dispute on a TRIPS provision involving the interpretation of biodiversity protection issues.[48] Indeed, I agree with the reasoning of Pauwelyn who argues for indirect applicability:

[46] The FAO International Treaty on Plant Genetic Resources for Food and Agriculture (ITPGRFA) (November 3, 2001), http://ext-ftp.fao.org/ag/cgrfa/it/ITPGRe.pdf.

[47] Boisson de Chazournes and Mbengue, "A propos du principe du soutien mutuel", 855.

[48] *Ibid.*, 857. See in favor of the indirect applicability Boisson de Chazournes and Mbengue, "Trade, Environment and Biotechnology: On Coexistence and Coherence", in T. Cottier and D. Wüger (eds.), *Genetic Engineering and the World Trade System* (Cambridge University Press, 2008).

> [A]ll international law binding on both parties to a dispute may, in principle, be part of the applicable law before a WTO panel. Non-WTO law is part of this applicable law, and may, in particular, provide a defense against violation of WTO law (for example, an environmental agreement binding between the disputing parties may, depending on the relevant conflict rules, excuse a violation of GATT, independently of GATT Article XX). While non-WTO law to be referred to when interpreting WTO terms ought to be limited to law that reflects the common intentions of all WTO members, in my view, the applicable law in a particular dispute may also include law binding only between the two disputing parties.[49]

[49] J. Pauwelyn, "Bridging Fragmentation and Unity: International Law as a Universe of Inter-connected Islands" (2004) 25 *Michigan Journal of International Law* 910.

Part II

The protection of genetic resources in intellectual property law

2 The TRIPS Agreement and the patent protection of genetic resources

This chapter examines the international and regional law obligations related to the patenting of life forms. The creation and the evolution of the international patent system has led to Article 27 in the TRIPS Agreement, thus internationalizing the practice of patenting life forms.[1]

Because the matter of the protection of biodiversity and TK in the IP system is strictly connected with the interests of developing countries, particular attention should be paid to some *development-oriented* principles of the TRIPS Agreement.

The following sections offer some interpretative paths that the judge, the policy maker, and the legal scholar can use to apply the relevant TRIPS provisions to any particular case at hand.[2] The objective is to explain some interpretative principles of the substantive TRIPS provisions that may help to achieve a fine-tuned balance between right-holders and users with regards to patent rights mandated by Article 27 of TRIPS. The achievement of this balance is central to the relationship between GR and TK provider countries (generally developing) and recipient countries (generally industrialized).

2.1 The general principles of the TRIPS Agreement

The TRIPS Agreement has the potential to contextualize IP law within the realm of general public international law, given its progressive universal adoption. The whole WTO-GATT legal architecture is based upon the intention of the international community to create a broader constitutional basis to regulate international trade. For instance, it has moved from the creation of negative obligations to the shaping of positive obligations: while GATT 1949 was based upon the ban on discrimination against foreign goods, WTO-GATT 1994 also includes positive

[1] T. Stewart (ed.), *The GATT Uruguay Round. A Negotiating History (1986–1992)* (Kluwer, The Hague, 1993).

[2] *Resource Book on TRIPS and Development* (Cambridge University Press, 2005) 118.

obligations like licensing and foreign direct investment. The TRIPS Agreement follows this pattern. For instance, it has moved from the negative obligation of National Treatment of foreigners in the Paris and Berne Conventions towards the concept of Most Favored Nation Treatment (MFN) of Article 4 of the TRIPS Agreement.[3] Whereas, under the concept of national treatment, States were obligated not to give less protection to foreigners than to their nationals, under the concept of MFN, States must grant to all the WTO Member States any further right of protection that they would have granted to a particular State with which they would have entered into an agreement.

2.1.1 Minimum standard of intellectual property protection

Article 1 of the TRIPS Agreement, entitled "Nature and Scope of Obligations," starts with a general statement giving effect to the obligations contained in the text. However, Members may implement more extensive protection than is required in the agreement, provided such protection does not contravene the "provisions of th[e] Agreement." The provisions include the national treatment rule in Article 3 and the most-favored-nation treatment rule in Article 4; rules requiring compliance with the relevant provisions of the IP Conventions concerned (Articles 1.3, 2, 9 and 35); minimum requirements for protection, scope and duration of IPRs (Articles 10 and 12 on copyright protection for computer programs and databases; Articles 25 26 on industrial designs; Articles 27, 28, 29, 31 and 33 on patents; and Articles 35, 36, 37.2 and 38 on layout-designs for integrated circuits); cooperation rules; rules imposing cooperation regarding implementation of TRIPS in general (Articles 66.2 and 67); and rules imposing cooperation regarding abuse of IPRs in the context of licensing practices in particular (Article 40(3) and (4)); rules on enforcement (Articles 41–61); rules pertaining to procedures for acquisition and maintenance of IPRs (Article 62); and dispute settlement rules (Article 64).

The TRIPS Agreement is, therefore, "a minimum rights agreement that leaves a fair amount of leeway to Member countries to implement its provisions within their own legal systems and practice and fine-tune the balance in the light of domestic public policy considerations."[4]

[3] M. Ricolfi, "Is there an Antitrust Antidote against IP Overprotection within TRIPS?" (2006) 10(2) *Marquette Intellectual Property Law Review* 331–32.

[4] *Protection of Intellectual Property under the TRIPS*, background paper submitted by the Secretariat of the WTO to Committee on Economic, Social and Cultural Rights at its 24th Session. Agenda Item 3: Implementation of the International Covenant on Economic, Social and Cultural Rights, UN Doc. E/C.12/2000/18, 5.

2.1.2 *The preamble*

The legal value of the TRIPS Agreement's preamble within the context of the WTO legal system sets forth the tone for the interpretation of the rest of the provisions of the treaty. It has been stated in the *Shrimp* case that the preamble of the WTO Agreement "does not only inform the GATT 1994, but also all other covered agreements."[5] Since the TRIPS Agreement forms part of the WTO Agreement as Annex 1C, the preamble of the WTO Agreement sets up general principles of conduct of the Member countries with respect to IP. This recital states that trade and economic relations:

> Should be conducted with a view to raising standards of living, ensuring full employment and a large and steadily growing volume of real income and effective demand, and expanding the production of and trade in goods and services, while allowing for the optimal use of the world's resources in accordance with the objective of sustainable development, seeking both to protect and preserve the environment and to enhance the means for doing so in a manner consistent with their respective needs and concerns at different levels of economic development.[6]

The general principle sets forth that Member countries should conduct their trade and economic relations in a manner consistent with their respective needs and economic, social and other concerns.

Recital 4 of the preamble to the TRIPS Agreement recognizes that IPRs are private rights. That is an implicit reference to Article 15.1(c) of the International Covenant on Economic, Social and Cultural Rights (ICESCR), which provides that "everyone has a right to benefit from the protection of the moral and material interests resulting from any scientific, literary and artistic production of which he is the author." Private rights are also reflected in Article 17.1 of the Universal Declaration on Human Rights (UDHR),[7] which provides that everyone has the right to own property alone as well as in association with others, and Article 17.2, which provides that no one shall be arbitrarily deprived of his property, as well as Article 27.2:[8] "the paragraphs clearly prescribe an individual right, and as such the

[5] *Appellate Body Report on United States – Import Prohibition on Certain Shrimp and Shrimp Products*, WT/DS58/AB/R (October 12, 1998) paragraph 129.

[6] WTO Agreement, Results of the Uruguay Round of Multilateral Trade Negotiations, The Legal Texts (WTO, Geneva, 1995), www.wto.org/english/tratop_e/envir_e/sust_dev_e.htm.

[7] Adopted by UN General Assembly Resolution 217(III) of December 1948.

[8] Article 27(2) of the *Universal Declaration on Human Rights* stipulates that "everyone has a right to the *protection* of the moral and material interests resulting from *any* scientific, literary or artistic product of which he is the author" (italics added).

paragraph is similar to a civil and political right."[9] This discussion shows how much IPRs are grounded in human rights law and that a total separation between the two in international law is counter-productive for both fields.

Recital 5 of the TRIPS Agreement recognizes "the underlying public policy objectives of national systems for the protection of IP, including development and technological objectives […]." It can be argued that the provisions of the TRIPS Agreement need to be interpreted in accordance with the public policy objectives as they vary according to each country's need. It is, therefore, within the power of a Member to strike the right balance.[10]

Recital 6 of TRIPS addresses the specific concerns which arise in connection with least developed country Members: "*recognizing* also the special needs of the LDC Members in respect of maximum flexibility in the domestic implementation of laws and regulations in order to enable them to create a sound and viable technological base."

In this context, it is worth mentioning the so-called transitional provisions, which allow DCs to comply with the minimum standard of protection in 2005 and LDCs in 2016.[11] In other words, poor countries have been given the right to postpone the implementation of patents in their national law until the end of the transition period. However, the practice of certain rich countries – first and foremost the US – in pressing for immediate implementation of pharmaceutical patents with retroactive effect (the so-called *pipeline solution*) seems to be undermining these provisions.[12] Moreover, there are special transitional provisions pursuant to Article 70.8 for pharmaceutical patents. The so-called *mailbox* system is set up to store pending patent applications that, if granted upon the implementation of the system after the transitional period, will be retroactively effective from the date of mailbox application.[13]

[9] G. Melander, "Article 27", in Eide Asbjorn *et al.* (eds.), *The Universal Declaration of Human Rights: A Commentary* (Scandinavian University Press, Oslo and Oxford University Press, 1992) 429–32, 431.

[10] Ricolfi, "Is there an Antitrust Antidote?", 15–16.

[11] Articles 65 and 66 of the *TRIPS Agreement* (Part IV). The transitional period applies automatically. Concerning LDCs, they have been granted the possibility of further extension upon duly motivated request. *Doha Declaration on the TRIPS Agreement and Public Health*, paragraph 7 (November 14, 2001), www.wto.org/english/tratop_e/TRIPS_e/public_health_e.htm#declaration.

[12] C. Correa, "Intellectual Property Rights, the WTO and Developing Countries" (Zed Books Ltd, Zed Books, Malaysia, Penang, London, New York, 2000) 10.

[13] J. Curci and M. Vittori, "Improving Access to Life-Saving Patented Drugs – Between Compulsory Licensing and Differential Pricing' (2004) 7 *The Journal of World Intellectual Property* 739. The scope of these rights was not clarified either in the Agreement or in the finding of the WTO Panel which dealt with the WTO Dispute Resolution Body, *Report of the Panel on India – Patent Protection for Pharmaceutical and Agricultural Chemical Products,*

2.1.3 *Objectives*

Articles 7 and 8 of the TRIPS Agreement state specific objectives and principles that could provide a solid legal framework for a more "development-oriented" interpretation of the substantive rules applicable to patents. Article 7 reads:

> The protection and enforcement of International Property rights should contribute to the promotion of technological innovation and to the transfer and dissemination of technology, to the mutual advantage of producers and users of technological knowledge and in a manner conducive to social and economic welfare, and to a balance of rights and obligations.

It is clear that the first objective of Article 7 is the protection of IPRs, that must not be exclusively intended to promote "*technological innovation*" but also "*the transfer and dissemination of technology.*" This aspect is undoubtedly of great importance to DCs especially in the light of the *teleological interpretation* of the TRIPS provisions.[14] According to Article 8, Member States are given, under certain circumstances, the possibility to implement IP rules taking into account their public interests. The article goes even further, allowing countries to adopt appropriate measures in order to circumvent IPR holder abuses.

The second objective is "the mutual advantage of producers and users of technological knowledge." This objective implies a balance of rights between the producers (IPR holders) and of the users of IP protected products. That the exercise of IPRs should generate benefits for both the producers and users is in line with the goal of the general principle of Article 27 of the UDHR, that states that (1) everyone has the right to participate in the cultural life of the community, to enjoy the arts and to share in scientific advantage and its benefits; and (2) everyone has the right to the protection of the moral and material interests resulting from any scientific advancement of which he is the author. It is clear that paragraph (1) deals with the users of IP while paragraph (2) deals with the producers.

Article 7, stating that the exercise of IPRs should be conducive to social and economic welfare, echoes Article 4 of the ICESCR, which says:

> The State Parties to the Present Covenant recognize that, in the enjoyment of the rights provided by the State in conformity with the present Covenant, the State may subject such rights only to such limitations as are determined by law only so

WT/DS50/R (September 5, 1997). C. Correa, *Intellectual Property Rights and Use of Compulsory Licenses: Options for Developing Countries* (Southcentre, Geneva, 1999) www.southcentre.org/publications.

[14] N. A. Odman, "Using TRIPS to Make the Innovation Process Work" (2000) 3 *Journal of World Intellectual Property* 343.

far as this may be compatible with the nature of these rights and solely for the purpose of promoting the general welfare in a democratic society.

Article 7 allows a State to interpret the TRIPS provisions so that the protection and enforcement of IPRs does not run counter to the social and economic welfare of the State as a whole.

Articles 7 and 8 of TRIPS, if interpreted in light of Article 29(2) of the UDHR, may lead a State to restrict the exercise of IPRs in certain cases. The exercise of the legitimate interests of the international patent rights by the right-holder has to be balanced with the "legitimate interests" of the society at large. These provisions justify exceptions to patentability, set out in Article 27.2 of TRIPS Agreement in case the invention is contrary to "*ordre public* or morality" or causes "serious prejudice to the environment" (see section 6.3 below); when the patent provokes serious unbalances in favor of the patent holder, undermining human rights, e.g. the right to health as in the case of the patented life-saving drugs (see Article 31.f of TRIPS).[15] Howse writes that "the legitimate interests of the patent holder are not to be considered prior to other interests, but inherently in relation to those other interests, particularly those of users of technological knowledge."[16] In my view, this is the correct interpretation of Article 7 that states that the enforcement of IPRs serves the "*mutual advantage* of producers and users of technological knowledge" (italics added).

All these principles will be crucial in the interpretation of TRIPS Agreement provisions in light of other treaties that set forth the interaction between IPRs on the one hand and nutrition and preservation of the environment on the other hand. The possibility that these Articles 7 and 8 offer in modeling (even through restricting) certain exercises of abusive IPRs is central in our study, because, as Odman puts it: "the enforcement [of IPRs] may have a significant impact on the supply and price of these products in the countries of the Members."[17]

The case law is going in the direction of stating that of the objectives of TRIPS Agreement are not only the enforcement of minimum standards of IPRs, but also their protection in a manner conducive to the mutual

[15] Curci and Vittori, "Improving Access".

[16] R. Howse, "The Canadian Generic Medicines Panel: A Dangerous Precedent in Dangerous Times" (2000) 4 *The Journal of World Intellectual Property* 493, 502.

[17] Odman, "Using TRIPS", 348–49. WTO Panel Report on *United States – Section 110(5) of the US Copyright Act*, WT/DS160/R (June 15, 2000). *Canada – Term of Patent Protection*, WT/DS 170/AB/R (October 12, 2000). D. Gervais, *The TRIPS Agreement: Drafting History and Analysis* (Sweet and Maxwell, London, 1998) 64. *The Understanding on Rules and Procedures Governing the Settlement of Disputes* (1994), forms annex 2 of the WTO Agreement, www.wto.org/english/tratop_e/dispu_e/dsu_e.htm. *WTO Appellate Body Report on Canada – Measures Affecting the Export of Civilian Aircraft*, WT/DS70/AB/R paragraph 187 (August 2, 1999).

advantage of both producers and users and to strike a balance of obligations and rights and to contribute to social and economic welfare.

In light of these considerations, an overall evaluation of the *development-oriented* provisions analyzed thus far leads us to conclude that they provide DCs with a certain maneuvering space in which to carve out a patent system that serves their national development interests. This custom-made patent system, however, needs to be "consistent with the provisions of th[e] Agreement" (Article 8).

2.1.4 *The relation between the TRIPS Agreement and "any other relevant rule of international law"*

Among the various rules of interpretation set forth by the VCLT, Article 31.3 (c) has particular relevance: "3. There shall be taken into account, together with the context: [...] (c) any relevant rules of international law applicable in the relations between the parties."[18] While the concept of evolutive interpretation is particularly relevant in the fields of human rights and environmental agreements, it cannot be totally separated from Article 31.3(c) that sets forth the obligation of interpreting a treaty provision in light of the subsequent practice. It goes without saying that customary international law and general international law are also to be applied as long as a WTO rule does not provide otherwise.[19] This matter has particular relevance to our subject-matter as the TRIPS Agreement does not provide all the solutions to the relationship between TRIPS and other treaties.

The relevance of an international rule is to be decided on a case-by-case basis. Because many TRIPS Agreement provisions are intentionally vague, an evolutive interpretation is required to fully implement both the promotion of technological innovation and public interest according to Articles 7 and 8 respectively. A purely conservative interpretation of the TRIPS Agreement would be at the expense of developing nations. Relative terms like "special cases," "unreasonably prejudice," "limited

[18] Sir I. Sinclair, *The Vienna Convention on the Law of Treaties* 139 (2nd edn, Manchester University Press, Manchester, 1984). *Legal Consequences for States of the Continued Presence of South Africa in Namibia (South West Africa) notwithstanding Security Council Resolution 276 (1970)*, June 21, 1971, Advisory Opinion, *International Court of Justice Reports*, 16–31, paragraph 53 (1971). L. Boisson de Chazournes, "Qu'est-ce que la pratique en droit international?", *La pratique et le droit international*, Colloque de Genève de la Société française pour le droit international (Paris, Pedone 2004) 43. *Gabcikovo-Nagymaros Project (Hungary/Slovakia)*, September 25, 1997, *International Court of Justice Reports* 7–78 paragraph 140 (1997). G. Marceau, "A Call for Coherence in International Law: Praises for the Prohibition against 'Clinical Isolation' in WTO Dispute Settlement" (1999) 33 *Journal of World Trade* 87, 112.

[19] *Ibid.*

exceptions," "unreasonably conflict," "national emergency," and "cases of extreme urgency," like the imprecise obligations of soft law, justify the use of evolutive interpretation.

The application of evolutive interpretation was particularly decisive in the *Shrimp-Turtle* case where the Appellate Body made reference to several international instruments adopted by the UN[20] in order to interpret Article XX(g) of the GATT[21] and to come to the conclusion that sea turtles are "exhaustible natural resources."

It must also be noted that the exceptions under the TRIPS Agreement are phrased in evolutionary terms, which permits an interpreter, such as a WTO panel, to use any other relevant rules of international law to implement those exceptions or restrictions of rights. The analysis of Article 27(2) that stipulates that "Members may exclude from patentability inventions, the prevention within their territory of the commercial exploitation of which is necessary to protect *ordre public* or morality; including to protect human, animal or plant life, or health, or to avoid serious prejudice to the environment, provided that such exclusion is not made merely because the exploitation is prohibited by their law." Here, as the Appellate Body in the *Shrimp-Turtle* case did, countries can interpret the expression "the prejudice to the environment" in light of the relevant MEAs (see section 3.3.4 below).

By the same token, Article 31(b) of TRIPS allows for compulsory licensing (allowing others to use a patent without authorization of the patent holder) in case of a "national emergency," or other cases of extreme urgency or in cases of public non-commercial use. In order to define a "national emergency," the policy maker or the judge will naturally resort to the relevant international rules in the fields of health and human rights.

2.2 The patentability of biotechnology

2.2.1 The scope and utility of Article 27 of the TRIPS Agreement

Article 27 of the TRIPS Agreement was not created *ex nihilo*. It extended to all WTO Member States the practice of patenting life forms. In other words, it globalized the IP protection of biotechnological inventions that were cultivated in the US and the EU and other industrialized countries for some decades so that these industrialized countries could ensure a comparable level of protection worldwide. The law and practice of the US

[20] *Resolution on Assistance to Developing Countries*, adopted in conjunction with the *Convention on the Conservation of Migratory Species of Wild Animals* (June 23, 1979) 19 *ILM* 11, 15.

[21] *Appellate Body Report on United States – Import Prohibition of Certain Shrimp and Shrimp Products* (October 12, 1998), WT/DS58/AB/R paragraph 130–4.

and the EU biotechnological inventions have achieved patentability in the regions that have promoted it and can set forth some policy options for the WTO Member States that will have to implement Article 27.

IP protection originally concerned non-living matter. So it is hardly surprising that IP protection incorporated new features when it was called to extend to self-replicating materials, including PGRs. For the present purposes, I examine the form that IP protection has taken in the US and EU when applied to PGRs and will also consider certain IP-like regimes that have emerged in connection with PGRs. In doing so, both a positive and a normative approach will be adopted. I outline the essential features of the law as it stands in these legal systems and then present these features as possible paradigms and recommend or discourage their adoption by other legal systems, depending on the applicable circumstances and context.

Both biotechnological patents and plant varieties (via Plant Breeders' Rights – PBRs)[22] may protect innovation which incorporates, uses, or is based on pre-existing PGRs derived from nature or agriculture (that are simultaneously covered by the CBD). Hence, it is crucial to sketch the legal history of the requirements for IP protection for innovations based on GRs.

This section equally assesses the economic importance of the pre-existing PGRs that have been subsequently incorporated into the innovation. The question arises whether the pre-existence of the relevant resource may have an impact on the fulfillment of the applicable requirements for IP protection. It is submitted that equitable solutions for benefit sharing with the holders of GRs (on which certain biotechnological inventions are based) depend very much on this study.

The CBD-mandated obligation of sharing the benefits of the use of biodiversity have an impact upon innovative IP protected plants that use or incorporate PGRs, information, or even knowledge belonging to persons and entities other than the innovator (see chapter 5).

Here lies the possible integration into the patent system of non-IP matters such as the preservation of the environment and the protection of TK through sharing the benefits arising from its use. The classical patent law scholars set these matters outside the innovation process constituting the patented invention.[23] Therefore, these issues will be also referred to as "pre/post-IP" or "non-technical."

[22] P. Cullet *et al.*, "Intellectual Property Rights, Plant Genetic Resources and Traditional Knowledge", in T. Cottier and S. Biber-Klemm (eds.), *Rights to Plant Genetic Resources and Traditional Knowledge: Basic Issues and Perspectives* (CABI on behalf of Swiss Agency for Development and Cooperation and the World Trade Institute, London, 2006) 130–33.

[23] R. Schapira, "Biotechnology Patents in the United States" in S. Sterckx (ed.), *Biotechnology, Patents and Morality* (Aldershot, England, Burlington, VT, 1997) 171–72; S. Crespi, "Debate" *Biotechnology, Patents, and Morality*, ibid., 219–20. US Commissioner for Patents

The patentability of life forms has also been withstood by a second objection that has been forcefully raised by opponents and skeptics who argue that the granting of patents on new life forms may run the risk of benefiting inventions that jeopardize the basic principles of life and harm the environment. While this argument may stop short of proposing that patentability of life forms should be *per se* excluded, its advocates call for increased caution in the procedure which leads to biotech-patenting and in the standards it employs.

Environmental or ethical considerations are not *per se* part of the patent system that grants a monopoly right to all inventions complying with the patentability requirements. The integration of environmental and ethical issues in the patent system belong to the same category of problems found to integrate TK within the patent system (see Part III chapters 5–7). Environmental, ethical, and TK matters are considered by the classical patent system as non- pre- or post-IP matters. A review of environmental and ethical considerations is preparatory to the study of the defensive protection of TK, i.e. the study of how non-compliance with standards integrating these matters can lead to the ban of a certain patent application from patentability at the examination stage.

Laws and practices on patentability of biotechnology have spurred an "imitation effect"[24] in which the EU and other industrialized countries have integrated biotech patenting principles from the US. This expansion was driven by strong business interests and comparably weak opposition from other concerned parties. The legislative process in the EU tended to consider non-business concerns such as morality and equity with more emphasis, most recently in the context of the elaboration and adoption of the Biotech-Directive.[25]

The scope of patent protection was strongly debated during the TRIPS Agreement negotiations. Article 27 of the TRIPS Agreement provides that patents shall, subject to certain conditions, be available for any invention, whether a product or process, in all fields of technology, provided that they are new, involve an inventive step, and are capable of industrial application. The

Bruce Lehman, *Utility Examination Guidelines*, 60 Fed. Reg. 26,263, 1995; R. Merges, S. Menell and M.A. Lemley, *Intellectual Property in the New Technological Age* (Aspen publishers, Gaithesburg, New York 2000) 120–24; A. Gallochat, "Le brevet et l'éthique, ou le mélange des genres" (1993) 2 *Dossiers brevets* 18; P. Spada, "Liceità dell'invenzione brevettabile ed esorcismo dell'innovazione" (2000) 5 (1) *Rivista di diritto privato* 5.

[24] The expression "imitation effect" stems from the expression "term regulatory," contrary to the "race of laxity," see *Louis K. Liggett v. Lee*, 288 US, 517, 557–59 (1933). For a fuller explanation of these concepts, see M. Ricolfi, "Intellectual Property Rights and Legal Order" (2001) *Il diritto d'autore* 123–145, also available at *Global Jurist Advances*: Vol. 2: No. 1 www.bepress.com/gj/advances/vol2/iss1/art3.

[25] *Directive 98/44/EC of the European Parliament and of the Council of July 6, 1998* on the Legal Protection of Biotechnological Inventions, OJ L 213 of July 30, 1998, 13–21.

TRIPS Agreement is the first international treaty which makes it legal – and compulsory – to patent life. In other words, it is the first globally adopted treaty to make the patenting of life legal by requiring WTO Member States to provide patent protection for all fields of technology. Paragraphs 2 and 3 of Article 27 outline the inventions Member States may exclude from patent protection under specified conditions.

Article 27 of TRIPS states:

Patentable Subject-matter

1. Subject to the provisions of paragraphs 2 and 3, patents shall be available for any inventions, whether products or processes, in all fields of technology, provided that they are new, involve an inventive step and are capable of industrial application. Subject to paragraph 4 of Article 65, paragraph 8 of Article 70 and paragraph 3 of this Article, patents shall be available and patent rights enjoyable without discrimination as to the place of invention, the field of technology and whether products are imported or locally produced.
2. Members may exclude from patentability inventions, the prevention within their territory of the commercial exploitation of which is necessary to protect *ordre public* or morality, including to protect human, animal or plant life or health or to avoid serious prejudice to the environment, provided that such exclusion is not made merely because the exploitation is prohibited by their law.
3. Members may also exclude from patentability:
   (a) diagnostic, therapeutic and surgical methods for the treatment of humans or animals;
   (b) plants and animals other than micro-organisms, and essentially biological processes for the production of plants or animals other than non-biological and microbiological processes. However, Members shall provide for the protection of plant varieties either by patents or by an effective sui generis system or by any combination thereof. The provisions of this subparagraph shall be reviewed four years after the date of entry into force of the WTO Agreement.

The State's obligation to grant patents provides two exceptions. The first – particularly supported by the European States – is the exclusion of inventions from patentability where it is necessary to "protect *ordre public* or morality, including to protect human, animal or plant life or health or to avoid serious prejudice to the environment" (Article 27.2, see section 6.3 below). The second exception provides that Members are not obliged to grant patents on plants or animals (Article 27.3(b)). The prohibition of patents on plant and animal varieties contained in EPC has strongly influenced that phrasing.[26]

[26] S. D. Murphy, "Biotechnology and International Law" (2001) 41(1) *Harvard International Law Journal* 47. *The Convention on the Grant of European Patents*, for examples of provisions in the EPO similar to those in TRIPS and for a discussion on the complexity of the interpretation of patent provisions relating to genetically engineered plants and animals.

However, in the absence of international jurisprudence, the interpretation of this provision will be left to domestic patent law and the interpretation of pertinent judicial bodies. Thus, while providing some exceptions, TRIPS allows Member States to provide patents or a *sui generis* system of protection over living organisms. Article 30 provides the third type of limited exceptions to the exclusive rights conferred by patents when subject to certain qualifications.[27] Furthermore, Members may permit use of the patented invention by third parties without the authorization of the patent owner in certain circumstances (Article 31).

With regard to its interpretation and implementation, Article 27 contains one of the most contentious provisions underpinning the new multilateral trade system of WTO. According to a literal interpretation of this provision, four possible options of implementation are identified: (i) Member countries can allow patents on any invention in biotechnology by not excluding plants, animals, and biological processes; (ii) Member countries have the option to exclude plants, animals, and biological processes, but not exclude plant varieties, from patentability; (iii) Member countries have the option not to patent plant varieties, i.e. to exclude plant varieties from patentability and introduce a *sui generis* system, an IPR protection of its own kind for the protection of plant varieties; (iv) Member countries can also choose the US-like solution which boils down to a double protection system whereby plant varieties are not excluded from patentability and at the same time can enjoy *sui generis*-UPOV protection. It will be observed how the wording of Article 27 obliges Member States to provide some kind of IPR protection to almost all life forms. The patentability subject-matter of Article 27.3(b) can be visualized in the following table:[28]

WTO members must provide protection for:	WTO Members may exclude from patent protection:
Micro-organisms	Plants
Non-biological processes	Animals
Microbiological processes	Essentially biological processes for the production of plants or animals
Plant varieties (by an IP system which may be patents, a *sui generis* alternative, or a combination thereof)	Plant varieties

[27] Article 30 of *TRIPS*, Exceptions to Rights Conferred: "Members may provide limited exceptions to the exclusive rights conferred by a patent, provided that such exceptions do not unreasonably conflict with a normal exploitation of the patent and do not unreasonably prejudice the legitimate interests of the patent owner, taking account of the legitimate interests of third parties."

[28] G. Dutfield, "Sharing the Benefits of Biodiversity: Is There a Role for the Patent System?" (2002) 5 *Journal of World Intellectual Property* 899.

A closer look at Article 27.3(b) reveals that, although it allows countries to exclude plants and animals from patentability, TRIPS requires that all countries provide patent protection on micro-organisms and non-biological and microbiological processes. A serious disconnect exists between the patentable subject-matter in Article 27.3(b) and life forms that may be excluded from patent protection because a patentable subject-matter has no commonly accepted definition in international patent law. The patent systems that have started to patent inventions based on these life forms (US, Europe, and Japan) differ in their interpretation of the patentability of the subject-matter. So, depending on how it is defined, a plant cell can be considered a micro-organism even though it can grow into an entire tree. A patent on such a cell could extend to trees even if one cannot patent a plant variety. It has to be noted that in scientific practice the idea of fixed definition of "micro-organisms" is inherently flawed since scientific classification is continually evolving.[29]

Various constitutions affirm (e.g. Article XII sec. 2 of the Constitution of the Republic of the Philippines)[30] that the State is the owner of all "flora and fauna" and "with the exception of agricultural lands, all other natural resources shall not be alienated." In this case, allowing IPRs over fauna or flora can be seen as a form of alienation since IPRs are exclusive monopoly rights that give the IPR holder the right to prevent others, including citizens and possibly the government of the state, from using or producing the subject-matter of the patent.

It is also true that the language of Article 27.3(b) is open to wide interpretation. For most DCs, it is not clear how TRIPS distinguishes plants, animals, and micro-organisms which must be patented from those that are exempt; nor is it clear why essentially biological processes do not have to be patented, but microbiological and non-biological processes do. After all, a microbiological process is a type of biological process. For example, one common microbiological process uses an engineered gene to modify a biological product; more often than not the resulting product is new, involves an inventive step, is capable of industrial application, and

[29] M. Adcock and M. Lewelyn, *Microorganisms, Definition and Options under TRIPS* (Quaker United Nations, Geneva, November 23, 2000) (discussing the reluctance of the EPO to introduce a fixed definition because "it does not seem expedient to introduce such a definition as the rapid evolution in the field of microbiology would necessitate its frequent updating").

[30] Constitution of the Republic of the Philippines, Article XII, paragraph 2 (1987) www.chanrobles.com/philsupremelaw1.htm ("all lands of the public domain, waters, minerals, coal, petroleum, and other mineral oils, all forces of potential energy, fisheries, forests or timber, wildlife, flora and fauna, and other natural resources are owned by the State"); Constitución de la República Bolivariana de Venezuela, Article 124 www.constitucion.ve/constitucion.pdf (prohibiting the registration of patents over GRs).

is thus, apparently, a patentable invention under Article 27. It is unclear whether or not the fact that genetic engineering produces a life form is grounds for denying IP protection, since Article 27 excludes only "plants and animals other than micro-organisms," and genes are not whole plants or animals. Similarly, cell lines are derived from organisms using microbiological processes. So one can wonder at what point an engineered gene or cell does become a "plant" or "animal" rather than a patented biological component or the result of a patented "microbiological process." States may argue that the plant or animal that is the end product is not patentable, but the dividing line is not clear.[31]

Another approach States may take is to invoke *ordre public* or morality in order to deny such an IPR. Can this denial be justified under the TRIPS Agreement with respect to a gene used to create a vitamin-enriched food product, where there is no scientific basis for regarding the gene as harmful to human health or the environment? A tentative answer to this complex question will be provided in section 6.3.

It is evident that the inherently unclear language of Article 27 of TRIPS results from painstaking negotiations on a wide number of IP issues. It cannot provide precise guidance as to the detailed application of the treaty to biotechnological patents. At the same time it has certainly influenced the attitude of many DCs towards transnational biotechnology corporations.

2.3 Economic considerations on biotech-patents and their interaction with traditional knowledge[32]

Before entering into the analysis of the impact of Article 27 of TRIPS, it is crucial to briefly address some of the fundamental issues related to the economics of patents on life forms since policy considerations depend on the economic impact of patents.

The classic economic approach to justifying patents maintains that they are necessary to provide an incentive for innovations and for the diffusion and disclosure of these innovations. Without the protection provided by the patent system, inventors would have less incentive to disclose their inventions and would have incentives to keep them secret in order to preserve their economic benefits against competitors. With disclosure,

[31] D. Leskien and M. Flitner, "Intellectual Property Rights and Plant Genetic Resources: Options for a *Sui Generis* System" (1997) *Issues in Genetic Resources No. 6* (International Plant Genetic Resources Institute) 18–22.

[32] I owe special thanks to David Newell, for his basic relevant research on this matter while he was performing a short internship in Geneva under my direction, in the summer of 2005, during his Master in Public Policy studies at Brigham Young University.

patents help avoid the wasteful duplication of innovation efforts and instead channel resources towards unexplored areas of technology. Furthermore, patents provide instrumental incentives for the commercialization of innovative products and processes.

The application *ipso facto* of the justification of patentability of life forms is more complex, however.[33] The complexity of the debate increases when one approaches the patentability of PGRs involving an industrial corporate entity that transforms the PGRs and its related TK from another provider country.

The current form of IPRs was created by industrialized countries that traditionally employ formal and laboratory (or industrial) methods of scientific R&D of technologies. Other cultures do not take the same philosophical and moral view concerning property rights and IP as do industrialized societies. The lack of understanding between Northern, industrialized countries and least developed countries has led, in large part, to the exclusion of moral and philosophical views of holders of TK on the utilization of plants. Besides the treaties regulating this matter, equity, as a secondary source of international law, may have particular bearing upon the attempts at protecting TK in the context of the North–South or, in general, industrialized countries and DCs debate.[34] The concept of equity has been translated in the TRIPS Agreement through the objective of Article 7 of fostering a "balance of rights" by generating the welfare and incentives of holders of TK on the use of plants and industrialized corporations. Given the current situation, it appears that TK on the utilization of plants is at a distinct disadvantage in the market of IP. Such an approach will provide for long-term viability and sustainability of the international patent system in the field of biotechnology.

The following sections examine the considerations and the criteria for evaluating proposals that seek to create this balance, given current constructs and laws relating to international IPRs. The balance of rights is to reach economic welfare. Biotech-patents have particular economic justifications and implications on the international market that are worth examining before delving into the legal matters.

[33] J. Golden, "Biotechnology, Technology Policy, and Patentability: Natural Products and Invention in the American System" (2001) 50 *Emory Law Journal* 101.

[34] G. Abi-Saab, 'Cours général de droit international public' (1987) 207 *Le Recueil de cours de l'Académie de Droit International* 189–90 (1987) ; M. Rubino Sammartano, *International Arbitration Law and Practice* (Kluwer, The Hague, 2001) 457–58; G. T. Castillo, "Whose Ethics and which Equity?: Issues in the Conservation and Use of Genetic Resources for Sustainable Food Security", in International Plant Genetic Resources Institute (ed.), *Ethics and Equity in Conservation and Use of Genetic Resources for Sustainable Food Security*, Proceedings of a workshop to develop guidelines for the CGIAR (Foz do Iguaçu, Brazil, 1997) 19–31.

2.3.1 *The economic rationale of patent law*

Patents offer legal protection for inventions, i.e. for solutions to specific problems in a given field of technology. But patents are first and foremost a tool for achieving certain economic goals. After a patent is granted, the owner has a monopoly over the commercial exploitation of the idea/information for a limited period of time. The stated purpose of a patent is to stimulate innovation by offering higher monetary returns than the market otherwise might provide.[35] However, the incentive is usually secured by providing the author of the innovation a property right over it, i.e. the grant of a position of exclusivity whereby no person or entity may exploit the innovation itself without the prior authorization of the patent holder (who may be the author himself or his successors and assignees).

Another purpose of patent law is to promote the dissemination of the outcomes within the technological community. This incentive is secured by providing for (i) immediate disclosure of the innovation, by filing with a public record as provided by law; and (ii) the provision of a final term of the grant, after which the innovation falls in the public domain, i.e. may be exploited by anybody without authorization.[36] This justification is otherwise known as the "social contract" theory:

> society makes a contract with the inventor by which it agrees to grant him the exclusive use of the invention for a period and in return the inventor agrees to disclose technical information in order that it will later be available to society.[37]

The classical IP scholarship has crafted each protection according to the principle of "allocative efficiency" – according to which the long-term benefits flowing to society from the protection granted to a particular class of creators or innovators outweigh the (mainly short-term) costs imposed by the monopolistic structure of the grant itself.[38] The "mainstream legal literature" has also applied such a standard IP question to the field of biotechnology as well.[39] This scholarship is "conventionally analytical, cost- and benefits-oriented."[40]

[35] F. Abbott, T. Cottier and F. Gurry (eds.), *The Intellectual Property System: Commentary and Materials* (Kluwer, The Hague, 1999) Part I, 25.

[36] R. Mazzoleni and R. R. Nelson, "Economic Theories about the Benefits and Costs of Patents" (1998) 32 *Journal of Economic Issues* 1033 and 1040, that refers to a disclosure function and dissemination function.

[37] T. Penrose, *The Economics of the International Patent System* (The Johns Hopkins Press, Baltimore, 1951) 32.

[38] P. Torremans and J. Holyoak, *Intellectual Property Law* (Oxford University Press, 2006) 17, 20.

[39] Cullet *et al.*, 'Intellectual Property Rights, Plant Genetic Resources and Traditional Knowledge', 119–25.

[40] M. Ricolfi, "Biotechnology, Patents and Epistemic Approaches" (2002) *Journal of Biolaw & Business* 77.

Patent rights on inventions are considered as a "natural right." Moreover, certain pragmatic considerations justify this legal monopoly because such rights promote the public interest. Pragmatic justification rests upon the "exchange-for-secrets" or the "monopoly-profit-incentive" concepts: an inventor discloses the results of his efforts to the public; in exchange, he is granted a monopoly-like privilege for a limited period of time with respect to the commercial use of his invention. In this way, the State provides the inventor with a competitive advantage in consideration of the disclosure of the inventor's intellectual achievement.

Arguments against the patentability of biotechnology often fail to consider the negative effects of exclusion of this science from patent protection. If biotech inventions are excluded from patentability as some interest groups would like, this does not mean by itself exclusion from legal protection. A technical solution, which happens not to be patentable, may still be protected *de facto* but also *de jure* as secret know-how and under trade secret law.[41] A ban on patenting may, therefore, fail to discourage innovative activity in the relevant field while inducing innovators to seek the alternative protection provided by trade secret law, which is hardly a desirable outcome since it would decrease rather than increase public scrutiny of novel technologies, and especially those that concern the building blocks of life.

The investment flow in biotechnology is also a reason for the patent protection. Companies indeed invest in R&D only if they know that there are legal means that will prevent the newly generated knowledge from being disseminated to competitors and into the public domain. Economists define this problem as the "incomplete appropriability of knowledge."[42] Patents provide a solution since they allow the investing firm to appropriate a return on its investment in R&D by protecting its invention against unauthorized duplication.

Yet, a certain level of public control over the patentability practice of biotech science is needed from the economic and ethical points of view (see section 6.3 below)

2.3.2 *Inherent economic hardships in including traditional knowledge in the economics of the intellectual property system*

In seeking to establish a better balance for the current IPRs system, the scope of policy effects, both normative and descriptive, should be

[41] Merges *et al.*, *Intellectual Property in the New Technological Age*, 27–53.

[42] K. J. Arrow, "The Rate and Direction of Inventive Activity: Economic and Social Factors", in *Economic Welfare and the Allocation of Resources for Invention*, (Princeton University Press, National Bureau Economic Research, Princeton, New Jersey, 1962) 609.

addressed. It should be noted that this approach to establishing criteria will hold global economic welfare as the ultimate criteria, as opposed to pursuing the economic welfare of either TK holders or corporations in industrialized countries, thus providing for the long-range viability and sustainability of a given solution. A solution that favors the economic welfare of one group over that of another, as it has arguably been the situation to date, will not provide the moral justification necessary for viability and sustainability. Descriptively speaking, the best solutions should seek to maximize the economic welfare of all nations as well as that of their internal groups of ethnicities, races, social divisions, and other social groups in all economic sectors.

The complications unique to TK regarding the use of rights in PGRs are threefold.

First, a solution to the problem of exploitation of TK by corporations in industrialized countries has to be found for those seeking to maximize global economic efficiency and welfare. Solutions that are external to market forces are necessary to correct for the market failure that has put TK on the use of plants at such a distinct disadvantage. This in and of itself frames the debate in a manageable form; yet there are additional factors that complicate the process.

The second issue is that the inherent nature of TK and the cultures surrounding it do not incorporate concepts of IP in the same way as do the cultures that created the very constructs that govern current international IPRs. This dilemma creates what could potentially be a deadlock. A satisfactory solution to the current imbalance will be one that not only satisfies the need of adapting the current international IPRs system to potentially provide TK with property protection, but also provides for the inherent problems associated with the mere existence of a property protection system; a system not directly compatible with the cultures and norms of TK. Such a solution must provide for some degree of cross-cultural compatibility, while recognizing that perfect compatibility is probably not achievable.

The third complication in deriving a solution for TK in the IP marketplace is a consideration of the market power of industrial corporations from the North (i.e. the US). The inherent market power, which is a function of the comparatively vast resources of industrial parties, must be somehow brought into check if TK holders are to gain the "leg up" they need in order to start and gain initial market power in the IP market place. Without such protection, even providing for the potential leveling of the playing field through constructs and administration will not sufficiently compensate for the lack of resources that prevents TK from engaging in the IP marketplace. This third complication must be given due note by any potential balancing solution.

In light of these general considerations on the economic underpinnings of patent law on biotechnological innovation made on the basis of TK related to PGRs, the major questions that arise are (i) how TK holders fit into the economic rationale of patents, and (ii) whether the application and adjustment of the international patent system in the case of TK holders are justified. As has been noted above, the scope, or in other words, the breadth and the height of patents granted, must be taken into consideration when weighing the economic costs and benefits of issuing a patent.

There are no current policy measures to modulate the height and the breadth of patent protection. In the economy of the relations between technologically patented knowledge and informal TK, a too broad biotech-patent that uses the TK and the GR held by a community in a provider country will limit the ability of TK holders to use their own knowledge and even to innovate on the basis of their own GR. The claims of the patent may be so broad as to exclude the community from developing a similar product on the basis of that GR and TK.

This product could potentially be cheaper and more accessible to populations local to the TK holders as well as to the global population, because of the generally lower labor costs in the areas where indigenous groups live. Thus, both regional and global economic welfare could be adversely affected by biotech patents that are too broad. The monopoly power granted to patent holders should be limited in such a way as to encourage TK holders to participate in creating imitations or closely related solutions that biotechnology presents. If the monopoly power granted to a patent holder is too broad, the deadweight loss problem associated with all monopoly power in the market place will be exacerbated. This process will create inefficiency in the production of biotechnological products to the detriment of the general economic welfare. The vast biological diversity present in the environments from which TK holders derive their knowledge and resources underscores the need not to exclude TK holders from practicing in the global or even regional and local marketplaces due to their great potential for contribution to global economic welfare.

In addition, the scope of patents' height constitutes another source of the problem. Patents whose scope is very high prevent improvements or modifications of the patented technology. This impediment has implications for those who seek to improve upon existing technologies. Again, the problem here is the inefficiency caused by the deadweight loss associated with the monopoly power granted by patents. If the patents' scope is too high, then TK holders will have little incentive to improve upon their TK. If the scope of patents is limited in height, then there will be greater incentives and possibilities for TK holders to improve upon existing technology. This limitation of the patent's height comes at the risk of de-motivating the original innovators from creating and patenting new technologies. An optimal solution will balance the incentives of original innovators to innovate while still motivating TK holders to improve upon existing technologies.

When taking into consideration both the height and the breadth of the scope of patents in biotech, two solutions have been proposed to minimize the impact of monopoly-induced deadweight loss from patent granting. The first is to issue patents that are low and broad in scope. There are patents that are easy to improve upon but difficult to imitate. Because of the limits on imitation, the original patent owner will be able to recoup resources expended on producing the new technology and thus be motivated to keep innovating. Those who seek to improve upon existing technologies will have a freer hand in doing so than if the patent had been broad in scope. The second approach to optimizing economic benefits is to issue patents that are high and narrow in scope. These are patents that are easy to imitate but difficult to improve upon. Because of the limits on improving the current technology, the original innovator will be able to recoup investments in producing the new technology and thus continue innovating. On the other hand, those seeking to create innovative solutions that perform functions that are similar to those of existing, patented technologies will have a freer hand in doing so, thus stimulating innovative imitation of technologies and decreasing the costs to society.

2.4 Conclusion

This chapter has observed how the TRIPS Agreement has opened up new commercial opportunities through IP protection of biotechnology in all countries. In spite of its incomplete construction and its inherent vagueness, Article 27 of TRIPS can have powerful effects on international commerce. However, these powerful effects are fraught with controversies aggravated by the fact that many genetic "inventions" claimed in the North derive from TK from the South with allegations of misappropriation. The industrialized countries' companies have sophisticated technologies – such as genetic engineering – to extract value from biodiversity. Through patent protection on life forms, major transnational corporations, e.g. Monsanto/Car, can take genes from, for instance, a butterfly in the fields, forests, or coastal waters of DCs, then manipulate them in their labs and obtain patent protection thereon. Consequently, the mere transformation of resources in the laboratories of industrialized countries has resulted in a public outcry from DCs because corporations are paid royalties based on the DCs' preserved resources and TK. The economic considerations on biotech-patents and their interaction with TK have marked the necessity to rebalance the rights and obligations of the partners in this transnational trade system.

The internationalization of the patentability of GRs that the TRIPS Agreement has generated has brought about an unbalance of rights and obligations both of non-State actors (transnational corporations, indigenous and local communities) and of States. Therefore, the international community has adopted other international legal instruments to counterbalance the potential negative consequences of unequal benefit sharing. The potential controversial relationship between the TRIPS Agreement and other relevant treaties on GRs and TK will be addressed in the following chapter.

3 The relationship between the TRIPS Agreement and treaties protecting genetic resources and traditional knowledge

This section analyzes the legal impact of the TRIPS Agreement on the implementation of other related international legal instruments in the field of environmental law and human rights. In order for States to implement all the applicable international law, these treaties need to be analyzed and interpreted. Accordingly, one cannot study the field of IP in clinical isolation from other fields of international law as well as diplomatic and international political relations. The very existence of Article 27 of TRIPS has evoked strong criticism by NGOs and indigenous communities in DCs that try to resist the alleged practice of "biopiracy" and market control over biodiversity through industrialized countries' use of the patent system. This campaign has resulted in a diplomatic strategy to modify the international norms of patentability contained in the TRIPS Agreement and other patent treaties.[1] In sum, the rhetoric used by TK holders and some DCs is that they are having to pay for IP protected products, including royalties, that are based on the use of their own resources and knowledge.

The differences between the IP and environmental legal regimes dealing with this subject-matter are wide. A patent under TRIPS requires novelty, whereas community rights under the CBD are founded on pre-existing rights to biodiversity and associated knowledge. IPRs on biodiversity-related "inventions" are therefore dependent upon the prior "rights" of communities. TK holders argue that these latter rights are undermined by the very existence of the rights detailed in TRIPS and inspired by "myopic" industrial interests. They also assert that the implementation of TRIPS will systematically negate the wider historical contribution made by communities in DCs to the planet's biodiversity. Through the adoption of Article 27 of the TRIPS Agreement, Northern countries have embarked on a rapid and spectacular race to engage Southern countries in international obligations on IP

[1] According to WIPO, citizens and corporations of industrialized countries hold 95 percent of the patents in Africa, almost 85 percent of those in Latin America, and 70 percent of those in Asia. WIPO, data set IP/STAT/1994/B, released in November 1996.

protection for biotechnology when they themselves have yet to define clear guidelines within their own IP systems to assure protection of biodiversity. States have preferred to deal with the protection of the rights of TK holders in fora that have not yet comprised WTO or TRIPS Agreements. Various treaties in the field of international environmental law are creating an alternative system of protection of PGRs that, in some aspects, may conflict with TRIPS mandated obligations: hence, the necessity of exploring the major options for the revision of Article 27 within the TRIPS Council of WTO to render it more compatible with MEAs dealing with the same subject-matter. This legal analysis attempts to convey an overall message of extreme caution when dealing with the delicate question of monopolized private ownership of the building blocks of life.

3.1 The impact of the TRIPS Agreement on CBD obligations

This section provides an overview of the legal relationship between TRIPS and non-IP treaties that present provisions that can affect IPRs. Vice versa, States' obligations of environmental law are highly affected by the exercise of IPRs. This section analyzes some fundamental legal and political alleged conflicts arising from the implementation of the CBD and the TRIPS Agreement. They form with ITPGRFA a triad of treaties that weaves the network of obligations of States and international organizations.

3.1.1 Principles of the CBD

The Rio de Janeiro CBD aims at setting up an international framework for the preservation and utilization of the world's biological resources. The CBD[2] is a result of prolonged international pressure on nations to respond to the destruction of, and unequal profits from, the biodiversity of the South. After years of debate, the CBD was completed in 1992.[3] It came into force in 1993, and today 188 States have ratified it.[4]

The CBD represents an important watershed in international efforts to promote biodiversity conservation. In the first place, the Convention binds signatories to a number of basic principles regarding how, by whom, and for whose benefit biodiversity should be conserved.[5] The CBD codifies the well-established principle of international law that States have a sovereign right over their territory, including their natural resources.[6] Before the CBD

[2] CBD. [3] *Ibid.* [4] *Ibid.*

[5] F. McConnel, *The Biodiversity Convention: A Negotiating History* (Kluwer, The Hague, 1996).

[6] C. M. Correa, *Sovereign and Property Rights over Plant Genetic Resources, FAO Background Study Paper No. 2* (Commission on Plant Genetic Resources – First Extraordinary Session, Rome, 1994).

codified the principle of a State's sovereignty over its biological resources, most States affirmed this principle in their constitutions, typically stating that the State owns all "flora and fauna" and that all other natural resources, with the exception of agricultural lands, shall not be alienated. It is maintained that if States allow IPRs over flora or fauna, this is a form of alienation because IPRs by their nature are exclusive monopoly rights that prevent others from producing the patented flora or fauna. Before the CBD's adoption, many questioned whether biological resources were included in the legal category of "heritage of mankind," (see sections 3.3.1 and 3.3.2 below) or whether States lacked the ability to exercise sovereignty over biological resources and to subject GRs to private property rights.[7]

The shift to the ideas propounded by the CBD came from an increasing commercial interest in biological and GRs and a desire to subject such resources to private property claims, namely IP. Much of the movement came in the form of PBRs and patents gave their owners an exclusive right to control any commercial use of these resources.[8] Amidst the global pressure to privatize biological resources, the CBD stands as an important watershed in international efforts to promote biodiversity conservation. For instance, the Convention binds signatories to a number of basic principles regarding how, by whom, and for whose benefit biodiversity must be conserved.[9]

Article 1 of the CBD states its overall objectives. These objectives include first, the "conservation of biological diversity"; second, the sustainable use of biological diversity components; and, finally, the "fair and equitable sharing of the benefits arising out of the utilization of genetic resources."[10] The CBD recognizes the sovereign rights of States over their biological resources in Articles 3 through 15.[11] Article 3 recognizes that "States have the sovereign right to exploit their own resources," including, under Article 2, biological and genetic resources of actual or potential value.[12] Article 8(j) requires Contracting States to:

respect, preserve and maintain *knowledge, innovations and practices of indigenous and local communities embodying traditional lifestyles* relevant for the conservation and sustainable use of biological diversity and promote their wider application with the approval and involvement of the holders of such knowledge, innovations and

[7] C. Joyner, "Legal Implications of the Concept of the Common Heritage of Mankind" (1986) 35 *International and Comparative Law Quarterly* 190.

[8] B. L. Kagedan, *The Biodiversity Convention, Intellectual Property Rights, and Ownership of Genetic Resources: International Developments*, prepared for Industry Canada, International Property Policy Directorate (Industry Canada, Ottawa, January, 1996); D. Janzen, "A New Lease on Life", in W. Reid *et al.* (eds.), *Biodiversity Prospecting: Using Genetic Resources for Sustainable Development* (World Resources Institute, Washington DC, 1993).

[9] F. McConnel, *The Biodiversity Convention. A Negotiating History* (Kluwer, The Hague, 1996).

[10] CBD, 823. [11] *Ibid.*, 824–29. [12] *Ibid.*, 823–25.

practices and *encourage the equitable sharing of the benefits arising from the utilization of such knowledge, innovations and practices.*[13] (italics added)

Article 15 discusses the details of regulating access to GRs through increased transparency in the patent application process (e.g. including the submission of information about the origin of the PGR). The first paragraph gives States sovereign rights over their resources and confers on them the "authority to determine access to [their] genetic resources."[14] Paragraph 4 allows access to GRs, subject to "mutually agreed terms," while paragraph 5 specifies that the same access "shall be subject to PIC of the Contracting Party providing such resources."[15]

Additionally, Article 15 states that the transfer of technology is an invaluable instrument for the effective implementation of the CBD,[16] and lists a set of rights conferred on provider States. However, the CBD also provides symmetric obligations on the recipient State; for instance, paragraph 7 of Article 15 provides that each contracting party:

shall take legislative, administrative or policy measures, as appropriate, and in accordance with Articles 16 and 19 [...] *with the aim of sharing in a fair and equitable way the results of research and development and the benefits arising from the commercial and other utilization of genetic resources* with the Contracting Party providing such resources. Such sharing shall be upon mutually agreed terms.[17] (italics added)

This provision establishes access to the biological resources of the provider country on a *quid pro quo* basis, and it lists a set of rights conferred on provider States. Consequently, it contains the potential to foster transfer of technology from the industrialized countries to DCs.

Finally, paragraph 5 of Article 16 asserts that IPRs must not conflict with the conservation and sustainable use of biodiversity.[18] Therefore, the CBD not only gives rights to provider States, but also regulates the transfer and interaction between provider and recipient States.

3.1.2 General considerations on the legal relationship between TRIPS and the CBD obligations

Access to GRs – from which genetically engineered products are developed – is becoming one of the most critical areas of debate between industrialized and DCs.

Many DCs regard the relationship between TRIPS and the CBD as one of opposing principles. On the one side stands the principle of economic growth propounded by the TRIPS Agreement. On the other side is the principle of sustainable development served by the CBD. Industrialized

[13] *Ibid.*, 826. [14] *Ibid.*, 828. [15] *Ibid.* [16] *Ibid.* [17] *Ibid.* [18] *Ibid.*, 829.

countries seek to globalize and harmonize IPRs because such rights will assist innovations in getting to the market. They argue that economic growth will result from improving dynamic efficiency through stronger IPRs. Pushing markets towards the high "technology fix," however, stands in stark contrast to the kind of economy advocated by committed environmentalists who believe that States should regulate development to take into account environmental costs and effects.

From a strictly international law perspective,[19] the alleged conflict between TRIPS and the CBD does not exist, as will be observed throughout this study. The alleged conflict between these two treaties is rather spurred on by moral and rhetorical assumptions. For instance, one assumption is that the patent regime is a Western form of IPR that is totally unsuitable to the majority of societies in the South that have accepted TRIPS by acceding to the WTO. Another assumption is that private rights are completely alien to indigenous communities because the vast majority of their farmers, who manage biodiversity at the local level, are accustomed to collective rather than private rights.

The CBD is intended to strengthen DCs' capacities to conserve and use biological diversity on a long-term basis by reserving all rights over those resources to the DCs and by including the right to enjoy the benefits of their resource base. Southern hemisphere countries feel consistently exploited because of the structural imbalances between countries rich in biological diversity and those with strong technological and legal infrastructures.

Conversely, TRIPS intends to provide private property rights over products and processes regardless of whether they are related to biodiversity. This intention is wholly aligned with the interests of many multinational companies – and through pressure by the multinationals, TRIPS' intended results have overwhelmingly been achieved.

The legal debate over the CBD and TRIPS has been highly politicized; it is indeed alleged that, as a consequence of the implementation of both international treaties, an acrimonious legal conflict may arise from inconsistent provisions contained therein.

Besides the first-glance fundamental political contradictions between CBD (which recognizes sovereign rights on GRs) and TRIPS (which obliges States to grant private rights to the same), there are a myriad of legal inconsistencies between IPRs applied to life forms under the TRIPS

[19] J. Curci, 'The New Challenges to the International Patentability of Biotechnology: Legal Relations Between the WTO Treaty on Trade-Related Aspects of Intellectual Property Rights and the Convention on Biological Diversity' (2005) 2 *International Law and Management Review* 16 ff.

Agreement and the obligations of CBD. Four types of areas of inconsistency between TRIPS Agreement and the pre-existing CBD obligations should be identified: (i) access to GRs, (ii) fair and equitable sharing of benefits from using GRs, (iii) respect for TK held by the indigenous communities, and (iv) transfer of technology.

These points of apparent conflict should be understood bearing in mind that contracting parties to the CBD have an obligation to cooperate and ensure that IPRs are "supportive of and do not run counter to [the CBD's] objectives."[20] Moreover, Article 22 of the CBD states that its provisions will not affect countries' rights and obligations to "any existing international agreement, except where the exercise of those rights and obligations would cause a serious damage or threat to biological diversity."[21] Article 22 continues stating that this harmonization process between TRIPS and CBD is mainly "subject to national legislation and international law" and stands as a basis for countering the runaway march of IPR regimes.[22]

When a conflict exists between two treaties dealing with the same subject-matter, the applicable rule is *lex posterior derogat legi anteriori* (the later law prevails over the earlier), enshrined in Article 30.2 of the VCLT.[23] Under this rule, TRIPS will prevail since it came into force after the CBD. However, on a *prima facie* reading of the treaties and interpreting the treaties narrowly (*stricto sensu*), the subject-matter of the CBD and TRIPS differ fundamentally; therefore, States should fully and simultaneously implement both of them. For instance, although both Article 27 of TRIPS and some of the provisions of the CBD deal with the utilization of biological resources, they do so to achieve two different objectives that are not necessarily mutually exclusive.[24]

Although TRIPS single-mindedly deals with IP and does not address environmental issues *per se*, some provisions regulate the same object and have the same purpose as CBD provisions. In order for both treaties to be fully applied and acceptable so that all countries will ratify them, certain provisions contained in both treaties need to come into harmonization.

Maljean-Dubois defines the controversial relationship between these two international instruments as an apparent conflict (namely an *emboîtement* or *désarticulation*) rather than an incompatibility; he posits that a relationship of

[20] CBD, 829. [21] *Ibid.*, 832. [22] *Ibid.*, 829.

[23] Article 30 of the VCLT states: "When a treaty specifies that it is subject to, or that it is not to be considered as incompatible with, an earlier or later treaty, the provisions of that other treaty prevail."

[24] C. Noiville, "Biodiversité et propriété intellectuelle. L'impossible conciliation?", in F.-D. Vivien (ed.), *Biodiversité et appropriation: les droits de proprieté en question* (Elsevier, Paris, 2002).

complementarities has yet to be developed.[25] Such complementarities, however, can be realized through adequate interpretation of all the obligations under both and through further legislative work to harmonize the two treaties for the benefit of the international community.

In spite of the fundamental political contradictions that seem to exist between CBD and TRIPS, legally speaking, inconsistencies between IPRs applied to life forms under TRIPS and the obligations of CBD are multi-faceted. The inconsistencies particularly reveal themselves in the following fields: the access to, and fair and equitable sharing of, benefits from the use of GRs; the respect for TK held by the indigenous communities, and the transfer of technology.

The alleged inconsistencies between TRIPS and the CBD reside at the crossroads between the opposing perspectives of North and South. This complex debate is intensified by equating the opposition between the North and South as the opposition between the "haves" and the "have-nots."[26] Taking a more optimistic approach, one can find a viable solution potentially able to resolve the North-South conflict by realizing that each side has something to give and something to gain by cooperation: industrialized countries have technology and biotechnological manufacturing capabilities, and DCs are richly endowed with biological diversity which is lacking in most industrialized countries.

The private property regime on biological diversity established by TRIPS may undermine the implementation of the benefit-sharing provisions of the CBD that require the knowledge or material holder's PIC for the use of GRs and TK.[27] TRIPS does not require transparent demonstration of PIC and is therefore inconsistent with the CBD in that regard. Without such a PIC obligation in TRIPS, private entities from countries (generally industrialized ones) that use GRs in innovative processes will limit their efforts to seek and exploit benefit sharing with the countries of origin (generally developing ones). Industrialized countries generally believe that the best way in which both treaties can be complied with is through leaving the parties (such as private entities and provider countries) to freely negotiate contractual solutions of benefit sharing and PIC.

The aim of TRIPS, to homogenize national IP regimes, may jeopardize a country's freedom to choose the way it wants to deal with the use and protection of its biodiversity and the related TK. The process of the SPLT

[25] S. Maljean-Dubois, "Biodiversité, biotechnologies, biosécurité: le droit international désarticulé" (2000) 127(4) *Journal du Droit International* 966–67.

[26] J. F. Badimboli Atibasay, 'The International Legal Regime for Biotechnology Patenting: An Appraisal from the Standpoint of Developing Countries' (2001) 31 *Revue générale de droit* 291–325.

[27] CBD.

at WIPO goes even further in the details of the fixation of patent legal concepts. This issue most conspicuously arises when companies or consumers acquire genes from a company in a State that manipulates, patents, and sells a genetically modified biological product rather than acquiring the genes from the State that asserts rights to them: rights to sovereign control to benefit sharing, or to PIC through the CBD. Communities of DCs have risen up against this kind of "piracy" of indigenous and local community knowledge through the imposition of IPRs on life forms and related knowledge.[28]

The well-known phenomenon of "bioimperialism"[29] or "biopiracy" describes the way in which industrialized countries "conquer" biological resources illegitimately. This strong terminology has a history. Industrialized countries have accused DCs of pursuing "intellectual piracy," and after the adoption of TRIPS, DCs have accused industrialized countries of "biopiracy." DCs coined this term as part of a counter-attack strategy to describe the misappropriation of GRs by private entities in the North. These DCs felt they were no more intellectual pirates than are corporations that acquire resources and TK from their countries and use these TK and GRs in their R&D programs by acquiring patents and other IPRs without compensating the provider countries and communities.[30] However, such rhetoric did little to prevent the legalization of this alleged "biocolonialism" or "conquest." Through TRIPS, the South has an obligation to grant patents, trademarks, and trade secrets without any compensation to the local communities that preserved and bred the biological resources.[31]

It is argued that IPRs can prevent countries from realizing the full and practical meaning of the CBD articles regarding national sovereignty over their natural resources and the rights of their local and indigenous communities.[32] This prevention frustrates the ultimate goal of fairly distributing the benefits arising from the use of GRs situated in the contracting parties' territories.[33]

[28] RAFI, *Enola Bean Patent Challenged* (January 5, 2001), www.etcgroup.org/documents/news_enolabean.pdf.

[29] K. Aoki, "Neocolonialism, Anticommons Property, and Biopiracy in the (Not-So-Brave) New World Order of International Intellectual Property Protection" (1998) 6 *Indiana Journal for Global Legal Studies* 11.

[30] P. Mooney, "Why I Call It Biopiracy", in H. Svarstad and S. S. Dhillion (eds.), *Responding to Bioprospecting: From Biodiversity in the South to Medicines in the North* (Spartacus Press, Oslo, 2000) 37–43.

[31] T. Cottier, "The Protection of Genetic Resources and Traditional Knowledge: Towards More Specific Rights and Obligations in World Trade Law' (1998) 1 *Journal of International Economic Law* 555 (this author states that some 90 percent of genetic information and TK are found in DCs).

[32] CBD, 825–26.

[33] CBD, 828–30 (CBD Articles 1, 15.7, and 19.2 present the relevant provisions on benefit sharing relating to the conservation of TK and the conservation of biodiversity contained

I have observed that there are many differences between TRIPS and the CBD that complicate the integration of the two agreements to address the concerns of biopiracy or biocolonialism. Under TRIPS, the patent right is a private right held by a non-State actor; the rights to control biodiversity are public rights that reside in the State. Thus rights and obligations under TRIPS and the CBD are different types of rights owned by different types of entities but are applied to the same subject-matter. The TRIPS Agreement led provider countries to dramatically change the way in which their GRs are accessed by private entities.

3.1.2.1 The impact of TRIPS on access to genetic resources

The TRIPS preamble defines IPRs as private rights. Because these rights are subject to the general WTO principle of national treatment (which requires States to give citizens of other nations the same privileges it grants its own people), the implementation of TRIPS Article 27.3(b) will affect private and individual property rights globally. The global scope of these rights may destabilize the national sovereignty espoused by the CBD, which aims to recognize the inherent rights of indigenous and local communities.[34] Although Article 15.1 of the CBD recognizes "the sovereign rights of States over their national resources" and the national governments' ability to determine access to GRs, the provision does not refer to the question of the ownership of these resources. The CBD simply submits access to GRs to the PIC of the party on mutually agreed terms aimed at sharing the benefits arising from the utilization of such resources.[35] Not only can firms find GRs within the territory of States, but also within germplasm and seed banks.

The CBD, in dealing with "the *ex situ* conservation of components of biological diversity,"[36] leaves legal issues on the ownership of biological resources held in trust in gene banks unanswered. Therefore, biopiracy has benefited from a loophole in the legal status of materials held by gene banks like the Consultative Group on International Agricultural Research (CGIAR). The FAO ITPGRFA[37] continues to try to close this loophole through the creation of the Multilateral System for PGRs. According

in Articles 8(j) and 10(c)). For some socio-economic consideration, K.A. Goldman, "Compensation for Use of Biological Resources Under the Convention on Biological Diversity: Compatibility of Conservation Measures and Competitiveness of the Biotechnology Industry" (1994) 25 *Law and Policy International Business* 695; S. Prakash, "Towards a Synergy Between Biodiversity and Intellectual Property Rights" (1999) 2 *Journal of World Intellectual Property* 821.

[34] CBD, 829. [35] CBD, 827–8.

[36] *Ibid.*, 826. *Ex situ* means "outside the place," i.e. "conservation of a plant outside of its original or natural habitat, such as in a gene bank or greenhouse." CGIAR, *Future Harvest Center Glossary*, www.futureharvest.org/about/glossary.shtml#e.

[37] *The FAO International Treaty on Plant Genetic Resources for Food and Agriculture (ITPGRFA)* (November 3, 2001), http://ext-ftp.fao.org/ag/cgrfa/it/ITPGRe.pdf.

to CBD Article 15.3,[38] national authorities should provide for the acquisition, conservation, storage, and management of these *ex situ*[39] collections.[40] As opposed to the ITPGRFA, the CBD does nothing to centralize the governance of genetic material stored in gene banks around the world.

I think that, from a legal standpoint, there is no conflict between the affirmation of sovereign rights over States' GRs recognized in Articles 3 through 15 of the CBD and the recognition of private rights over the same resources in Article 27 of the TRIPS Agreement. This is so because under TRIPS the same States that adopted the CBD have expressly exercised their sovereign rights to grant IP protections to private parties for inventions based upon their own GRs in accordance with the stated conditions of patentability. Rather, conflicts between the CBD and TRIPS may arise from the conditions/requirements of patentability of innovation based upon GRs protected by the sovereign States (as discussed at section 6.2 below). TRIPS does not currently include PIC and benefit sharing and this may be found in conflict with Article 15 of the CBD.

3.1.2.2 The impact of TRIPS on the protection of traditional knowledge

Section 1.1.4 has described how TK is created, preserved, and disseminated. TK includes mental inventories of local biological resources, animal breeds, local plants, crops, and tree species. It may also include such information as which trees and plants grow well together and which plants are indicator plants, plants that indicate soil salinity or that flower at the beginning of the rains. It also includes practices and technologies such as seed treatment, storage methods, and tools used for planting and harvesting. TK also encompasses belief systems that play a fundamental role in a peoples' livelihood and in maintaining their health and the environment.

TK is a blend of knowledge and experience integrated with a coherent world-view and value system. Therefore, TK is usually collective in nature and considered the property of the entire community. Because ownership and property rights under modern legal systems are foreign to most traditional, knowledge-based communities, many conclude that TK is *res*

[38] CBD, 828 (Article 15.3 states, "For the purpose of this Convention, the genetic resources being provided by a Contracting Party, as referred to in this Article and Articles 16 and 19, are only those that are provided by Contracting Parties that are countries of origin of such resources or by the Parties that have acquired the genetic resources in accordance with this Convention").

[39] *Ibid.*, 826.

[40] An in-depth analysis of the interaction among IPRs, the CBD, and the FAO ITPGRFA falls outside the scope of the present study.

nullius – the property of nobody – until it is discovered by explorers, corporate scientists, governments, and so on. This legal approach to TK does not take into account the fact that customary laws recognize forms of ownership separate from those designated by IP law.[41] Traditional communities view TK ownership as a responsibility rather than as an exclusive property right.

The TRIPS Agreement does not guarantee any protection of TK. TK is a fundamental source for the sustainable management of biological diversity and for the development of new and socially beneficial products through, for instance, long-term selective breeding of food crops or knowledge of medicinal plants.[42] For this reason, the CBD established Article 8(j) for the preservation of TK.

The conservation of biological resources implies enormous responsibilities. While TRIPS does not allocate these responsibilities to those who will benefit from ownership rights in these resources, under the CBD States, local communities and indigenous people are primarily responsible for the preservation of biological resources. TRIPS, instead, allocates these responsibilities to IPR holders, which only cultivates their monopoly by effectively suspending national or community sovereignty over local GRs. Consequently, governments and communities have little means of regulating access to or demanding a share of benefits in their own GRs to which they no longer own the rights.

Ultimately, the impact of IPRs on TK is not yet clear. On the one hand, it can be argued that IPRs will eventually encourage investment in the preservation of these practices. On the other hand, it can be claimed that the lack of protection of TK leads to its misappropriation by IP holders. This discussion becomes more complicated when many leaders of indigenous communities assert that the commoditization of TK is inherently unjust on religious grounds. Consequently, even IP protection of TK is considered inappropriate *ab initio*.

It can be affirmed that the TRIPS Agreement is completed by the objectives of the CBD to protect TK. Part III of this book demonstrates how IP law can still undergo an interpretation and even transformation to allow the objective of benefit sharing by using IPRs and/or *sui generis* systems apt to protect TK and to compensate its holders when their preserved GR is the basis of a patent.

[41] C. Correa, *Traditional Knowledge and Intellectual Property: Issues and Options Surrounding the Protection of Traditional Knowledge* 3 (Discussion Paper, Quaker UN Office, Geneva, 2001), www.quno.org/geneva/pdf/economic/Discussion/Traditional-Knowledge-IP-English.pdf.

[42] G. Dutfield, "The Public and Private Domains: Intellectual Property Rights in Traditional Knowledge" (2000) 21(3) *Science Communication* 276–78.

Table 2. *Synopsis of the points of conflicts between CBD and TRIPS according to the DCs standpoint.*[43]

CBD (1992)	TRIPS (1994)	CONFLICT
Nation States have sovereign public rights over their biological resources *(Preamble, Article 15.1)*.	Biological resources should be subject to private IPRs. Compulsory licensing, in the national interest, should be restricted *(Article 27, Article 21)*.	National sovereignty implies that countries have the right to prohibit private IPRs on life forms (biological resources) to maintain sovereign control. TRIPS requires the provision of IPRs on micro-organisms, non-biological and microbiological process, as well as patents and/or *sui generis* protection on plant varieties.
The use or exploitation of TK, innovations, and practices relevant to the use of biodiversity must give rise to equitably shared benefits *(Preamble, Article 18.4, Article 8.j)*.	States must provide patents for all fields of technology, including biological resources. There is no mechanism for sharing benefits deriving from the patent between a patent holder in one country and the donor of material in another country where the invention is derived *(Article 27.1)*.	The CBD gives DCs a legal basis to demand benefit sharing. TRIPS does not mention any authority for such a right for DCs.
Access to biological resources requires the PIC of the country of origin. It also requires the "approval and involvement" of local communities *(Article 15.5)*	There is no provision requiring PIC for access to biological resources which may subsequently be protected by IPR.	The CBD gives States legal authority to diminish the incidence of "biopiracy" by requiring PIC. TRIPS does not mention this authority with the risk to promote biopiracy.
States should promote the conservation and sustainable use of biodiversity as a common concern of humankind, taking into account all rights over biological resources.	The safeguarding of public health and nutrition, and the public interest in general.	CBD places the public interest and common good over private property and vested interests. TRIPS grants private rights over the same subject-matter.

[43] This synopsis has been inspired by the one effectuated by GRAIN; I have modified it in accordance to illustrate the relations between the two conventions, www.fao.org/ag/cgrfa/itpgr.htm.

3.2 The impact of UPOV on the freedom of exchange of seeds

UPOV has been promoted by some industrialized countries as the benchmark of the "effective *sui generis* system" for PVP required by Article 27.3 (b) of TRIPS Agreement. The purpose of the UPOV Convention is to ensure that the Member States of the Union acknowledge the achievements of breeders of new plant varieties by making available to them exclusive property rights, PBRs, on the basis of a set of uniform and clearly defined principles. These principles include the stability, uniformity, and distinctiveness of the new plant variety giving rise to the right.

Until recently most countries allowed farmers and other traditional breeders to be exempted from the provisions of such rights, as long as they did not indulge in branded commercial transactions of the varieties. Now, however, after an amendment in 1991 and the subsequent harmonization of the principles established in the Convention, UPOV itself has tightened the monopolistic nature of plant variety breeder rights by substantially removing the exemptions for farmers.

3.3 Some international intellectual property aspects of the FAO ITPGRFA

This section assesses the implications of the FAO ITPGRFA of 2001. Together with TRIPS and the CBD, it builds an international legal framework regulating access and exploitation of PGRs.

First and foremost, this section deals with an area that is within the purview of the FAO rather than of UNEP. Moreover, it would appear that the way the issue of access to GRs is dealt with in the IUPGRs (adopted in 1983 at an FAO Conference), on which the ITPGRFA builds, is to a certain extent at variance with the CBD approach.

A major difference between the CBD and ITPGRFA is that the ultimate goal of the CBD framework is to leave it to the interested parties to make their own contracts on matter of access to GRs and the exploitation thereof. Indeed, GRs and IPRs are seen as property rights, the value of which may be optimally assessed and exploited through market mechanisms.[44] The IUPGRs took a different approach just a decade later in the somewhat specialized field of agricultural PGRs. In fact, the IUPGR was originally predicated on the rather different assumption that PGRs are

[44] S. D. Murphy, "Biotechnology and International Law" (2001) 42 *Harvard International Law Journal* 77, who appropriately brings our attention to the several market failures which may jeopardize the workability of the resulting scheme.

part of the "common heritage of humankind,"[45] so that, while they should be preserved through international conservation efforts, they should also be freely exchanged throughout the world.

3.3.1 Legal nature of farmers' rights

The recognition of farmers' rights has been progressively viewed by the international community as a moral duty rather than as a positive legal obligation or even an economic incentive.

Farmers' rights are intertwined with IPRs. Article 9 of the ITPGRFA describes farmers' rights both as "positive rights," which are something resembling IPRs and as "negative rights," which consist of limitations on or exceptions to the IPRs of others. The positive rights are more clearly stated in paragraph 9.1, which outlines the underlying principle of equity, and paragraph 9.2, which delegates the protection of farmers' rights to national legislations. The Article reads as follows:

9.1. The Contracting Parties recognize the enormous contribution that the local and indigenous communities and farmers of all regions in the world, particularly those in the centres of origin and crop diversity, have made and will continue to make for the conservation and development of plant GRs which constitute the basis for food and agri-arts production throughout the world.

9.2 The Contracting Parties agree that the responsibility for realizing Farmers' Rights, as they relate to Plant Genetic Resources for Food and Agriculture, rests with national governments. In accordance with their needs and priorities, each Party should, as appropriate, and subject to its national legislation, take measures to protect and promote Farmers' Rights, including

- (a) **protection of Traditional Knowledge** relevant to plant genetic resources for food and agriculture;
- (b) the right to **equitably participate in sharing benefits** arising from the utilization of plant genetic resources for food and agriculture;

[45] UN General Assembly Official Records 25th Session, Resolution 2749 (XXV), Declaration of Principles Governing the Seabed and the Ocean Floor, and the Subsoil Thereof, Beyond the Limits of National Jurisdiction, Su 28 A/8028 (1970); then this concept was incorporated at the Report of the Conference of FAO, 25th Sess., Resolution 4/89, Agreed Interpretation of the International Undertaking (1989); Report of the Conference of FAO, 25th Session, Doc. C89/REP, Doc. C89/REP, and Res. 5/89, Farmers' Rights (November 29, 1989); and finally in the FAO ITPGRFA, with its Multilateral System that facilitates access to a list of PGRs.

(c) the right to **participate in making decisions**, at the national level, on matters related to the conservation and sustainable use of plant genetic resources for food and agriculture.

9.3 Nothing in this Article shall be interpreted to limit any rights that farmers have to save, use, exchange and sell farm-saved seed/propagating material, subject to national law and as appropriate.

Until the adoption of this article, negative farmers' rights were considered in fora like the CBD and WIPO and have been occasionally considered in the context of their broader mandates. UPOV does not address farmers' rights, positive or negative, *per se*. However, given the implications of Article 9.2, the future of farmers' rights depends very much upon how nations who are parties to the ITPGRFA interpret and implement its provisions. As a result, even after the FAO treaty upgrade of the International Undertaking, the content of "farmers' rights" remains somewhat indeterminate since its content is vague.

Moreover, there is little consensus on the exact rights to be conferred.[46] They could not be shaped in the form of IPR for two main reasons: (i) the subject-matter of the ITPGRFA is not IP; (ii) the purpose of farmers' rights protection differs significantly from that of patents, copyrights, and PBRs.[47] These IP protections cover well-defined subject-matter, whereas the concept of farmers' rights apply to less definable incremental contributions to the innovation process.[48] Farmer's rights remain, nevertheless, a starting point in the future for effective protection of indigenous agricultural knowledge that is likely to develop. Farmers' rights include know-how about informal plant breeding relating to plants, plant varieties, crops, landraces, traditional PGRs for food, and agriculture, along with their wild and weedy relatives both *in situ* and *ex situ*.[49]

The scope of the rights of farmers should vary with the type of contribution of the farmer and the species in question. For instance, the scope of farmers' rights in techniques for the inclusion of the wild or weedy plant relatives should be very limited. Wild or weedy relatives are part of each country's natural resources – just like wild animals for fur or minerals in

[46] M. Girsberger, *Biodiversity and the Concept of Farmers' Rights in International Law* (Peter Lang, Berne, 1999) 172.

[47] H. El-Saghir, J. Mwijukye and G. Issahaque, "Plant Varieties, Biodiversity and Developing Countries", *Collection of Papers of the LL.M and Post-Graduate Specialization Course on Intellectual Property* (WIPO Worldwide Academy and University of Turin, 2003) 535.

[48] M. Blakeney, *Protection of Plant Varieties and Farmer's Right*, International Seminar on the Role of International Property in the Field of Biodiversity and Traditional Knowledge, jointly organized by the Brazilian Institute of Industrial Property and the European Commission (Manaus/Amazonas/Brazile, September 9–11, 2001) paragraph 4.2.

[49] Plant resources that are covered by the FAO ITPGRFA may also be patentable, thus becoming unavailable to Members of the FAO treaty.

the mines; it may be argued that farmers did not plant or develop those resources – they exist in nature.

Farmers as a collective group may have had something to do with the development and selection of landraces,[50] farmer's know-how, plant varieties and traditional crops, and thus they represent a man-made resource and theoretically should be protected by farmers' rights of wider scope. Even wild species may deserve protection in national law, providing benefit sharing arising from the further innovation based upon them. The wording of Article 9 can be submitted to a wide interpretation. The real obstacle to defining the scope of farmers' rights is ensuring that a certain improvement to a landrace can be assigned to a specific farmers' group (especially in relation to the landraces conserved *ex situ*).[51]

In sum, Article 9 combines with the CBD to indicate that international law does not prohibit the assignment of rights to wild or weedy species to the individual States, the collective groups of farmers, or communities. This right may conflict with Article 27 of TRIPS.

3.3.2 *Farmers' rights as positive or negative rights*

Farmers' rights set out in Article 9 of the ITPGRFA can be construed either as positive or negative rights. The former view supports the notion of farmers' rights as IPRs (exclusive rights), whereas the latter favors the implementation of an alternative legal form of protection such as systems of compensation and immunization in case of utilization of the landrace by third parties.

The implementation of farmers' rights as positive rights has been elaborated by Cottier who calls them traditional IPRs.[52] They may encompass, *inter alia*, farmers' rights to TK or know-how relating to plant and animal GRs (grassroots innovations). According to this view, farmers' rights may be suitably referred to as IPRs since the knowledge and information concerned, while in the public domain, has been part of the traditional heritage of specific communities and individuals: "it has been intellectual and mental, and it should become a legal property in the future" by removing this intangible knowledge from the public domain.[53]

[50] M. Halewood *et al.*, "Farmers, Landraces and Property Rights: Challenges to Allocating *Sui Generis* Intellectual Property Rights to Communities over their Varieties", in T. Cottier and S. Biber-Klemm (eds.), *Rights to Plant Genetic Resources and Traditional Knowledge* (CABI, London, 2006) 176.

[51] *Ibid.*, 193–99.

[52] T. Cottier and M. Panizzon, "A New Generation of IPRs for the Protection of Traditional Knowledge in Plant Genetic Resources for Food, Agricultural and Pharmaceutical Uses", ibid., 203–35. Dutfield, "The Public and Private Domains", 287.

[53] Cottier, "The Protection of Genetic Resources", 1841.

TK does not rely on the patent-like concept of novelty.[54] States that thoughtfully and carefully wish to craft national legislation to protect positive farmers' rights should start by seeking an interface between "traditional resources rights" and IP law.[55] Section 7.1.2 below will be dedicated to Cottier's model when compared to other proposals including Reichman's.

On the other hand, the implementation of farmers' rights as negative rights is based on the clear distinction between farmers' rights and PBRs: the latter are IPRs protecting plant varieties while the negative farmers' rights are exceptions to such IPRs. If farmers' rights were realized as IPRs, they would include a number of more or less typical characteristics of IPRs, such as (i) a *right to exclusiveness* in using, selling, or reproducing the protected subject-matter, and the exclusive right to compensation; (ii) *individual ownership*, a right to exclusiveness in using, selling, or reproducing the protected subject-matter, and the exclusive right to compensation. Girsberger, a major proponent of this negative right's solution, points out that farmers' rights are collective by definition (e.g. they are assigned to the collective interests of all involved in conserving crop germplasm without being innovations assigned to specific farmers), and thus are not compatible with the individual rights and ownership that define standard IPRs.[56] Additionally, the IPR option can cause substantial expenses (in the form of administrative bodies and procedures, litigation, and scientific investigations) that can consume the compensations achieved through farmers' rights, especially if the demand for the use of farmers' right protected subject-matter is not too large.

In Girsberger's view, farmers' rights are intended as incentives for the conservation of biodiversity, so they should merely include the right to compensation.[57] On this line of reasoning, Girsberger's thesis concludes that geographical indications (GIs), patents, undisclosed information, and PBRs are not suitable to protect the subject-matter of farmers' rights mainly because these IPRs are aimed at protecting modern innovations.[58]

Girsberger discusses in depth the possible content of farmers' rights,[59] identifies possible criteria for protection,[60] and identifies those farmers

[54] *Ibid.*, 1836.

[55] T. Cottier, "The Protection of Intellectual Property Rights: A Requirement for Technology Cooperation, Foreign Investment and Equitable Returns in Biotechnology Prospecting, Biotechnologie für Entwicklungsländer?", in *Chancen und Risiken der Biotechnologie bei Landwirtschaftlichen Nutzpflanzen* (Schweizerisches Zentrum für Internationale Landwirtschaft ed., 1995) 65.

[56] Girsberger, *Biodiversity*, 255–59.

[57] The recognition of this right should also be enshrined in TRIPS Article 27.3(b) during its current revision, Girsberger, *Biodiversity*, 255–59.

[58] *Ibid.*, 321. [59] *Ibid.*, 206. [60] *Ibid.*, 215.

entitled to specific rights and obligations, for instance, the right to compensation. Holders of these rights can be individual farmers, indigenous groups or other rural communities, farmers involved in informal plant breeding, State entities, and the international community.[61]

In sum, while Cottier favors implementation of farmers' rights through what are called "positive rights," Girsberger addresses potential forms of implementation of "negative rights." The question arises whether a State can implement Cottier's and Girsberger's proposals at the same time. A positive answer is possible. On the one hand, a specific national legislative enactment may render farmers' rights under Article 9 ITPGRFA as exceptions to exclusive rights generated by a *sui generis* system of PVP (Article 27.3(b) of the TRIPS Agreement as an alternative to UPOV 1991), with the effect of reducing the breadth of the claim of BPRs. This is the implementation of farmers' rights as positive rights. This same farmers' right can simultaneously become a Cottier-modeled Traditional Intellectual Property Right (TIPR) that can be subject to an exclusive right, if it matches the requirements (to be set in the national legislation).[62]

Some authors in industrialized countries may regard indigenous agricultural knowledge systems as similar to general scientific information in that they are part of public or community knowledge.[63] This means that there would not be any legal protection or assignability possible at the international level. Farmer's rights under Article 9 of the ITPGRFA do not add much to the current legal status of TK. Innovations based on TK do not reach a high degree of ingenuity so to match novelty and non-obviousness, inventive step and industrial application requirements and achieve protection by a patent. Because TIPRs do not require any of these restrictive requirements, they are more appropriate to protect grain-sized innovations.

An example clarifies the possible effects of implementing farmers' rights. A group of farmers in the Peruvian Altiplano grow a landrace of potatoes characterized by a unique purple flesh. These potatoes are then purchased by a Peruvian governmental conservation agency at an open market. The same potato is then transferred to a Mexican potato breeder who uses it as a parent in several crosses. Many years and generations later, this Mexican

[61] *Ibid.*, 227.

[62] Cottier and Panizzon, "A New Generation of IPRs", 223–25.

[63] J. Philips, "The Diminishing Domain" (1997) 8 *European Intellectual Property Review* 429–30. *Contra* see M. S. Swaminathan, *Farmers' and Breeders' Rights for India – Farmers' Rights and Plant Genetic Resources: Recognition and Reward; A Dialogue* (Macmillan India Ltd, Madras, 1995) 246–47; L. E. Ewens, "Seed Wars: Biotechnology, Intellectual Property, and the Quest for High Yield Seeds" (2000) 23(2) *Boston College International and Comparative Law Review*, 285–310 www.infoeagle.bc.edu/bc_org/avp/law/lwsch/journals/bciclr/23_2/23_2_TOC.htm

breeder obtains a purple potato variety that can be used to make potato chips, and he obtains a patent or a UPOV-like plant variety (potatoes are clonally propagated). Subsequently this purple potato chip is picked up by a large food company like Kraft, and it becomes one of their best sellers. At this juncture, the Peruvian farmer seek to exercise these newly acquired farmers' rights, or so-called "abstract rights."

Regarding the effects of the implementation of farmers' rights under ITPGRFA, Article 9, the national legislation of Mexico could allow the Peruvian community to continue to innovate using the potatoes with purple flesh, regardless of any exclusive IPR based upon that PGR. In the Girsberger negative-right approach, the right of the group of farmers in Peru to use the potato would constitute an exception to the exclusive right held by the Mexican potato breeder and Kraft. The difference between holding and not holding this right is that any other traditional farmers' group would be prevented by the right held by the Mexican potato breeder and Kraft from breeding the same purple potato variety. Even before Article 9 of the ITPGRFA was enacted, the patent did not cover any of the elements existing in nature, but, in this illustration, Article 9 grants a general special protection to the farmers that hold TK related to the PGR. The exercise of this type of implementation of farmers' right is without prejudice to other international systems of benefit sharing (e.g. the PGRs belongs to the Multilateral System, see sections 3.3.3.1. and 3.3.3.2 below).

If national legislation provides for a system of positive rights as envisaged by TIPRs, the Peruvian farmers' right to the potatoes will also have the quality of exclusiveness.[64] The fact of holding an exclusive right upon some of the use of the relevant PGRs endows the group of farmers of the Peruvian Altiplano with more bargaining power in a contract negotiation setting.

The implementation of farmers' rights contained in Article 9 of the ITPGRFA will help DCs identify and implement such traditional agricultural systems, moving the world toward a more equitable implementation of Article 27.3(b) of the TRIPS Agreement and spurring disclosure of agricultural information that will benefit all. Many questions raised by the ITPGRFA and its relationship to other international instruments still need to be addressed. Meanwhile, it is safe to say that the implementation of the FAO Treaty will play a crucial role in moving toward an interpretation that harmonizes CBD and TRIPS (see chapter 5) and that allows for the creation of an ABS regime in the provider country.

[64] See the difference that Cottier makes between assignable and non-assignable TK subject-matter in view of legal protection, Cottier and Panizzon, *A New Generation of IPRs*, 217–19.

3.3.3 *The access- and benefit-sharing regime of the ITPGRFA and intellectual property rights*

The ITPGRFA established two ABS regimes for PGRs for food and agriculture (PGRFA): one for the plant genetic resources for food and agriculture (PGRFA) listed in an annex to the agreement, protected under the Multilateral System, and the other covering all other species that are governed by the national legislation.[65] States have a fundamental obligation to facilitate access to contracting parties through their national systems.

3.3.3.1 The Multilateral System[66]

The Multilateral System is an *ex situ* conservation arrangement operated by FAO, providing for facilitated and transparent access to PGRs from around the world for research and breeding for agricultural development. Farmers from around the world may deposit PGRs such as seeds and plants into the *ex situ* conservation bank.

Article 11(1) establishes a Multilateral System under which, Contracting Parties – and therefore also recipient States, in perfect harmony with the CBD – agree to shape their IPRs in conformity with the mandate of the Treaty (see e.g. Article 12.3(d)) and in particular to provide for benefit sharing in the form of monetary and other benefits (see e.g. Article 13.2(d)).

Article 12.3 provides that "access" to the resources included in the Multilateral System "shall be provided in accordance" with a list of conditions. In this regard, (d) provides that "recipients shall not claim any IP or other rights that limit the facilitated access to resources for food and agriculture, or their genetic parts or components, *in the form received from* the Multilateral System." This article is a compromise resulting from numerous discussions between industrialized and DCs. While the ITPGRFA does not conflict with the TRIPS Agreement *per se* (which allows for WTO members to exclude plant varieties from patentability), it would conflict with any bilateral "TRIPS-plus" agreements calling for protection over isolated gene material derived from any PGRs bank. By specifying that IPRs will not be claimed over materials "in the form received" by the Multilateral System, the ITPGRFA attempts to circumvent any potential conflicts between bilateral agreements while still providing for the various non-IPRs championed by the UPOV or TRIPS. Aside from this skirting of conflict, however, the language of 12.3(d)

65 Article 10.1 of FAO ITPGRFA.

66 Cottier and Panizzon, *A New Generation of IPRs*, 291–94.

remains deliberately ambiguous on invention versus discovery, i.e. it is not clear whether biological matter such as isolated and purified compounds or gene sequences are patentable. The language of Article 12.3(d) was not universally accepted at first. The admission of the obscure phrases "or their genetic parts or components" and "in the form," were hotly contested in the final round of negotiations. The US wanted the first phrase deleted and the second retained while DCs who opposed patent protection wanted the second phrase deleted and the first retained. In the end, both clauses were retained after the US lost a vote to have 12.3(d) omitted from the treaty altogether. The entire treaty was then unanimously adopted with only two abstentions by the US and Japan.

With regard to benefit sharing, Article 13.2(d)(ii) provides, *inter alia*, that "[t]he Contracting Parties agree that the standard Material Transfer Agreement" – the form and contents of which are mandated by Article 12.4 – "shall include a requirement that a recipient who commercializes a product that is a PGR for food and agriculture and that incorporates materials extracted from the Multilateral System, shall pay to the mechanism referred to in Article 19.3(f)."[67] This is intended to provide compensation to farmers for the contributions they have made and continue to make for the conservation and development of PGRs under Article 9 and to serve as "an equitable share of the benefits arising from the commercialization of that product."[68]

Before moving on to the logistics of the Multilateral System, it is worth noting that the ITPGRFA gives no clear answer to the vital question of when a resource is deemed a GR for purposes of the treaty. This ambiguity is largely due to the ITPGRFA's use of the CBD definition of "GRs" as its model. The CBD definition is utilitarian, focusing on the natural characteristics of a resource in nature. Genetic material, or Plant Genetic Material (PGM) in the case of the ITPGRFA, becomes a GR based on the use of that material for some purpose and not based on any intrinsic characteristic of the material. Therefore, the definition needs to be supplemented by provisions regarding the uses of genetic material that make that material a genetic resource. The only text within the ITPGRFA relevant to this concern is in Article 2 where the closing line of the *chapeau* states: "*These definitions are not intended to cover trade in commodities.*" This text suggests that the definitions provided are not self-sufficient since they

[67] Article 19.3(f) reads: "The functions of the Governing Body shall be to promote the full implementation of this Treaty, keeping in view its objectives, and, in particular, to: [...] (f) establish, as needed, an appropriate mechanism, such as a Trust Account, for receiving and utilizing financial resources that will accrue to it for purposes of implementing this Treaty."

[68] Article 13.2(d)(i) of FAO ITPGRFA.

must also exclude commodities. The text further emphasizes that the definition of genetic material has more to do with its various uses rather than its intrinsic characteristics.

After the PGRs are deposited in the FAO Multilateral System, they may be obtained for utilization and research. This access to PGRs includes not only the PGRs itself but also to all related data and information regarding the PGR. If a commercial product results from the use of a PGR in the Multilateral System, the ITPGRFA provides for payment of an equitable share of the monetary proceeds to the PGRs' provider. In this manner, the FAO is able to respect farmers' rights while providing for a mutually beneficial global access mechanism for both farmers and researchers.

While the Multilateral System is grounded in the principle of free exchange of PGRs, it recognizes the need for access control rules exercised by States.[69] For this reason, the CGRFA, the Multilateral System's regulatory body, makes its member States responsible for enforcing farmers' rights over the PGRs they provide to the Multilateral System.[70] Conditions for ABS will, therefore, be set out by the Member States in a material transfer agreement (MTA).

The MTA is an agreement whereby a country receiving genetic material from the Multilateral System agrees to share benefits received from the commercialization of that material with the provider country according to ITPGRFA, Article 12.4, which reads:

> facilitated access, in accordance with Articles 12.2 and 12.3 above, shall be provided pursuant to a standard material transfer agreement, which shall be adopted by the Governing Body and contain the provisions of Articles 12.3a, d and g, as well as the benefit-sharing provisions set forth in Article 13.2(d)(ii) and other relevant provisions of this Treaty [...]

In addition to sharing the benefits resulting from the commercialization of genetic material, the recipient countries have the responsibility to ensure that no claimed IPR will hinder the facilitated access to GRs.[71]

One of the most important provisions that connects the Multilateral System with IP issues is Article 12.3(d) that provides the facilitated access without the necessity of PIC. Article 13.2 provides for "the exchange of information and access to and transfer of technology, capacity-building, and the sharing of benefits arising from commercialization." The benefits arising from the commercial use of PGRFA shared under the Multilateral

[69] P. Cullett, *et al.*, "Intellectual Property Rights, Plant Genetic Resources and Traditional Knowledge", in T. Cottier and S. Biber-Klemm (eds.), *Rights to Plant Genetic Resources and Traditional Knowledge* (CABI, London, 2006) 112.

[70] www.fao.org/ag/cgrfa/itpgr.htm.

[71] Article 12.3(d) of FAO ITPGRFA.

System through a Trust Account should flow to the farmers in all countries, especially in DCs that conserve and sustainably use PGRFA.[72] Furthermore, in accordance with Article 12.3(h), access to listed PGRFA in *in situ* conditions is to be provided according to national legislation or according to standards set by the Governing Body of the Treaty.

The Governing Body of the Treaty drafts standard contracts for material transfer. Elements of these contracts are included in Article 12.4. The CGRFA has recently proposed a draft MTA contract that, while operating exclusively between parties to the MTA and not signatory parties to the PGRFA treaty, would nonetheless function within the context of the PGRFA treaty. For example, all relevant provisions of the ITPGRFA treaty are to be expressed as contractual rights and obligations of the parties to the MTA. The MTA ensures the implementation of 12.3(d) through the mandatory inclusion of 12.3(d) within its provisions, specifically stating that it should be "an obligation of the recipient."

MTA contracts will also include a dispute or arbitration mechanism separate from Articles 21 and 22 of the PGRFA treaty which provide for *compliance* and *settlement of disputes* respectively.[73] The MTA will ensure the application of 12.3(d) by including it as a mandatory clause within the MTA contract and Article 12.5 makes it enforceable under contract law.[74] It must be noted, however, that because the current draft of the MTA adopts the language of Article 12.3(d) verbatim, it thus incorporates the same ambiguities as discussed above. These ambiguities effectively entail that the terms and conditions of the MTA will vary depending on the jurisdiction in which particular parties to an MTA are found.

The MTA draft contract provides the most sophisticated example of an internationally agreed-upon system of ABS. The Multilateral System should be followed as a model for the management of other GRs that come under an international access regime. In particular, this treaty actualizes the concept of "common heritage" and can be compared to aspects of the treaty of the Seabed Authority provided by the UN Convention on the Law of the Sea.

[72] Article 13.3 of FAO ITPGRFA.

[73] Article 21 of FAO ITPGRFA. Article 22 reads: "If the parties concerned cannot reach agreement by negotiation, they may jointly seek the good offices of, or request mediation by, a third party."

[74] Article 12.5 of FAO ITPGRFA reads: "Contracting Parties shall ensure that an opportunity to seek recourse is available, consistent with applicable jurisdictional requirements, under their legal systems, in case of contractual disputes arising under such MTAs, recognizing that obligations arising under such MTAs rest exclusively with the parties to those MTAs."

3.3.3.2 Specifics of the relationship between intellectual property rights and the ITPGRFA Multilateral System

When a PGR is taken from the Multilateral System and is developed into a product, the question may arise as to when that product would become available without restriction for further research and breeding. When the product is patented, further considerations arise on the interaction among patents (TRIPS), plant variety (UPOV), and farmers' rights on PGRs in the Multilateral System (ITPGRFA). The Commission on Plant GRs for Food and Agriculture (CPGRFA) has stated that such products are available when they are in the public domain, or when protected by PVP (as in the case of UPOV or other *sui generis* systems), or by a patent system as long as they are made available through royalty-free licenses.[75] With regard to UPOV, the CPGRFA stipulates that authorization is not required where the protected materials are merely for further research and breeding.[76] It also allows for the use of the product.

There is also a question as to whether materials covered both by UPOV protections such as farmers' rights and by patents are free to be used without restrictions. The question of use of a patented variety under the UPOV system for research and breeding is not obvious.

UPOV 1991 does not allow for unrestricted use of a protected variety in further research and breeding and commercialization free of charge. Patents in TRIPS are a separate matter, and the permissibility of use of patented materials for research or commercialization of a product obtained by use of a patented product is very limited and varies from one legal system to another.[77] In both cases, however, the IPR is protected for a limited period of time, after which the material moves to the public domain.

The recipient countries are responsible to ensure that IPRs do not hinder the access to genetic material in the *form received* by the Multilateral System. Accordingly, the question arises as to what kind of right is necessary to protect GRs once they are no longer in the form received by the Multilateral System.

3.3.3.3 The consequences of the refusal of a State Party to the ITPGRFA and TRIPS to grant a patent on plant genetic resources acquired from the Multilateral System: a potential case for the WTO Dispute Settlement Mechanism

After having stated the general principles of the relationship among patents, PBRs and farmers' rights, this section will develop one specific

[75] *Second Meeting of the Commission on GRs for Food and Agriculture*, 4.

[76] *Ibid.*, 20.

[77] S. J. R. Bostyn, "One Patent a Day Keeps the Doctor Away? Patenting Human Genetic Information and Health Care" (2000) 7 *European Journal Health Law* 229, 248–49, quoted in Ricolfi, "Is there an Antitrust Antidote?", 357.

example in order to portray the legal conflicts arising from specific provisions of the ITPGRFA and TRIPS Agreement.

Again, a few remarks are in order here on the interrelations between the TRIPS Agreement, the CBD and ITPGRFA. Article 12.3(d) ITPGRFA[78] would seem to impose constraints on IP protection connected to PGRs in the Multilateral System. This restriction is much stronger than those mandated by the CBD. ITPGRFA bans IP claims on the GRs in the Multilateral System outright, whereas the CBD allows the corresponding IP claim, provided PIC and ABS obligations are complied with and the duty of information at the patenting stage is fulfilled.

The ITPGRFA sets two conditions precedent to the ban on patentability: (i) the prohibition concerns only IPRs which limit the facilitated access to the GR and (ii) the prohibition is triggered only to the extent that the limitation to access deriving from IP protection concerns the resource "*in the form received* from the Multilateral System." The former condition *may* imply that the ban does not concern plant varieties protection. Indeed, genetic material, however protected under UPOV or UPOV-like protection (Article 27.3(b) of TRIPS), is never restricted as far as downstream innovation is concerned, to the extent that the research exemption is provided in UPOV 1961 but not in UPOV 1991.[79] The latter may in turn open quite a large loophole in the field of patents. Indeed, genetically engineered plants are by definition different from the original plant on which recombinant DNA technology was applied so that patent protection never implies a limitation to access of the GR in its original, unaltered form – that is "in the form received from the Multilateral System." The restriction, then, only applies to the genetically altered resource.[80]

In this respect, the ITPGRFA ban on patenting of PGRs – in the form received by the Multilateral System – raises various legal problems of State compliance with international treaties in the field of IP and, in particular, with Article 27 of the TRIPS Agreement. An example can better illustrate this conflict. Let's suppose the US has granted a patent based on a plant in the Multilateral System. The US, which is a signatory but not a party to the ITPGRFA, requires a WTO Panel to enforce TRIPS Article 27.1 on a State that is a party to the ITPGRFA. Article 27.1 mandates that all States grant patents regardless of the field of technology. However, the ITPGRFA

[78] FAO, ITPGRFA.

[79] In UPOV 1991 and EU Regulation No. 2100/94, the notion of "essentially derived varieties" has arisen. However, nowadays downstream innovation is no longer unrestricted.

[80] R. Pavoni, "Accesso alle risorse fitogenetiche e diritti di proprietà intellettuale dopo il trattato dalla FAO del 2001" (2003) 3 *Comunità internazionale* 369, 382–83 (2003).

party cannot comply with this obligation without breaching Article 12.3(d) of the ITPGRFA that prohibits registration of the US patent on a Multilateral System PGR in the ITPGRFA party's national patent system. To extend the example further: suppose the US has granted a patent on a weed whose gene has been isolated and sequenced. The patent regards the biotechnological process as novel, non-obvious, and useful – the US requirements for obtaining a patent. The ITPGRFA party refuses the patent on the grounds that it is too closely related to a weed pre-existing in the Multilateral System. Indeed this is a case falling within the scope of Article 12.3(d) because the novelty of the innovation is not particularly high. The question arises whether this biotechnological invention can be granted a patent because its underlying process renders the form of the PGR different from the one received by the Multilateral System.

Furthermore, the question arises here as to whether one can patent the process related to the PGR without patenting the plant itself. The answer will partly depend on the scope of the claims of the patent. It goes without saying that any further use of the process related to that patent is going to be covered by the claims of this type of patent. It should be recalled that the problem consists in the fact that the US may request that a patent be granted on this invention, and that an ITPGRFA party State may refuse the patent on grounds of ITPGRFA Article 12.3(d).

It can be hypothesized that a State that is party to the ITPGRFA grants patents on plants as such (e.g., the ITPGRFA party is also a member of the EPC and of the EU biotech-Directive practice). The US can thus argue that the ITPGRFA party cannot discriminate as to the field of technology. A patent on the process based upon the weed in the Multilateral System should therefore be granted under TRIPS.

The conflict is blatant in this case: the ordinary meaning of the provision instituting the Multilateral System, Article 12, is in open conflict with the TRIPS Agreement because it places restrictions and burdens on patent holders of biotechnological inventions based on PGRs in the Multilateral System. Article 27.1 of TRIPS states that "patents shall be available and patent rights enjoyable without discrimination *as to* [...] *the field of technology*" (italics added). This is a clear case of international *de jure* discrimination. However, a recent WTO panel decision has rejected a similar claim of *de facto* discrimination by pharmaceutical patent holders.[81] Moreover, emerging legal doctrine is relaxing the prohibition on discrimination of Article 27.1 of TRIPS in case it conflicts with the "the ability of Members to target certain

[81] *WTO Panel Report on Canada – Patent Protection of Pharmaceutical Products*, WT/DS114/R paragraph 7.91 (March 17, 2000), www.wto.org/english/tratop_e/dispu_e/7428d.pdf and N. 439, *Canada – Term of Patent Protection.*

technological fields in dealing with the important national public policies referred to in Article 8(1)."[82]

I will examine the arguments that the two parties may put forth before a hypothetical WTO Panel. The first point the ITPGRFA party would argue is that the US, as a signatory of the ITPGRFA, has the minimum obligation not to undermine the purposes of the ITPGRFA provisions and their compliance by other contracting parties.[83] It would be very difficult to argue that the banning of the US patent in the ITPGRFA party national patent system is compliant with a clear-cut literal interpretation of Article 27.1 of the TRIPS Agreement. The State may still justify the special measures provided by Article 8[84] of the TRIPS Agreement (see section 2.1.1 above) in sectors of vital importance to its socio-economic and technological development that, in this case, involve a commonly agreed multilateral system created by the international community with PGRs considered as a common heritage of mankind. In sum, the request of the US implies the breach of an obligation *erga omnes* and not just towards the treaty State that cannot grant the patent.

In case the obligation under Article 27.1 of the TRIPS Agreement cannot be waived, the ITPGRFA party can still invoke the application of compulsory licenses provided by Articles 30, 31, and 40.1, when interpreted teleologically in light of (in order to fulfill the purposes of) Article 7 of the TRIPS Agreement (see section 2.1.3 above).

Compulsory licenses in this case can be justified by the concept of mutual supportiveness between WTO and other relevant treaties, the measures that the ITPGRFA compliant State can take for the sake of "transfer and dissemination of technology" (Article 8 of the TRIPS Agreement), and to counter the "abuse of IPRs" (Article 40.1 of TRIPS Agreement). Here the question arises whether, notwithstanding the issuance of a compulsory license, the ITPGRFA signatory State that grants a patent in these circumstances violates Article 12 of the ITPGRFA. As a matter of fact, the circulation and

[82] Ricolfi, "Is there an Antitrust Antidote?", 345–47.

[83] This principle is enshrined in Article 18 of the VCLT (obligation not to defeat the object and purpose of a treaty prior to its entry into force): "A State is obliged to refrain from acts which would defeat the object and purpose of a treaty when: (a) it has signed the treaty or has exchanged instruments constituting the treaty subject to ratification, acceptance or approval, until it shall have made its intention clear not to become a party to the treaty; or (b) it has expressed its consent to be bound by the treaty, pending the entry into force of the treaty and provided that such entry into force is not unduly delayed."

[84] Article 8.1, TRIPS: "Members may, in formulating or amending their laws and regulations, adopt measures necessary to protect public health and nutrition, and to promote the public interest in sectors of vital importance to their socio-economic and technological development, provided that such measures are consistent with the provisions of this Agreement."

exploitation without restraint of PGRs that are in the Multilateral System is considered vital for sustainable development of many DCs. In other words, Article 8.1 can be interpreted in such a manner that the national legislation compliant with the ITPGRFA Multilateral System may ban abuses of patent rights.

The IPGRFA Party cannot invoke the patentability exceptions of *ordre public* or "the serious prejudice to the environment" of Article 27.2 of the TRIPS Agreement because granting the patent would not harm the environment. Furthermore, it is hard to see how the patenting of a Multilateral System PGR would be contrary to *ordre public*. The patenting act is only contrary to a multilateral agreement, the *lex specialis* of Article 12.3(d) of ITPGRFA.

With regard to the specific ban on patents, the US may rebut that ITPGRFA is not a WTO constitutive treaty, thus the WTO Panel may only apply its own law that is limited to WTO treaties, in this case only the TRIPS Agreement. The US can also rebut because the measure applied by the ITPGRFA cannot be based upon Article 8 of TRIPS since it clearly states that the measure has to be "consistent with the provisions" of the TRIPS Agreement. Banning patents on that ground is not consistent with Article 27.1 of TRIPS.

The WTO should apply the rules of interpretations of treaties as outlined in section 2.1.4 above. Since both provisions regard patent rights, the rule of *lex posterior derogat priori* is not applicable in this case since the US is party only to TRIPS and not to ITPGRFA. The WTO Panel should decide as follows:

1) WTO cannot directly apply the ITPGRFA by stating what the obligations of the US vis-à-vis the ITPGRFA contracting parties are. In other words, the WTO does not have jurisdiction to state that the US should not require that country to breach Article 12.3(d) of the ITPGRFA that excludes patents on the PGR as received in the Multilateral System.
2) Because Article 12.3(d) of the ITPGRFA is relevant to IP law, this provision is to be taken into account in the interpretation of the TRIPS Agreement. The WTO appellate body has in the past considered both WTO jurisprudence[85] and "relevant rules of international law" in deciding disputes.[86] For example, in the *Shrimp-Turtle* case, the

[85] H. J. Jackson, *The World Trading System: Law and Policy of International Economic Relations* (Massachusetts Institute of Technology Press, Cambridge, Mass., 1997) 121–22; A. Chua, "Precedent and Principles of WTO Panel Jurisprudence" (1998) 2(16) *Berkeley Journal of International Law* 171–95, 183; Panel Report on *India – Patent Protection for Pharmaceutical and Chemical Agricultural Products* (complaint by the US), WT/DS50/R paragraph 7.19 (September 5, 1997).

[86] O. Cattaneo, "Interpretation of the TRIPS Agreement: Considerations for the WTO Panels and Appellate Body" (2000) 3 *The Journal of World Intellectual Property* 627–81, 675, referring to the *North Sea Continental Shelf*, paragraph 73, 43.

Appellate Body made reference to several international instruments adopted by the UN in order to interpret Article XX(g) of the GATT and to come to the conclusion that sea turtles are "exhaustible natural resources."[87] The Appellate Body referred to conventions like the UN Convention on the Law of the Sea,[88] the CBD and the Resolution on Assistance to DCs adopted in conjunction with the Convention on the Conservation of Migratory Species of Wild Animals.[89] In light of the ICJ advisory opinion on *Namibia*, the Appellate Body upheld the view that "an international instrument has to be interpreted and applied within the framework of the entire legal system prevailing at the time of interpretation."[90] It follows, therefore, that the WTO may, on a case by case basis, consider other relevant international agreements while implementing the provisions of the TRIPS Agreement.[91]

3) The measures contemplated by Articles 8(1), 8(2) and 40(2) are to be implemented as long as they are "consistent with the provisions" of Article 27.1 of the TRIPS Agreement. Any limitation to the rights of the patent holder cannot be used "as a pretext to undermine the protection of IPRs as guaranteed by the TRIPS Agreement."[92] In light of this principle, eliminating all patent protection for this invention is not compliant with Article 27 of the TRIPS Agreement. Article 8 of TRIPS allows limitations to the rights conferred since "Article 27 does not prohibit *bona fide* exceptions to deal with problems that may exist only in certain product areas."[93] The teleological interpretation of the TRIPS Agreement would preclude requiring states to patent PGRs made available in the Multilateral System because such PGRs are, to a certain degree, already part of the public domain. In this context, the minimum standard of protection for the promotion of inventiveness that the TRIPS Agreement seeks to accomplish is not suitable where the patent sought is not properly enforceable within a State that is party to

87 *Shrimp-Turtle* case, paragraphs 130–4. In this case the Appellate Body employed the evolutive method in interpreting the phrase "exhaustible natural resources" appearing in Article XX(g) of GATT.

88 *United Nations Convention on the Law of the Sea.*

89 Final Act of the Conference to conclude a *Convention on the Conservation of Migratory Species of Wild Animals* (June 23, 1979) 19 *ILM* 15 (1980).

90 *Shrimp*, paragraph 130.

91 G. Marceau, "A Call for Coherence in International Law: Praises for the Prohibition Against 'Clinical Isolation' in WTO Dispute Settlement" (1999) 33 *Journal of World Trade* 87–152 especially 123.

92 And consistent with the finding of *WTO Panel Report US – Sec. 211 Omnibus Appropriations Act of 1998*, WT/DS176/R paragraph 8.57 (August 6, 2001); Ricolfi, "Is there an Antitrust Antidote?"

93 Panel Report, *Canada – Term of Patent Protection.*

the ITPGRFA. Additionally, it is widely recognized that any interpretation of the TRIPS Agreement must be broad and allow for great deference to national law. WTO panels must therefore exercise judicial restraint when passing judgment on national patent systems.[94]

4) The Panel considers its role as the interpreter of WTO Treaty provisions in light of relevant international law. It should therefore require the ITPGRFA party to allow registration of the patent. In this case there should be no exception to patentability. No "special measure" envisaged by Articles 7 and 8 can sufficiently justify the total ban on patentability in order to comply with another treaty like the ITPGRFA.[95] I hasten to add, however, that the problem in applying the special measures provided by Article 7 or 8 is that the ban in this case does not concern a particular sector but the origin of the PGR which is the Multilateral System of the ITPGRFA. Ultimately, one may wonder whether an expansion of the potentially increasing scope of the patentability exception of *ordre public* can serve in this case to justify the ban on patentability of a PGR in the Multilateral System. The EU or US patent systems' classic and very restrictive interpretation of the concept of *ordre public* would certainly not include such a broad interpretation. This is the reason why section 6.3 below strives to expand the possible use of the patentability exception of *ordre public* moving from a classic to a more radical interpretation in the new and more complex context of international trade.

In sum, ITPGRFA seeks to provide a mechanism whereby access to PGRs may be both facilitated and protected under farmers' rights. Within the ITPGRFA there are provisions that conflict with a literal interpretation of Article 27 of the TRIPS Agreement. Many TRIPS provisions leave a large margin of interpretation to the States. Much of this interpretation hinges on whether to stress IPRs, which will favor the interests of the IPR holder, or public policy, which in turn benefits the user.[96] The fine-tuned balance of IP protection within a national IP system is achieved through the application of all the international norms at hand interpreted in light of the concept of mutual supportiveness. However, in the particular case that has been sketched, there is no mutual supportiveness possible between Article 12.3(d) of the ITPGRFA and Article 27 of the TRIPS.

[94] J. Jackson, "Dispute Settlement and the WTO. Emerging Problems" (1998) 1(3) *Journal of International Economic Law* 342.

[95] Ricolfi, "Is there an Antitrust Antidote?," 347. [96] *Ibid.*

3.3.4 *Overview of the status of traditional knowledge in international human rights law*

Certain international human rights conventions contain some provisions that relate to the possible IP protection of TK.[97] The very definition of "indigenous people and local community" and TK, as stated in Article 8(j) of the CBD, depends on their evolutionary interpretation in light of international human rights instruments. The normative value of these human rights instruments is essential to this determination.

Whereas at the national level States sometimes provide for recognition of indigenous peoples through registration, at the international law level there is no agreed definition of what indigenous people or local communities are in the context of holders of IPRs regarding their TK.[98]

3.3.4.1 International Labor Organization Convention No. 169

The first international treaty within the UN system on the legal protection of indigenous peoples is the ILO Convention No. 169.[99] This Convention, although adopted within the ILO, concerns the human rights of indigenous and tribal peoples, and includes some rules concerning labor, social security, and health matters. Its main objective is to assert the relationship between indigenous and tribal peoples and their land; it does not directly address the issue of protection of TK.

The importance of the Convention to this study is found in its definition of 'indigenous peoples' as:

> Peoples in independent countries who are regarded as indigenous on account of their descent from the populations which inhabited the country, or a geographical region

[97] D. Posey, "International Agreements for Protecting Indigenous Knowledge", in V. Sanchez and C. Juma (eds.), *Biodiplomacy: Genetic Resources and International Relations* (ACTS, Nairobi, 1994) 125.

[98] It suffices to mention the attempts at producing a definition by the UN Special Rapporteur Martinez Cobo found in UN Doc. E/CN.4TSub.2/1986/7/Add.1 S.5 and by the World Bank Operational Directive, where it is admitted that the varied and changing contexts in which indigenous peoples are found do not make it possible to outline a definition that can capture their diversity. T. P. Stoll and A. Von Hahn, "Indigenous Peoples, Indigenous Knowledge and Indigenous Resources in International Law", in S. Von Lewinski (ed.), *Indigenous Heritage and Intellectual Property* (Kluwer, The Hague, 2nd edn, 2008) 5–28.

[99] ILO Convention No. 107 of June 26, 1957, 328 *UNTS* 247. ILO Convention No. 169 of June 27, 1989, http://ilolex.ilo.ch:1567/scriptslconvde.pl?CI69. The committee proposed the adoption, in 1957, of the Convention Concerning the Protection and Integration of Indigenous and Other Tribal and Semi-Tribal Populations in Independent Countries. This Convention, commonly referred to as Convention 107, essentially dealt with measures to integrate indigenous people into modern production systems. This Convention was revised in June 1989 as International Labour Organization Convention, Convention Concerning Indigenous and Tribal Peoples in Independent Countries (June 7, 1989), www.cwis.org. The revised Convention eschews the goal of promoting the assimilation of indigenous and tribal peoples.

to which the country belongs, at the time of conquest or colonization or the establishment of present State boundaries and who, irrespective of their legal status, retain some or all of their own social, economic, cultural and political institutions.[100]

This definition is important because indigenous people are identified in relation to four vital factors: time, geographical space, persistence, and territorial occupation by outside population.

As regards the specific rights granted to indigenous people, a certain number of principles can be found in the ILO Convention.[101]

Article 2(2b) provides that governments shall have the responsibility of developing measures for "promoting the full realization of the social, economic and cultural rights of these peoples with respect for their social and cultural identity, their customs and traditions and their institutions." Article 5(a) provides that "the social, cultural, religious and spiritual values and practices of these peoples shall be recognized and protected, and due account shall be taken of the nature of the problems which face them as groups and individuals."

Particularly important is Article 13(1), which attempts to explicitly recognize collective rights of indigenous peoples when it states that "governments shall respect the special importance of the cultures and spiritual values of the peoples concerned, of their relationship with the lands or territories, or both as applicable, which they occupy or otherwise use, and *in particular the collective aspects of this relationship*" (italics added).

Besides the fact that this convention has not been ratified by a large number of States, these provisions are far from creating a legal basis for IP protection of TK held by indigenous peoples. These provisions offer a clear example of a soft law *negotium* introduced in a hard law *instrumentum*.[102]

However, at least at the conceptual level, this Convention lays the foundations of the inter-relationships between cultural heritage law, land rights, and cultural rights of indigenous peoples in their own traditions, including TK.

[100] *Ibid.* Article 1(b).

[101] *Ibid.* Article 3(1) of the Convention provides for a non-discrimination clause regarding male and female members of the indigenous peoples; this is a provision that can be found in legal instruments concerning the appropriate sharing of benefits from the exploitation of TK. Some general human rights principles of this ILO Convention have found their way into national laws on TK.

[102] *Negotium* refers to the actual content of the international instrument, that is, the rules and obligations indicating behavioral expectations or "obligations of good will," such as the obligation to act in an appropriate manner or the obligation to consult, negotiate and co-operate. When States have to negotiate on a certain subject-matter it is generally easier that they reach a consensus on a *negotium* in soft law rather than in a hard law *instrumentum*. And this is so even if the wording of the *negotium* is so precise that some may define it as hard law. Consequently, soft law instruments are proliferating in international law.

3.3.4.2 The Universal Declaration on Human Rights and the Covenants

Article 27(2) of the UDHR states that "everyone has the right to the protection of the moral and material interests resulting from any scientific, literary or artistic production of which he is the author." Read together with Article 15(1)(c) of the ICESCR, which states that "to benefit from the protection of the moral and material interests resulting from any scientific, literary or artistic production of which he is the author", no particular necessity of protecting indigenous knowledge is indicated. As a consequence, national or international IP law can hardly constitute a breach of either article.

Since the subject-matter of the rights of an indigenous community is collectively shared, collective rights of indigenous and local communities "are excluded from the human rights standard on IP, which is an individual right."[103] It is also true that indigenous groups often do not have a concept of individual private ownership of property. The lack of individual assignability of the right of ownership is the first obstacle to recognizing a specific human right over collectively owned TK. The justiciability of human rights has been based on the individual right vis-à-vis the State: "This emphasis may limit the utility of Western concepts in helping indigenous peoples maintain their identity and rights in the face of pressure to assimilate and yield to the 'modern' world."[104] The legal exercise and enforcement of a collective right is entrenched with major complexities stemming from the applicability of inner-tribe customary laws, which, in turn, are based on structures composed by decision makers, elders, shamans, and group healers.

Despite these problems of definition, these provisions contain a "soft legal basis" for indigenous and local peoples to be entitled to benefits arising from the use of their knowledge and resources. What is more, this concept can be used to buttress other international legal instruments that have the same kind of mandated obligation, such as the CBD.

3.3.4.3 Draft UN Declaration on the Rights of Indigenous Peoples

The Working Group on the Draft Declaration on the Rights of Indigenous Peoples has been working on a precise definition of what an indigenous people is and which internationally protected right it should enjoy. The Draft Declaration on the Rights of Indigenous Peoples is a step forward in

103 Stoll and Von Hahn, "Indigenous Peoples", 18–19.

104 J. R. Axt, M. L. Corn, M. Lee and D. M. Ackerman, *Biotechnology, Indigenous Peoples and Intellectual Property Rights, Congressional Research Service* 27 (The Library of Congress, Washington DC, 1993).

international law for the rights of indigenous peoples in matters of "(a) the definition of indigenous peoples; (b) self-determination; (c) ownership of land and resources; (d) establishment of distinct political and economic institutions; and (e) national and territorial integrity."[105] It sets forth human-rights principles which have an impact on TK and biodiversity. It not only accepts the rights of self-determination of indigenous peoples in Article 3 but it also confers the "collective right to live in freedom, peace and security as distinct peoples" (Article 7.2), and a right to the full recognition of their laws, traditions, and customs (Article 27).

More specific provisions can be found in Article 12.1 of the Draft Declaration:

> Indigenous peoples have the right to manifest, practice, develop and teach their spiritual and religious traditions, customs and ceremonies; the right to maintain, protect, and have access in privacy to their religious and cultural sites; the right to the use and control of their ceremonial objects; and the right to the repatriation of their human remains.

The former version of the Declaration called on States to abstain from removing indigenous peoples from their lands or territories (Article 29), to respect their traditions and indigenous knowledge, and to restore and protect the environment (Article 28).

Article 31 builds upon the concepts developed in the ILO Convention by stating that:[106]

1. Indigenous peoples have the right to maintain, control, protect and develop their cultural heritage, traditional knowledge and traditional cultural expressions, as well as the manifestations of their sciences, technologies and cultures, including human and genetic resources, seeds, medicines, knowledge of the properties of fauna and flora, oral traditions, literatures, designs, sports and traditional games and visual and performing arts. *They also have the right to maintain, control, protect and develop their intellectual property over such cultural heritage, traditional knowledge, and traditional cultural expressions.* (italics added)
2. In conjunction with indigenous peoples, States shall take effective measures to recognize and protect the exercise of these rights.

Article 23 of the Draft Declaration grants:

> the right to determine and develop priorities and strategies for exercising their right to development. In particular, indigenous peoples have the right to be actively involved in developing and determining health, housing and other

[105] L.18/Rev.1, www.ohchr.org/english/issues/indigenous/groups/groups-02.htm.
[106] Stoll and Von Hahn, "Indigenous Peoples", 22–23.

economic and social programmes affecting them and, as far as possible, to administer such programmes through their own institutions.

According to Article 25 "indigenous peoples have the right to maintain and strengthen their distinctive spiritual relationship with their traditionally owned or otherwise occupied and used lands, territories, waters and coastal seas and other resources." Various authors state that the reference to "other resources" could be broadly interpreted to cover TK and traditional resources.[107] Instead of totally leaving the implementation of Article 31 to the discretion of States, Article 11.2 gives some guidelines, though not very precise, as to the matter of:

> redress through effective mechanisms, which may include restitution, developed in conjunction with indigenous peoples, with respect to their cultural, intellectual, religious and spiritual property taken without their free, prior and informed consent or in violation of their laws, traditions and customs.

The affirmation of certain rights of indigenous peoples over land in very broad terms has posed major obstacles towards wide acceptance by the majority of States. The text was adopted by the first session of the Human Rights Council on June 29, 2006.[108] The third committee of the General Assembly adopted a resolution to hold further consultation and adopt a declaration during the 61st Session of the General Assembly ending in September 2007.[109]

The *instrumentum* of the General Assembly Resolution is soft and non-binding but its *negotium* may have a normative value of customary law especially with regard to Article 31 that specifies that indigenous people have the "right to maintain, control, protect and develop their intellectual property over such cultural heritage, traditional knowledge, and traditional cultural expressions." The choice of verbs reveals a willingness on the part of States to protect the inherent rights of these people to the point of openly mentioning IP which includes the exclusive right. In an objectivist perspective, the precision and specifity of this *opinio juris* may (for some even instantaneously) transform the content of the provision from soft law of the resolution into a customary norm that renders TK as an IP subject-matter held by the indigenous people. It is certainly premature to interpret this draft resolution as as an expression of the *opinio juris communis* (international common consent)

[107] *Ibid.*, S. K. Verma, "Protecting Traditional Knowledge. Is a Sui Generis System an Answer?" (2004) 7(6) *Journal of World Intellectual Property* 765.
[108] UN Doc. Assembly/AU/ Dec. 141 (VIII).
[109] Working Group of the Commission on Human Rights to Elaborate a Draft Declaration in accordance with § 5 of the UN General Assembly Resolution 49/214 (December 23, 1994).

with the aim of rendering these principles customary norms. Nevertheless, this trend may crystallize[110] customary principles that have strong relevance to the relationships between TRIPs Agreement and the CBD (see also section 6.1.1).

3.4 Conclusion of Part II

Part II has demonstrated how the TRIPS Agreement has a strong impact on the commercial exploitation of GR-related TK. Contracting parties to both TRIPS (that regulates the private property innovation in this field) and the CBD, UPOV, IUTPGRFA (regulating the protection, preservation, and conservation of GR and TK upon which the formal inventions are based) encounter hurdles implementing both bodies of treaties in their national laws. Because TRIPS only recognizes a patent in the case of novelty, industrial application, and non-obviousness, and the CBD provides pre-existing rights to GRs and related TK. Prior informed consent and mechanisms of benefit sharing are still to be incorporated in the international IP system. However, it is possible to come to the conclusion that all the provisions of TRIPS Agreement and the CBD can be simultaneously applied. Hence, they do not contain *per se* conflicting norms. The TRIPS Agreement and Article 12.3(d) of the ITPGRFA are in serious conflict in the reciprocal relationship between countries where one is party to both treaties and the other is party only to the TRIPS Agreement.

The study of conventions stemming from international environmental law and international IP law has also demonstrated that there is no international agreement on a precise definition of the status and the rights of indigenous peoples and local communities with regard to IPRs on GR and related TK. Yet, defining the rights of indigenous peoples and local communities has been the intention of various international human rights and environmental policy bodies for several years. The maintenance of traditional lifestyles and their contribution to the resources valuable for humankind at large is generally promoted by all these documents and declarations.

From all these international legal instruments the conclusion can be drawn that the obligation to include the protection of GR related

[110] The term "crystallization" is borrowed from the world of chemistry. This term describes the opposite process from dissolution, whereby a solid is dissolved in a solution. Crystallization is used to describe the process whereby the solute parts slowly from the solution in crystal form. Whereas precipitation is an instant process, as it is the result of a chemical reaction, crystallization is a slow one since it results from evaporation of the solvent; see also the explanation given by G. Abi-Saab, "Cours général de droit international public" (1987) 207 *Le Recueil de cours de l'Académie de droit international* 160–169, 171.

TK in the IP system lies more on the shoulders of the international community as a whole than upon each individual State. This obligation consists in a duty to negotiate an international regime that protects TK from misappropriation for commercial purposes. This is a prospective obligation rather than one that is currently incumbent upon States. All these findings reaffirm *a fortiori* the rights of States over their GRs and TK vis-à-vis foreign entities that access and commercially exploit those resources.

Part III

The protection of traditional knowledge in the international patent system

Part III

The protection of traditional knowledge in the international patent system

Part III focuses on the IP methods of redistribution of benefits with TK holders in the international trade context and analyzes the proposals that have been put forward by governments, the international legal doctrine, the recommendations of international organizations, statements made by States, and NGO-sponsored studies in order to make the IP system more supportive of the benefit-sharing treaty provisions.

In this regard, it will be necessary to outline a few preliminary relevant conceptual distinctions along with a clearer definition of the legal concept of TK so as to present some viable options including some CBD principles in the implementation of the TRIPS Agreement (see section 4.2 below).

The main existing options within the narrow boundaries of IP laws to internationally protect TK attached to GRs are both "defensive" and "positive/offensive".

Among the defensive protection mechanisms, the introduction of a certificate of origin in the patent examination system can help trace more easily acts of misappropriation of GRs. This can encourage compliance with the obligations under the CBD, i.e. the duty of bio-prospecting companies to negotiate an agreement with the provider country concerning the conditions of GRs' utilization (see section 5.2 below). PIC implementation can occur through the introduction of disclosure requirements and certification about the GRs issued by the country of origin during the application procedure at the Patent Office of the recipient State. Apparently a simple requirement, the submission of certificates of origin is surrounded by difficulties and inconsistent interpretations both within WIPO and the TRIPS Council. This proposal can also be facilitated by the compilation of TK in databases (see section 6.2.3 below). If DCs let patent examiners consult their TK databases, they can be of paramount importance to prevent the mistaken granting of patents whose claims overlap existing TK. This TK can be considered prior art able to destroy novelty (see section 6.2 below).

In section 6.3 the exception to patentability of *ordre public* and morality of Article 27.2 of the TRIPS Agreement will also be thoroughly examined with a view to modulating the patent system and making it more compliant with environmental concerns. Although this section is more concerned with the relationships between TRIPS and CBD than TK protection in the IP system, it also explores the possibilities in which the patentability expectations of *ordre public* and morality can include claims concerning misappropriation of TK.

With regard to the positive protection of TK related to PGRs, a *sui generis* system of PVP, as imposed by Article 27.3(b) on all WTO Members, should be studied first. At this juncture, the analysis moves into the more delicate issue of applying certain TRIPS provisions to the needs and expectations of local communities. This matter will be considered through the implementation of the "effective *sui generis* system" mandated by Article 27.3(b) of TRIPS, i.e. the interaction between patent rights and the PVP conceived in the DCs for plant varieties. Incidentally, this section sketches the methods of matching this obligation with the exigency of defending "farmers' rights." This will require me to construe systems on the basis of the major instruments, the subject of analysis up to this point.

4 Towards clearer legal definitions

4.1 An intellectual property approach to the concept of traditional knowledge

TK, by gaining importance, has become the new buzzword for IP law. In an attempt to define TK with the view of eventual protection in mind, the WIPO IGC on IPGRTKF of the WIPO uses the term to refer to "tradition-based literary, artistic or scientific works; performances; inventions; scientific discoveries; designs; marks, names and symbols; undisclosed information; and, all other tradition-based innovations and creations resulting from intellectual activity in the industrial, scientific, literary or artistic fields."[1]

The notion "tradition-based" refers to knowledge systems, creations, innovations, and cultural expressions "which have generally been transmitted from generation to generation; are generally regarded as pertaining to a particular people or its territory; have generally been developed in a non-systematic way; and, are constantly evolving in response to a changing environment."[2] TK in everyday life is governed by a series of holistic and dynamic local and indigenous customary laws and where there is no distinction between the sacred and secular. This holistic outlook on life accounts for the way these communities operate with a sacred notion of the inherent "one-ness" between man and the natural order, where all living things are interrelated and interdependent. There follows from this view a pervasive respect for the environment and an acute sensitivity to its agricultural and medicinal qualities. Such communities have evolved complex relationships with their surroundings, often expressed through totemic relationships with various species and religious ceremonies involving the celebration of the human–nature interaction.

[1] *Intellectual Property Needs and Expectations*; T. Taubman, 'Genetic Resources', in S. Von Lewinski, *Indigenous Heritage and Intellectual Property* (Kluwer, The Hague, 2nd edn, 2008) 192–216.

[2] *Intergovernmental Committee on Intellectual Property and Genetic Resources, Traditional Knowledge and Folklore Elements of a* Sui Generis *System for the Protection of Traditional Knowledge* (March 29, 2002) WIPO/GRTKF/IC/3/8.

Many traditional communities do not view their TK as property owned by an individual or group. Rather, the possession of a story or medicinal knowledge pertinent to the customs of the community carries with it the responsibility to implement that knowledge with utmost respect for the other community members, animals, plants, and places with which the story or medicine may be associated. Thus, for many traditional communities, their TK entails a bundle of relationships and obligations rather than a bundle of economic rights as under the common law property system. The notion that such elements of TK can be "owned," and with it the possibility that other responsibilities and relationships pertaining to that knowledge could be negated, is incomprehensible.

Categories of TK include the following subject-matters: agricultural knowledge; scientific knowledge; technical knowledge; ecological knowledge; medicinal knowledge, including related medicines and remedies; biodiversity-related knowledge; "expressions of folklore"[3] in the form of music, dance, song, handicrafts, designs, stories, and artwork; elements of languages, such as names, GIs, and symbols; and, movable cultural properties. Excluded from this description of TK would be items not resulting from intellectual activity in the industrial, scientific, literary, or artistic fields, such as human remains, languages in general and "heritage" in the broad sense.[4]

The working definition of TK within the CBD is "the knowledge, innovations and practices of indigenous and local communities embodying traditional lifestyles" as well as "indigenous and local technologies."[5] But these formal and concise statements indicate what TK contains, more than what TK is. A precise and internationally accepted legal definition of TK has not yet been conceived. Meanwhile, the WIPO IGC on IPGRTKF has not deemed any definition necessary for identifying the legal elements of a mechanism for TK's protection. As a matter of fact,

> most patent laws, for example, do not precisely define the concept of an 'invention'; equally, international harmonization and standard-setting in patent law have proceeded without specific or authoritative international definitions of this fundamental concept – although what constitutes an 'invention' has strong elements of

[3] Benin Law on the Protection of Copyright of March 15, 1984 quoted in *Intergovernmental Committee on Intellectual Property and Genetic Resources, Traditional Knowledge and Folklore* (June 13–21, 2002), www.wipo.int/documents/en/meetings/2002/igc/doc/grtkfic3_9.doc.

[4] WIPO, *Intellectual Property Needs and Expectations of Traditional Knowledge Holders: WIPO Fact Finding Mission on Intellectual Property and Traditional Knowledge* (WIPO, Geneva, 2001) 50; Taubman, "Genetic Resources", in Von Lewinski, *Indigenous Heritage*, 192–216.

[5] Articles 8(j) and 18.4 of the CBD. Moreover, the UN has made a significant contribution in this domain in F. Abbott, T. Cottier and F. Gurry (eds.), *The Intellectual Property System: Commentary and Materials* (Kluwer, The Hague, 1999) 506.

harmony in practice, significant differences continue to apply at the national level after some 120 years of progressive international harmonization.[6]

The WIPO IGC on IPGRTKF Secretariat has however decided to break the holistic working concept of TK into two separate categories with corresponding legal tracks: (i) TK related to biodiversity, i.e. genetic (or, more generally, biological) resources such as traditional medicinal know-how, traditional agricultural practices, and local/indigenous planting materials, also called "technical TK"[7] and (ii) TK related to the arts such as handicrafts and expressions of folklore or culture destined to the development of a system duly adapted to the characteristics of expressions of folklore. This distinction appears to have various conceptual advantages in the attempt to make TK subject-matter more suitable for protection under existing IP systems.

Notwithstanding their differences, "GR-related TK" and "cultural expressions and folklore" are really two sides of the same coin. They also pose similar challenges to the existing IP system in that (i) neither of them would qualify for IP protection because they belong to the public domain; (ii) it is very difficult to identify the right-holder since this knowledge is collective, i.e. has been developed within a community.

This study's approach to GR-related TK follows three provisos:

(i) It follows the *summa divisio* propounded by the WIPO IGC on IPGRTKF in order to study possibilities of using and adapting IPRs to accommodate the protection of TK to the greatest possible extent. In other words, my main efforts are aimed at studying how to protect various types of TK subject-matter in the classic or existing IP system. This study will only marginally take into account the new types of IPR or the customary rules capturing the holistic nature of TK for legal protection.

(ii) The use of the term *TK* is a contraction of *GR* or *biodiversity related TK*. This use follows the definition of the CBD. It suffices to consider the lack of an internationally agreed definition of the terms "local" or "indigenous" associated with the "innovations and practices" to imagine how difficult it is to reach a precise definition of this subject-matter so as to fulfill TK holders' objectives in the realm of IP. One can, for instance, wonder whether they comprise aboriginal peoples, or what is the difference between the knowledge held by indigenous and local communities.[8]

[6] WIPO/GRTKF/IC/3/8, 5.

[7] *Traditional Knowledge – Operational Terms and Definitions, Intergovernmental Committee on Intellectual Property and Genetic Resources, Traditional Knowledge and Folklore*, Third Session, WIPO/GRTKF/IC/3/9 (June 13 to 21, 2002).

[8] T. Taubman and M. Leistner, "Analysis of Different Areas of Indigenous Resources", in Von Lewinski, *Indigenous Heritage*, 76–79.

(iii) It must be clear by now that my use of the term TK is more restricted than some of the non-IP binding and soft-law instruments in various fields of international law and policy. This is so because my aim is to include TK in the IP system. I use the more restricted term for the specific purpose of integrating GR-related TK in the *lex lata* of IP. This definition of TK confines itself to the knowledge attached to the GR that constitutes the basis of an IP-protectable innovation. It also shows what the value of TK is and what is the condition to qualify it as such. The term TK is usually associated with "heritage" of a given civilization, thus including "everything that belongs to the distinct identity of a people and which is theirs to share, if they wish, with other people. It includes all things which international law regards as the creative production of human thought and craftsmanship, such as songs, stories, scientific knowledge, and artworks. It also includes inheritances from the past and from nature, such as human remains, the natural features of the landscape, and naturally occurring species of plants and animals with which a people has long been connected."[9] Delving into the analysis of such approach to TK in all its aspects in connection with IPRs would let us fall into the holistic approach that I have excluded from this study. Indeed these aspects of the concept of TK are neither directly protectable nor enforceable through internationally agreed-upon IPRs. For purposes of the present study, such soft-law definitions are then useful only to determine whether a certain knowledge can be qualified as "traditional" or not (see the more precise distinction made at section 4.2.4).

It can be argued that this approach to TK is too minimalistic, thus running counter to the current diplomatic efforts in various fora to include all types of needs and expectations of TK holders. At any rate, this chosen approach does not intend to undermine national, regional, or international efforts to protect TK by new *sui generis* IPRs, even by the adoption of a binding instrument. Nevertheless, it is highly unlikely that industrialized States – who have encouraged the globalization of IPRs – may in turn accept the possibility of granting protection to a new generation of IPRs for this type of knowledge. The compatibility of the maximalist or holistic approach to TK, through *sui generis* IP systems, is fraught with intricate and difficult questions of basic definitions and enforcement that will largely lie outside the scope of this research.

The scope of research encompasses all the commercially used GR-related TK present in any human community, hence also in industrialized societies, and not only in industrialized countries where knowingly "indigenous" or "aboriginal" groups are to be found such as in the US, Canada,

[9] *Report of the UN Economic and Social Council on the Cultural and Intellectual Property of Indigenous Peoples* (E/CN.4/Sub.2/1993/28 (July 28, 1993).

Australia, New Zealand, Finland, Norway, etc. As a matter of fact, the Portuguese decree-law is a first example of definition of TK among EU Member States for the purpose of protection:

> TK comprises all intangible elements associated with the commercial or industrial utilization of local varieties and other autochthonous material developed in a non-systematic manner by local populations, either collectively or individually, which form part of the cultural and spiritual traditions of those populations. That includes, but is not limited to, knowledge of methods, processes, products and designations with applications in agriculture, food and industrial activities in general, including traditional crafts, commerce and services, informally associated with the use and preservation of local varieties and other spontaneously occurring autochthonous material covered by this Decree.[10]

4.2 A few relevant analytical distinctions on biodiversity and related traditional knowledge[11]

Besides the effort of the WIPO IGC in this field and the personal scholarly efforts of the head of its Secretariat, Anthony Taubman,[12] Ricolfi has suggested a few further analytical distinctions that should be taken into account in the international and national law-making process. Outlined in this way, they serve to clarify the subsets of problems in this complex area.[13]

4.2.1 The distinction between provider country rules and recipient country rules

The term "country providing genetic resources" is defined in Article 2 of the CBD as "the country supplying GRs collected from *in situ* sources, including populations of both wild and domesticated species, or taken from *ex situ* sources, which may or may not have originated in that country." Furthermore, Article 15.3 of the CBD states that "[f]or the purpose of this

[10] *Document Submitted by Portugal Ministry of Agriculture, Rural Development and Fisheries*, Decree-Law n. 118/2002 stated in WIPO/GRTKF/IC/8/13 www.wipo.int/edocs/mdocs/tk/en/wipo_grtkf_ic_8/wipo_grtkf_ic_8_13.doc

[11] This section is mainly inspired by a joint conceptual work directed by Professor Ricolfi, University of Torino, for the Istituto Agrononico per l'Oltremare of the Italian Foreign Ministry. I will summarize, quote, and systematize his findings and add a few additional comments; M. Ricolfi, "Intellectual Property and Biodiversity: A Review of Legal and Conceptual Issues and of Policy Options" (2004) *Atti del Seminario*, Istituto Agronomico per l'Oltremare Firenze, 40, http://brasile.iao.florence.it/documenti/ricolfi.pdf.

[12] Taubman and Leistner, "Analysis of Different Areas of Indigenous Resources", in Von Lewinski, *Indigenous Heritage*, 59–180. Taubman, 'Genetic Resources', *ibid.*, 181–292.

[13] Ricolfi, "Intellectual Property and Biodiversity," 37. S. Bragdon and H. Bragdon, "Major Legal Regimes Affecting Plant Genetic Resources (PGR): The Convention on Biological Diversity (CBD), the International Undertaking (IU) and the TRIPS Agreement", in T. Cottier and P. Mavroidis (eds.), *Intellectual Property: Trade, Competition and Sustainable Development* (University of Michigan Press, 2003) vol. 3, 448–49.

Convention, the GRs being provided by a Contracting Party, as referred to in this Article [...] are only those that are provided by Contracting Parties that are countries of origin of such resources or by the Parties that have acquired the GRs in accordance with this Convention." The term "Contracting Party providing genetic resources" is also used in Article 15.7 of the CBD and in paragraphs 16(d)(iii) and 24 of the Bonn Guidelines.[14]

In this regard, Ricolfi observes that:

> biodiversity may be preserved, with or without human intervention, in one given country, while subsequent technological contributions may be applied to a GR (usually in laboratory conditions) in another country. If this is the case, then it may generally be said that a transfer (of genetic material, of genetic information or of other items, as the case may be) has taken place. In this perspective, rules concerning the first country may be visualized as *provider country rules*; rules concerning the second country as *recipient country rules*.[15] (italics added)

4.2.2 *The distinction between genetic resources for food or for drugs as a seed or as a food*

Depending on the point of observation, the GR can be considered as information, as a food, or as a drug. If it is considered as information, at an abstract level, the resource is researched to identify its DNA sequences, the portions of its code which account for relevant properties (e.g. resistance to saline agents, nutritional properties, and active principles accounting for therapeutic effects). Still on the information level, the resource should be considered as self-replicating, so that the first generation is liable to generate subsequent generations. In the case of a food GR, the resource may be considered a seed (if it is intended to generate subsequent generations of the same resource) or food (if it is intended for human consumption). The economic value of the information and the TK can be different, and this distinction is very relevant for drawing up contracts on benefit sharing (see section 5.2 below).

4.2.3 *The distinction between the place of geographical origin and the country of initial origin*[16]

The place of geographical origin does not always coincide with the country of initial origin. The "place of geographical origin" is referred to as the

[14] *Bonn Guidelines on Access to Genetic Resources and Fair and Equitable Sharing of the Benefits Arising out of their Utilization*, COP of the CBD Decision VI/24 (April 7–19 2002), www.biodiv.org/doc/meeting.aspx?mtg=Cop-06.

[15] Ricolfi, "Intellectual Property and Biodiversity", 40.

[16] S. Klemm, "Origins and Allocation of Traditional Knowledge and Traditional PGRFA: Basic Questions", in T. Cottier and S. Biber-Klemm (eds.), *Rights to Plant Genetic Resources and Traditional Knowledge* (CABI, 2006) 165–72.

country of source (see section 6.1 below), and the "country of initial origin" is referred to as the country of origin. There is substantive evidence that many PGRs have widely and even transoceanically traveled the earth both because of natural agents and human migration.[17]

It is widely believed that the practice of using a given PGR as food usually stretches over a more or less wide area and goes back substantial periods of time so that it may become problematic to univocally identify a specific place from which the resource derives. In these cases, it may be easy to identify the place from which a specific physical resource has been derived (place of geographical origin), but this does not mean that the same is identical to the country of initial origin.

The first step is the precise determination of the country of origin of a given GR under Article 2 of the CBD[18] or of the center of origin under Article 2 of the ITPGRFA[19] by domestic provider country legislation.

Ricolfi observes that PGRs "may be intended for a variety of uses, ranging from human consumption as food to therapeutic uses (e.g. use of the active principle identified in a specific PGR to cure a specific disease). There are reasons to keep the two situations separate, at least for some analytical purposes and in particular to identify, assess or generate the appropriate rules."[20]

In the case of a PGR used as a traditional drug, it is usually a resource containing an active principle. This type of PGR appears to be found only in a much more restricted ("discrete") area than a PGR used as food. A PGR used as a drug may have resided in the same place for a long time either unknown or known to a restricted number of members of the local population. As a matter of fact, it would appear that knowledge of therapeutic properties of PGRs does not tend to travel as extensively as knowledge of their nutritional properties.

The methods of identification of the country of origin should be enshrined within provider country domestic legislation. It is possible to spell out a link between this matter of the country of origin of PGRs and copyright law. In copyright law, it has been convincingly argued[21] that when determining the ownership regime for copyright protected works, it

[17] J. Sorensen and C. Johannessen, "Biological Evidence for Pre-Columbian Transoceanic Voyages", in V. Mair (ed.), *Contacts and Exchange in the Ancient World* (University of Hawaii Press, 2006) 238–98.

[18] Article 2 of the CBD states: "'Country of origin of genetic resources' means the country which possesses those genetic resources in *in-situ* conditions."

[19] Article 2 of the *FAO ITPGRFA* states: "'*Ex situ* collection' means a collection of plant genetic resources for food and agriculture maintained outside their natural habitat."

[20] Ricolfi, "Intellectual Property and Biodiversity," 3.

[21] L. C. Torremans, "The Law Applicable to Copyright: Which Rights Are Created and Who Owns Them?" (2001) 188 *Revue international du droit d'auteur* 37, 71.

is preferable to adopt a uniform solution by exclusively referring to the country of origin of the work instead of determining it in accordance with a variety of criteria by the laws of the multiple jurisdictions where the work may be exploited. In the same manner each provider country should take all the necessary measures to ensure that a certain GR has its origin in its country.

In the law-making process or in the establishment of benefit-sharing contracts, another distinction that needs to be taken into account is the one between ownership *in situ* and *ex situ*, with the "possibility that different ownership rules apply to *ex situ* as opposed to *in situ* resources."[22]

4.2.4 *The distinction among the genetic resource, traditional knowledge referring to it, and technology applied to it*

The distinction between the resource itself provided by nature and the human ingenuity applied to it (whether technology or TK) may be appropriate to define the scope of applicable international legal instruments and the negotiating position of various countries that receive or provide technology.

GRs *per se*, "pristine and untouched" from time immemorial, may prove to be valuable for the application of human ingenuity expressed in the form of TK or laboratory technology. The application of human ingenuity over TK related to a GR occurs through the selection or the breeding of varieties that are the most apt to the intended use or by applying laboratory technology and even biotechnology. The research applied on TK can consist of the identification of its health or agricultural properties (see the full application of this distinction in chapter 5).

This analysis is mainly interested in cases where the information contained in a GR is used in a laboratory for further innovation. This information contained in the GR may have been preserved by informal human contribution, that we call TK, as a sort of added value to the GR *per se*.

On the one hand, GRs *per se* may be protected through pre-IP or non-IP regimes, i.e. a legal regime that precedes the acquisition of the IPR based upon the GR *per se*. This regime will call for the compliance with a regime of access conditions including benefit sharing, before the application of an IPR. Indeed, the CBD as well as the ITPRGFA place GRs in their wild state or naturally occurring in the mandated access regime (e.g. the Multilateral System in section 3.3.3.1 or domestic ABS in chapter 5). On the other hand, TK calls for an IP or quasi-IP regime.

[22] Ricolfi, "Intellectual Property and Biodiversity," 37.

CBD covers both plant and animal GRs as such and contains some provisions for the protection of TK concerning the resource itself. One of the main differences between TK related to GRs and the GRs *per se* is the *term of protection*. TK is a product of human ingenuity as much as it is of technological innovation. Hence, any types of protection granted for it must provide for a term of protection. Ricolfi observes that GRs are valuable because they "serve as a kind of insurance against the risks deriving from genetic erosion – e.g. of diseases affecting the main crops. They exist in their untouched or pristine form just because they have been preserved for a very long time, either because they have been let alone by local populations or, if there has been human intervention, this kind of action has not translated into tampering with their genetic make up."[23] Given these facts, "a final term for access and benefit sharing does not make sense in connection with a regime concerning the GR *per se*."[24] The existence of a term of protection determines the fact that the regime applicable to TK is akin to IP (or quasi-IP), whereas the one applicable to GRs *per se* is pre-IP, because it deals with rules concerning ABS relating to resources to which no human ingenuity has been applied at the time of access and until the time technological innovation acts upon it.

The CBD, the FAO ITPGRFA, and the TRIPS Agreement mandate a type of protection of TK, though it is not clearly specified. In this context the relationship between GRs and TK raises a number of questions to which the following chapters seek answers: what is the connection between the rules pertaining to such knowledge and the access regime concerning GRs which may or may not be associated with such knowledge? Should the two sets of rules be linked to each other or be kept separate? Is the adoption of the one a prerequisite for the adoption of the other? Just to indicate what direction this line of reasoning could take, is it possible under the present international framework to adopt rules so that TK is protected under an IP (or quasi-IP) regime, while GRs are subject to a non-IP (or pre-IP) regime?

4.2.5 *The distinction among types of protectable knowledge*

Apart from traditionally held customary laws implicit in the normal functioning of a community, there are several types of TK that deserve special attention.[25] The most important of these include TK as it relates to (i) the

[23] *Ibid.*, 40. [24] *Ibid.*, 30.

[25] Taubman and Leistner, "Analysis of Different Areas of Indigenous Resources", in Von Lewinski, *Indigenous Heritage*, 89–90 (2008).

environment, (ii) agricultural knowledge, (iii) medicinal knowledge, and (iv) technology.

Environmental TK typically refers to that knowledge that a community derives from its interaction with the terrain, meteorology, ecosystem, and biodiversity of a particular area. This knowledge is also known as traditional ecological knowledge (TEK), that is to say, those aspects of TK that are specifically directed at the maintenance and conservation of the environment and its limited natural resources. Both agricultural and medicinal knowledge fall under the umbrella of traditional ecological knowledge.

Traditional agricultural knowledge (TAK), a subset of TEK, is particularly pertinent to this study in that it concerns all knowledge relating to plant and animal species as well as breeding techniques. TAK also extends to knowledge of soil types, ground preparation, pest control, crop rotation, animal husbandry, harvesting and storage techniques, etc. Farmers applying TAK usually follow long-standing farming and land use practices designed to conserve the biodiversity of the area, regulate the exploitation of natural resources, as well as provide for other local benefits as stipulated by the customary laws and the TEK of the region.

Traditional medicinal knowledge (TMK) was first officially recognized as a source of primary health care by the World Health Organization (WHO) in the Primary Health Care Declaration of Alma-Ata (1978) and has been globally addressed since 1976 by the traditional medicine programme of the WHO. That program defined TMK as "the sum total of all the knowledge and practices, whether explicable or not, used in diagnosis, prevention and elimination of physical mental or social imbalance and relying exclusively on practical experience and observation handed down from generation to generation, whether verbally or in writing."[26] In this context, TMK is not just knowledge of plant- and animal-based remedies, but also the knowledge and performance of healing rituals used in their application. An important distinction must be made between traditional and indigenous medicine in South Asia and China. On the one hand, you have a system of codified "traditional medicine" while on the other you have an "indigenous," non-codified system of medicinal know-how.

A qualifying element of these types of TK can be defined as "traditional technology." Technology in relation with TK should not be confused with laboratory innovation technology that Western cultures are familiar with. By traditional technology is meant those inventions created by tradition-based communities to better manage their relationships with their

[26] *General Guidelines for Methodologies on Research and Evaluation of Traditional Medecine* (2000)WHO/EDM/TRM/2000.1, www.whqlibdoc.who.int/hq/2000/WHO_EDM_TRM_2000.1.pdf.

respective environments. These technologies range from the simple yet remarkably efficient portable tool-kit developed by indigenous peoples in Australia, to complex irrigation systems such as the famed rice terraces of Ifugao in the Philippines.

The above list is not an exhaustive list of all the types of TK. TK is also found in sacred rituals, folklore, and language. Because such educative elements are common mechanisms for propagating cultural values, not surprisingly much of a community's TK is preserved within these activities. However, because they are at best tangential to the scope of this study, they are given only passing mention. The epistemological status of the distinctions drawn so far is mainly operational and not objective. Their main purpose is to help us in identifying and generating rules apt to protect TK.

4.2.6 *Ownership/sovereignty of/over the material and ownership/sovereignty of/over the information*

Any ABS domestic provider-country legislation should draw a distinction between the regime concerning untouched or pristine GRs on the one hand and those that have come down in the present form as a consequence of human intervention on the other hand. In connection with the latter, legislation should in turn:

> distinguish between ownership of the specific material and physical specimen of the resource on the one hand and ownership of the information content which a given specimen has in common with the other specimens of the same taxonomic description and that distinguishes it from specimens of a different description. Indeed, individual ownership of the former (the individual resource specimen) is compatible with collective or communal ownership of the information incorporated in it.[27]

The provider country has to set up an "access right,"[28] i.e. the authority may prevent access to its GRs by any person or entity, except if in compliance with a set of predetermined rules.

These rules can be grounded in public law in accordance with the State sovereignty prerogatives. The State can alternatively establish a property right (*in rem*) over the GRs present in its territory.[29] In this case the State

[27] Ricolfi, "Intellectual Property and Biodiversity", 40. Article 4(1) of the *Act of the Umbria Region* in Italy of September 4, 2001, n. 25.

[28] T. Heide, "Copyright in the EU and the U.S.: What 'Access Right'?" (2001) *European Intellectual Property Review* 469.

[29] T. W. Merrill and H. E. Smith, "What Happened to Property in Law and Economics?" (2001) 111 *Yale Law Journal* 357.

would establish an exclusive intangible right over the GR[30] – thereby inevitably raising the question of identifying its holders. It does not seem inevitable that the corresponding property right, or "intangible right," is conceived as IP.

There are two options that can be followed by an ABS regime that protects the PGRs *per se*, as Ricolfi notes: (i) it may protect the PGRs *per se*, independently from the technical innovation or the related TK; (ii) it may also protect the DNA information that PGRs carry in them.

Here the question arises whether we are in the presence of an IPR or a non-IPR. Ricolfi solves the problem by stating that it should be identified as a "pre-IPR" to the extent this property right may turn out to be valuable insofar as TK concerns it or technology innovation is applied to it, i.e., it is not an IPR *per se.*[31]

Here the complexities arise as regards the country of origin and the country of source (see section 4.2.3 above).[32] The problem of original sovereignty of PGRs *per se* is very complex and is one of the reasons industrialized countries do not want to participate in a clear internationally binding mechanism. The protection may indeed open a Pandora's box of intersecting claims over PGRs (mainly among developing provider countries) while at the same time PGRs are used by formal innovators in industrialized countries. The creation of a clearing house mechanism is an attempt to solve this problem.[33]

4.2.7 *Difference between commercial and research exploitation*

States or country authorities are free to set forth different regimes of access to GRs depending on their commercial or research exploitation. Adequate consideration should be paid to the needs of researchers, through, for instance, a smoother access regime "in terms of timing for authorization; of waiver of rights to compensation and the like."[34] A monitoring mechanism could be set in place to avoid the problem of an access originally intended for research purposes.[35]

[30] B. Sherman, "Regulating Access: Intellectual Property Law and Biodiscovery" (2003) 13 *European Intellectual Property Review* 301.
[31] Ricolfi, "Intellectual Property and Biodiversity", 31–32. [32] *Ibid.*, 32.
[33] S. Biber-Klemm and J. Curci, "Clearing House Mechanisms", in Cottier and Biber-Klemm (eds.), *Rights to Plant Genetic Resources*, 269 ff.
[34] *Ibid.*, 38. [35] *Ibid.*, 38.

5 The construction of an access- and benefit-sharing regime and intellectual property issues: criteria and options

This section explores an ABS regime in accordance with the mandated provisions of the relevant international legal instruments and the major policy options that States and the international community can follow.

While in other fields of IP law, one can avail oneself of comparable experiences, I do not know of any analogy in establishing an access regime. For instance, when shaping new types of protection like the neighboring rights of phonogram producers, of software, or of the *sui generis* protection of databases, one would resort to copyright and other *sui generis* regimes as paradigms to follow or to depart from according to the circumstances. In contrast, some of the possible suggestions aiming at protecting TK and GRs originally stem from IP law, e.g. the CBD and the ITPGFRA legal regimes.[1] Here resides the difficulty in spelling out an IP system inspired by non-IP standard concepts and non-IP parallel implementation regimes.[2]

In this endeavor, the concept of *mutual supportiveness*[3] plays a pivotal role as do all the other rules of interpretation of applicable treaties and the rules for solving conflicting provisions.

The bilateral correspondence of the efforts on the part of biodiversity recipient and provider countries is absolutely indispensable for ensuring that the system is functional. These efforts consist of adopting and enforcing rules aimed at guaranteeing compliance with biodiversity provider rules regarding access in connection with the grant of IP protected

[1] The first is already implemented by various countries; the latter is still at the preparatory stage.

[2] J. O'Hagan and C. McAndrew, "Restricting International Trade in the National Artistic Patrimony: Economic Rationale and Policy Instruments" (2001) 10 *International Journal of Cultural Property* 32; M. D. Birnhack, "The Dead Sea Scrolls Case: Who is an Author?" (2001) 23(3) *European Intellectual Property Review* 128; and recent legislation providing for protection of technological measures, as seen in the analysis by T. Heide, "Copyright in the EU and the US: what 'Access Right'" (2001) *European Intellectual Property Review* 469.

[3] R. Pavoni, "Accesso alle risorse fitogenetiche e diritti di proprietà intellectuale dopo il trattato dalla FAO del 2001" (2003) 3 *Communità internazionale* 369, 382–83.

innovations incorporating GRs originating from provider countries. If access rules in provider countries imply a process that is slow, unpredictable in its extension over time, or implies a consent by a multitude of authorities, or no authorities exist, the whole mechanism becomes unworkable.[4] Not only State authorities but also indigenous peoples and research institutions to foreign private companies, NGOs and bureaucracies must make a sound public choice analysis: this part of the study is therefore devoted to examining criteria and options that represent common denominators for the sound choices operated by provider countries.

Section 5.1 will study the generalities of the CBD mandated regime, that combines the two elements of ABS and the contractual solution. Section 5.2 will observe the fragility of the contractual solution and then proposes ways in which the national ABS regime can include mandatory contractual provisions in the relations between recipient industrial parties and providing countries and/or indigenous communities.

5.1 The CBD mandated access- and benefit-sharing regime

Various countries (mainly biodiversity provider countries) are currently drafting legislation intending to control access to GRs. This section outlines the main choices that stand before provider countries as they shape ABS legislation in accordance with the CBD. The construction of an ABS regime can be divided in different options inspired by the State practice and legal doctrine. This section is mainly inspired by Article 15 (on the connection between the provider and recipient entities) and Article 8(j) (on national implementation).

A very important aspect of the CBD-mandated protection is the participation of the TK stakeholders. In order to achieve the objective of the sustainable use of the resources, ABS provider domestic legislation should clarify the role of indigenous peoples in the benefit-sharing process. The "Guidelines on Access to GRs and Fair and Equitable Sharing of the Benefits Arising out of their Utilization" (Bonn Guidelines)[5] provided a detailed guide for the development of ABS regimes in accordance with Article 8(j) of the CBD (No. 1 and 9). One of the objectives should be to "contribute to the development [...] of mechanisms and ABS regimes that recognize the protection of TK [...] in accordance with domestic laws and relevant international instruments." The Guidelines encourage the participation of the TK stakeholders in negotiations on the implementation

[4] M. Heller, "The Tragedy of the Anticommons: Property in the Transition from Marx to Markets" (1998) 111 *Harvard Law Review* 621–88.

[5] Bonn Guidelines.

of ABS legislation at the national and regional level. Moreover, States should provide the pertinent information and capacity-building as support for negotiation (Nos. 17–21).[6]

The access to GRs can be regulated through a State authority with the objectives of preservation and the promotion of trade in the informational values of TK and GRs. Two main options are here available in this regard: (i) the protection of TK can be a prerequisite for the implementation of, and in a legislative instrument separate from an ABS domestic provider country legislation, or (ii) TK and ABS matters can be combined in the same piece of legislation.[7]

5.2 An access- and benefit-sharing regime and the contractual solution

The CBD encourages mutual agreements between bioprospecting companies and provider countries with the hope of, or better, with the expectation of attaining equitable benefit sharing. It more precisely dictates that States shall have the "authority to determine access to their GRs" (Article 15), and, in addition, it contains the duty of negotiation of agreements between bioprospecting companies and provider States as a condition for accessing biological resources (see Article 8(j)).

While property and tort law may protect the holder of the resource against any third party and therefore also against those third parties with whom the holder has had no prior dealing, contractual mechanisms apply only to the parties to the agreement. Contractual devices may nevertheless be expanded also to encompass parties that did not have prior dealings, especially when the law provides for an affirmative duty to negotiate.

This section suggests two complementary characteristics: (i) provider countries can include in an access regime a series of provisions protecting the interests of the weaker contractual party, i.e. local communities (see Article 15 of the CBD); (ii) the international

[6] Some of the Contracting Parties are concerned about the voluntary character of the Guidelines. It is feared that due to their voluntary nature, they remain ineffective (Communication from Brazil, *Review of Article 27.3(b)* IP/C/W/228, November 24, 2000) www.docsonline.wto.org/DDFDocuments/t/IP/C/W228.doc. Accordingly, the COP explicitly decided to keep the Guidelines under review, considering that they are but a first step of an evolutionary process (*Bonn Guidelines*). *Report of the World Summit on Sustainable Development, Johannesburg, South Africa* A/CONF.199/20, chapter I, resolution 2, annex (August 26 – September 4, 2002); *International Regime on Access to Genetic Resources and Benefit Sharing. Note by the Secretariat*, UN Doc. UNEP/CBD/MYPOW/6 (January 7, 2003).

[7] M. Ricolfi, "Intellectual Property and Biodiversity", (2004) *Atti del Seminario*, Istituto Agronomico per l'Oltremare Firenze, 38.

community formulates internationally agreed contractual provisions in order to strike a balance on sharing the benefits arising from the (IP) exploitation of PGRs.

It is a common opinion among DCs that the creation of legal conditions that favor and promote local exploitation of patents related to successful compounds would be a more effective measure in compliance with the objectives of TRIPS and the CBD than royalties or lump-sum payments for the transfer of genetic material. That system is the most effective application of the benefit-sharing obligation under the CBD: the DCs, often rich in biodiversity and poor in technology, release to industrialized countries their TK as embodied in the germplasm, and the industrialized countries, rich in technology and often poor in biodiversity, release to the DCs their technical and commercial knowledge as embodied in the newly developed biodiversity-based products and processes.

Biodiversity provider countries may enact laws to establish the production *in situ* of the biodiversity-based innovative products patented by a foreign company, yet, in doing so, they need to be in compliance with international obligations of international patent law. This can be achieved in three different types of statutory contractual provisions as indicated in section 5.2.2.1, the most effective of which is, in my opinion, the "local working of the patent" (see section 5.2.2.1.5).

This matter is approached with great hopes of seeing a transition from an era of confrontation to an era of *cooperation* between developing and industrialized countries. This is one of the areas of IP global issues that urges synergy and real interdependence between technologically advanced and biodiversity rich countries.

This section starts by considering the current legal frameworks of access to GRs and observes the main shortcomings of what I call the "contractual-freedom solution". Lessons can be learned from these experiences that can help shape a flexible series of contractual provisions to be adapted to various regional or national circumstances. In this context, I will outline the importance of transfer of technology to the provider country on the basis of legal logic and some armchair economic and legal reflection focused on the "local working requirement" of patents. All these efforts aim at achieving legal coherence between the TRIPS Agreement and the CBD.

Before delving into these matters, it is important to identify the party with whom a foreign company, interested in screening PGRs, should negotiate, according to the type of PGR that the company wishes to have access to. The following table offers a quick overview of the possible contractual partners according to the contractual subject-matter:

Types of Contract	State	TK holder	FAO Treaty Multilateral System
Contract between foreign company/institution and TK holder to obtain know-how license		YES	
MTA of any non-FAO Multilateral System PGR (see Treaty on PGRFA)	YES		
Material transfer and know-how license (or any IPR on TK)	YES	YES	
Material transfer of PGRs under Multilateral System of FAO Treaty on PGRFA			YES

I limit this analysis to contracts between provider country and foreign company. A deeper analysis of the contractual aspects of the ITPGRFA Multilateral System, i.e. the contract between an entity and the FAO for the utilization of PGRs of the Multilateral System, falls outside the scope of this study.

5.2.1 *The fragility of the contractual-freedom solution*

Contractual freedom is the solution that is proposed by most of the industrialized countries to the problem of benefit sharing on TK related to biodiversity between a private corporation and the provider State or the indigenous or local community.[8]

However, resorting to contractual freedom to negotiate benefit sharing in this field faces three strong limitations:

(i) Contracts or MTAs are negotiated and entered by a provider country and a foreign private entity. There is no international control over the fact that the private entity will enter into negotiation of a contract. The parties may voluntarily decide to comply with access legislation in the first place, but the contractual solution does not at all take care of the problem created by those firms and countries who do not want to comply with access legislation. It goes without saying that there is no control over the way in which the negotiations are handled.

(ii) The contractual parties create their rights by negotiation. Usually, economically powerful companies possess a stronger negotiating

[8] For instance, the US representative to the *Intergovernmental Committee on Intellectual Property and Genetic Resources, Traditional Knowledge and Folklore* (WIPO IGC on IPGRTKF), Intervention of the Delegation of the United States of America to the *WIPO Intergovernmental Committee on Intellectual Property and Genetic Resources, Traditional Knowledge and Folklore*, Third Session, Geneva, Agenda 5: Traditional Knowledge, WIPO/GRTKF/IC/3/7, 8, 9 (June 13 to 21, 2002).

power than provider governments themselves and, *a fortiori*, than the local communities. In these agreements, the interests of the weaker parties are unlikely to be duly protected.

(iii) When TK serves as a starting point for patented products, one very basic shortcoming is that in most cases, TEK is considered as know-how which cannot be qualified *per* se as an "inventive contribution" in the sense of patent law in order to claim "joint inventorship."[9]

When reading the provisions contained in the preamble of several of benefit-sharing agreements between bioprospecting companies and provider countries, one may receive the impression of goodwill to conform to the principles and the spirit of the CBD, especially regarding equitable relations between the parties. However, the substantive provisions hardly match the balanced articulation among the paradigms of bioprospecting/IPRs/sustainable exploitation. One of these examples is the agreement concluded between the US National Institutes of Health and the Government of the Philippines for the screening of specimens for important therapeutical relevance.

Four principal shortcomings recur in many contracts of this type:

(i) The contractual benefit sharing arising out of IPRs remains very uncertain or distant in the future. The main reason lies in statistical figures: generally, only one sample out of 10,000 or even 100,000 gathered specimens will yield the development of a marketable invention. Consequently, the benefit sharing at the contractual stage remains highly unpredictable. Thus, it is extremely unusual to find examples of contracts like the one between the Merck company and the Biodiversity National Institute of Costa Rica In Bio, in which Merck is under an obligation to transfer a certain amount of money for the simple right to gather samples.[10]

At the stage of sample gathering, there is indeed a deep difference between contracts for genetic modification of biological resources and those for direct valorization of biodiversity. In the latter types of contracts, the company or the bioprospector tends to acquire a great quantity of biological resources that will be used in a classical

[9] M. Blakeney, "Bioprospecting and the Protection of Traditional Medical Knowledge of Indigenous Peoples: An Australian Perspective" (1997) 6 *European Intellectual Property Review* 299–300. *Intergovernmental Committee on Intellectual Property and Genetic Resources, Traditional Knowledge and Folklore, Third Session Geneva, Seventh Session, Genetic Resources: Draft Intellectual Property Guidelines for Access and Equitable Benefit-Sharing* (November 1 to 5, 2004) WIPO/GRTKF/IC/7/9, 14.

[10] D. Posey and G. Dutfield, *Beyond Intellectual Property Rights: Towards Traditional Resource Rights for Indigenous Peoples and Local Communities* (International Development Research Centre, Ottawa, 1996) 44.

industrial process; for example, a company will buy seeds and plants or tree barks and use it to create a drug. In these contracts, IP issues are very marginal and are seldom addressed. Examples of these types of contracts include those concluded by Shaman Pharmaceuticals with the Consejo Aguarana/Huambisa in Peru.[11]

The parameters change (i.e. some benefits can flow to the provider-State entities) when the bioprospecting company seeks to obtain GRs in order to develop a search process through biotechnology. But another hindrance appears very soon: in this case once a molecule is synthesized or once the gene is cloned, the biotechnological company will not need to come back to the indigenous community to ask for more samples. The reproduction of the successful sample can be comfortably pursued in a lab without undertaking further sample-gathering activity. As a result, bioprospecting companies are hesitant to continue to remunerate the community or the provider State from which the sample was gathered. More equitable economic relations can be achieved through statutory provisions adopted by the country of origin. National legislation can provide for a contractually defined framework regulating the matter from the genesis of the bioprospecting operations.[12]

(ii) The fragility of the economic relationship inherent in such contracts is due to a second dilemma: indigenous communities will hardly be able to control the development of eventual inventions because the research on the samples is rarely done *in situ*. In other words, local people should invest enormous research efforts to know when and if their biological resource, once genetically modified, has been patented.

(iii) The lack of participation of the local community or of its State in the development process of the biodiversity-based product: to illustrate this shortcoming, I refer to the Agreement between the Peruvian Communities representing the Aguaruna and Huambisa Peoples and G. D. Searle & Co. of Monsanto Group, a US company, where the inherent potential of patenting the GR has endowed the "industrial party" (the licensee) with a much higher contractual power. Indeed, one long article in the agreement provides the sole duty on the licensee of the biodiversity material and industrial developer and owner of the patent on the new drug or seed to "grant back" to the biodiversity-providing licensor a non-exclusive license "for research

[11] *Ibid.*

[12] M.-A. Hermitte, *Mission sur la valorisation de la diversité biologique à Madagascar*, April 2000, www.panjuris.univ-paris1.fr (this document helped the Malagasy Government in drafting new law in conformity with TRIPS and the CBD).

use"[13] but "not for any commercial use." This provision means that the licence could not be conferred to and shared with a locally operating industry. Moreover, according to an article in the agreement, while the indigenous people are free to continue to make and sell their traditional products, it is clearly stated that "all products covered by such patent rights shall be conclusively deemed not to constitute traditional Aguaruna and Huambisa Medicinal Products and any and all methods covered by such patent rights shall be deemed not to constitute Traditional Aguaruna and Huambisa Methods." Thus, there is no provision for the local communities to participate in the industrial development of the new products. As a result, the biodiversity-related innovation based on the traditional germplasm, be it patented or not, will not be profitable to the community whose TEK has been crucial in the preservation of the successful compound.

(iv) Another problem relates to the fairness of the economic relationship set up by the international contract regarding the entities entitled to the compensation. As earlier observed, in the great majority of cases, the benefit-sharing provisions remunerate the resources themselves and very seldom the related know-how transferred by the indigenous communities.[14] The situation changes when a "trust fund" is especially instituted for the indigenous communities. The optimal solution is that the local communities must be either directly involved or even party to such agreements, which is rarely the case. In certain other contracts, the remuneration is uncertain because there is generally no obligation on the government to pay compensation from the benefits to the local populations.[15] In spite of the fact that there is an international mandate to seek fair and equitable sharing of the benefits from commercializing their GRs and related knowledge, contracts currently in existence are far from satisfactory. Moreover, even where a local

[13] C. McManis, *Recent Publications on Indigenous Knowledge Protection – New Directions in Indigenous Knowledge Protection* 71 (ATRIP Collection of Papers, 1999).

[14] In the contractual perspective the industrial party attributes to the biodiversity-providing local communities just a financial return, be it a lump sum and/or a royalty, from commercial exploitation of the new biodiversity based drug or food produce. Such US contractual models like the "Diversa-Yellowstone Cooperative Research and Development Agreement" and "INBio-Merck," both involving mere profit-sharing: reference in "the need and possible means of implementing the Convention on biodiversity into patents law" 388 (*AIPPI Yearbook* 2001/II, 28th Congress, Report of the US Delegation on Question 159).

[15] R.-J. Coombe, "Intellectual Property, Human Rights and Sovereignty: New Dilemmas in International Law Posed by the Recognition of Indigenous Knowledge and the Conservation of Biodiversity" (1998) 6 *Indiana Journal of Global Legal Studies* 59.

community is a direct party to the agreement, it will be much less able to take legal action against a company in another country in case of infringement. Therefore, legal advice at the early stages of the negotiation and the establishment of a "trust fund" can constitute preliminary warranties against such inequitable situations. For instance, local communities can be assisted by IP-knowledgeable NGOs to mediate and evaluate the terms and the implementation of such agreements.

In sum, entire freedom in the negotiation of contracts should not be seen as the best tool to enhance biodiversity valorization and preservation policies. Therefore, international standards should be elaborated to protect the weaker party and at the same time to avoid the forum shopping of bioprospecting companies which, when attracted by important compounds, may go where there is no national legislation protecting the weaker contractual party.

5.2.2 *Statutory contractual norms protecting the weaker party in license agreements*

The various aforementioned examples hint that the achievement of the goals more clearly defined by the CBD may be jeopardized under a "purely private" contractual perspective. It must not be forgotten that the substantial disproportion of contractual and economic power between industrialized countries' companies and DCs is robustly enhanced by the not always transparent diplomatic and political pressures imposed by industrialized countries' governments. The contractual-freedom solution may be criticized because it is undermined by this sort of not-so-private "third-party intervention" in so-called "private" agreements.

Drafting statutory contractual provisions in favor of TK holders is mainly inspired by the CBD that in Article 1, encourages, "each contracting Party to adopt legislative, administrative or policy measures aimed to achieve the fair sharing of results of R&D and the benefits arising from the commercial and other utilization of genetic resources." Note that this adoption must be "*in accordance*" with Articles 16 and 19 of the Convention. Article 16 obliges the developing and industrialized countries to cooperate, subject to national legislation and international law, in order to ensure that patents and other IPRs "*are supportive and do not run counter to [the Convention's] objectives.*" The implied objective is the "*equitable sharing of benefits.*" Article 19 provides that all parties to the Convention "*shall take all practical measures to promote and advance priority access on a fair and equitable basis by Contracting parties, especially DCs, to the results and benefits arising from biotechnologies based upon genetic resources provided by those Contracting parties.*"[16]

[16] Bonn Guidelines.

Moreover, in its objectives, TRIPS unequivocally states that members, "in *formulating or amending* their laws and regulations," should adopt "*measures* necessary to promote the public interest in sectors of vital importance to their *socio-economic and technological development*" (Article 8.1 and Article 7).

For all these reasons, in my view, States that are party to the CBD and TRIPS have an obligation to continue to cooperate in this direction, even to elaborate a certain number of internationally negotiated and agreed-upon standards of protection in favor of the weaker party, e.g. TK holders in the provider country. The international framework would thus prevent bioprospecting companies from forum shopping, i.e. they look for countries where there is no national legislation to protect the interests of TK holders that are indeed the weaker contractual party.[17] This process of drafting a Guide of Contractual Practices has already started within the WIPO IGC on IPGRTKF where four general and fundamental principles have been articulated:

> Principle 1: The IP-related rights and obligations set out in [the Guide of Contractual Practices] should recognize, promote and protect all forms of formal and informal human creativity and innovation, based on, or related to, the transferred GRs.
>
> Principle 2: The IP-related rights and obligations set out in [the Guide of Contractual Practices] should take into account sectorial characteristics of GRs and GR policy objectives and frameworks.
>
> Principle 3: The IP-related rights and obligations set out in [the Guide of Contractual Practices] should ensure the full and effective participation of all relevant stakeholders and address process issues related to contract negotiation and the development of IP clauses for ABS agreements, including in particular TK holders where TK is covered by the agreement.
>
> Principle 4: The IP-related rights and obligations set out in [the Guide of Contractual Practices] should distinguish between different kinds of use of GRs, including commercial, non-commercial and customary uses.[18]

These principles are to be further developed to cover more and more exhaustively areas of standard contract: legal jurisdictions and particular national laws; providers and recipients; GRs agreed or licensed uses of the

[17] The European Commission made a step forward by suggesting within the TRIPS Council that TRIPS and the CBD should be implemented "in a mutually supportive way," *Communication by the European Communities and their Member States on the Relationship Between the Convention on Biological Diversity and the TRIPS Agreement* (April 3, 2001), www.europa.eu.int/comm/trade/issues/sectoral/intell_property/wto_nego/index_en.htm.

[18] *Draft Intellectual Property Guidelines for Access and Equitable Benefit-Sharing*, WIPO/GRTKF/IC/7/9 (November 1 to 5, 2004). Besides the *Bonn Guidelines*, that were adopted by the Conference of Parties of the CBD, key elements of international law include the CBD and the FAO ITPGRFA.

genetic material and associated TK; time-frames; letters of intent; heads of agreement confidentiality; non-disclosure agreements; MTAs; licensing agreements; research agreements; R&D agreements; implications of joint ownership of IP; defining and sharing benefits from access; and dispute settlement issues.

5.2.2.1 Some contractual provisions in favor of the weaker party

In order to construct well-balanced contracts there is the possibility for the provider country to enact provisions in their statutory contract law to protect the weaker parties in transactions involving IP issues. The following indicated mandatory provisions are the ones that go beyond the usual provisions that protect the parties in license agreements.

For the sake of clarity, a distinction between licensing TK as know-how and MTAs for GRs can be drawn. Of course, the provisions for TK may also be applicable to MTAs.

Another important distinction needs to be maintained among the types of statutory provisions based on the degree of importance of such provisions for protecting TK holders:

Mandatory provisions that will apply whether or not they are present in the contract.

Relative mandatory provisions that may be derogated from in order to protect the interests of the weaker party.

Enabling provisions that apply where the contract fails to make any provision.

This differentiation also creates more flexibility, bearing in mind the importance of DCs' capacity to attract foreign companies in this business.

A note of caution should be emphasized here. Crafting an ABS regime is sensitive legal ground fraught with risks of "overprotection" that may eventually inhibit foreign companies from entering into the Kafkaesque meanders of national or regional access legislation and contract law. Indeed, GR-provider States may reap counter-productive outcomes if they shape complicated access legislation: no more benefits to be shared, which also means no financial resources for the preservation and valorization of biodiversity. All this kind of legislation and contract law leaves the door wide open for growing skepticism towards IP among their nationals. At the same time that many DCs are incorporating the CBD and TRIPS in their national legal systems, they should simultaneously find ways in which they can attract foreign companies to invest in bioprospecting. For this purpose, the legal regime of access to GRs has to be clear, certain, and simple, without overburdening private companies with too many hurdles. Any measures for protecting the weaker party in the provider State should be implemented after constructing a whole system supported by marketing infrastructures.

5.2.2.1.1 Provisions for license agreement on traditional know-how on the use of plants

This agreement between the TK holder (licensor) and the bioprospecting company/industrial party (the licensee) provides the licensee with exclusive or non-exclusive rights to execute or use a certain traditional technology without transfer of any tangible material. Know-how licenses can be an appropriate type of contract in the context of bioprospecting. They recognize indigenous and local communities' rights through a type of contract that is familiar to the commercial sector. However, such licenses may entail some legal complications that cannot be dealt with here. The fundamental provisions include the following:

- (i) Contracts must have a written form (mandatory).
- (ii) Any contractual license on an invalid IPR or know-how on TK on the use of plants is void because of the impossibility of execution of its object. However, no restitution of royalties can be invoked because of the *de facto* exercise of an exclusive right on the part of the licensee (mandatory).
- (iii) The licensor can transfer all the rights related to the traditional know-how, except the right to be mentioned as the inventor/innovator, i.e. moral right (mandatory).
- (iv) Diligence of the licensee in the exploitation of the technology; for instance, confidentiality in the use of the know-how (mandatory).
- (v) The licensee must indicate the eventual infringers to the TK holder (mandatory).
- (vi) Unless otherwise agreed, the licensee will report on the results of the exploitation of the licensed TK at least semi-annually (relative).
- (vii) The licensee communicates any improvements by the licensee to the licensor, and the licensee cannot exclude the licensor from the use of the improvements (mandatory).

5.2.2.1.2 Material transfer agreement between traditional knowledge holder (licensor) and industrial company

I have observed how the transfer of TK can be compared to the transfer of know-how. However, most of the time when TK is licensed, the contract will also involve the transfer of the GR. MTA is a term generally utilized for the licenses whose main object is the transfer of the biological material. In other cases, the term used is *license*, even if certain clauses are related to the *transfer* of tangible material.

According to the CBD obligations, the bioprospecting company/industrial party will have to directly negotiate with the State authority. In cases where the industrial party is interested in traditional know-how on the use of PGR, it may, according to the national legislation, also conclude a separate contract with the holders of the right in that regard.

This type of agreement is a standard license instrument in commercial and academic research partnerships. It is a contract that encompasses clauses adapted to the transfer of tangible biological material, including IPRs arising from it. It also defines the basic rights and responsibilities related to the specific materials transferred. The fundamental provisions include the following:

(i) The material to be transferred must be defined, including the descendants and the direct products (mandatory).

(ii) Transfer of possession without transfer of ownership (mandatory).

(iii) The use must be defined according to research purposes or commercial purposes (mandatory).

(iv) Payment of royalties.

Industrial companies usually prefer a royalty schedule that specifically rewards their own research investment in the development of proprietary mechanism-based assays. The sample collection is guided by traditional uses, and the information is provided with the sample. The royalty schedule is a complex matter with various approaches according to the field of research. Broadly speaking, provider States may require a royalty schedule arrangement after the compound has been chemically analyzed and preliminary data on extraction shows a certain degree of efficacy in the field in which it is applied. At this time, more information regarding the compound is added, and this can result in the negotiation of better royalty rates. Because of this possibility of higher rates, the relevant provision should require flexibility to allow further negotiations. The royalty structure must be based on (a) relative contributions of the parties to invention and development; (b) information provided with the samples; (c) novelty or rarity of sample organisms.

(v) TK not protected by any IPR must be rewarded *per se* in the royalty structure of agreements as intellectual contributions to an invention (enabling).

This relationship between the development of the compound and the royalty negotiation can be particularly important for the *in situ* conservation of knowledge and the plants to which it relates.

(vi) When the provider State considers TK as sacred, TK must not be disclosed (enabling).

If the provider State wants to lower any type of risk regarding payment of royalties, it can require an advance monetary payment in the form of a trust fund, lump-sum or milestone payments, per sample fees, payment for re-supply of sample, or in-kind contributions of equipment, training medicines, etc. These advance monetary payments may also include capacity building such as training, equipment, and infrastructure development. Other less tangible and long-term benefits may include research on diseases or on other fields important to the provider country (enabling).

(vii) Confidentiality in the use of the PGR (mandatory).

(viii) No transfer to third parties (mandatory).

(ix) Communication of the results of the research (mandatory).

(x) Indication of the source/origin of the material in all the publications (mandatory).

(xi) Determining the regime of modification on the original material. In technical terms this is called "grant-back" or "reach through" (relative).

(xii) Return or destruction of the material at the end of the contract (enabling).

(xiii) Waiver on responsibility of environmental impact of the derived products. This provision prevents the industrial company from suing the provider State for harms caused by the genetically altered GR (mandatory).

(xiv) Solutions to work the patent locally (see section 5.2.2.1.5 below) (mandatory).

It is a commonly shared view among DCs that creating legal conditions for favoring and promoting local exploitation of patents related to the successful compounds would be a more effective measure in compliance with the objectives of TRIPS and the CBD than royalties or lump-sum payments for the transfer of genetic material.[19]

These provisions can foster and realize the most effective application of the benefit-sharing obligation under the CBD: the DCs, rich in biodiversity and poor in technology, release to industrialized countries their TK as embodied in the germplasm, and the industrialized countries, rich in technology and poor in biodiversity, release to the DCs their technical and commercial knowledge as embodied in the newly developed biodiversity-based products and processes.

Biodiversity provider States may enact laws to establish the production *in situ* of the biodiversity-based innovative products patented by a foreign company. The industrial party will very rarely voluntarily work the patent locally. Two alternatives can serve this purpose: (a) the provider State may issue a compulsory license, (b) it may require that the industrial party works the patent locally as a patent requirement. The two options are examined in the following separate sub-sections.

[19] G. Ghidini, "Equitable Sharing of Benefits of Biodiversity-Based Innovation: Some Reflection under a Neem-Tree", *ATRIP Congress* (New Delhi, 2002).

5.2.2.1.3 Compulsory licenses

Should the patent owner refuse to produce locally, the biodiversity provider country may issue a compulsory license to domestically located industries for the production and sale of products covered by their patents. This compulsory license is in compliance with Article 31 of TRIPS, and it must be non-exclusive, non-discriminatory, and based on fair terms. Moreover, according to Article 31(f) such license should focus fundamentally on the supply of the local market – thus excluding, if not marginally, any export activities of the licensee. Article 31 of TRIPS provides for the imposition of compulsory licenses only when related to specific goals and upon specific conditions. This measure to achieve the goal of the local working of the patent is the expression of a confrontational relationship between the two parties, because issuing a compulsory license severely interferes with the fundamental patentee right to freely choose how and where to exploit its monopoly. Such a measure provides "a formidable means of coercion over patentees."[20]

5.2.2.1.4 Plea for the implementation of the local working requirement as an international patent law obligation

The problem of benefit sharing seems to be polarized in two extreme solutions: on the one side, the provider country will attempt to adopt the extreme measure of compulsory licenses and, on the other side, the patentee will merely offer financial compensation for the germplasm or for any other GR of the provider country. Legal thinking needs to be stretched in order to suggest some solutions to the "biocolonialist" pattern whereby the DC exports its raw materials and the industrialized country returns its finished goods to the provider country to be sold at unaffordably high prices.

A compromise solution can be found between the safeguards of the companies from industrialized countries and the interests of the biodiversity-provider States. This compromise can be found if the provider country's IP norms provide for the obligation of the patentee to *locally work* the patent. Of course, this solution is better than the straightforward imposition of a compulsory license under TRIPS Article 31 (on other use of the patent without authorization of the right-holder) and Article 8.1 (on the adoption of measures necessary to protect special interests of the State) because it is less intrusive on the exclusive rights of the patentee.

The *local working requirement* in the provider country represents a very effective way to implement both the "benefit sharing" obligation under

[20] M. Ricolfi, "Is There an Antitrust Antidote, against IP overprotection within TRIPS?" (2006) 10(2) *Marquette Intellectual Property Law Review* 343.

Article 19 of the CBD and the "mutual advantage" objectives of the TRIPS Agreement. However, this is a controversial measure that is far from gathering the unanimity of the doctrine which can be outlined as follows:[21]

Economic justifications for the local working requirement. This solution guarantees a patentee's competitive advantage since it avoids the problem of the provider country issuing a compulsory license to an unrelated third party who can even be a competitor of the patent holder. Instead, in this case, the patentee can directly choose the local licensees/co-venturers, whereas, of course, the patentee may not do so when a compulsory license is issued. Indeed, the patentee can secure a slow spill-over of industrial and commercial know-how with the consequent reduction of the patentee's time-lead natural to the patent.[22]

Through the ensured local working of the patent, the provider State gains the transfer of technical know-how and the flow of financial resources that in many cases may exceed those resulting from a blunt compulsory licence to a third unrelated party. This means growth in the provider State's own capacity for R&D even after the patent right has elapsed. Domestic know-how can then be developed using the State's own biodiversity-based products.

Local requirement in Article 5.A.2 of the Paris Convention and the repeal of WTO law. The legal basis supporting this proposal is Article 5.A.2 of the Paris Convention which has been incorporated into the TRIPS Agreement (see Article 2.1). On this matter, the Paris Convention states that "Each country of the Union shall have the right to take legislative measures providing for the grant of compulsory licenses to prevent *the abuses* which might result from the exercise of the exclusive rights conferred by the patent, *for example, failure to work*" (italics added). In light of this provision, the default of local working can be seen as an abuse *per se*.[23]

The local working requirement of patents had been a long-established requirement for patent protection at the national level until the adoption of Article 27.1 of the TRIPS Agreement. However, in a WTO dispute brought by the US against Brazil, the US challenged the legitimacy of such local working requirements on the grounds that it had been repealed by Articles 27.1 and 28 of the TRIPS Agreement. The US withdrew the complaint before the Panel was formed and rendered the decision.[24]

[21] *Ibid.*, 343–9.

[22] G. Van Overwalle, "Belgium Goes Its Own Way on Biodiversity and Patents" (2002) 24 *European Intellectual Property Review* 235–6.

[23] Article 2(1) of TRIPS, says that Members "shall comply" with the substantive provisions of the Paris Convention.

[24] *Brazil – Measures Affecting Patent Protection* (*United States v. Brazil*), Request for the Establishment of a Panel by the United States, January 9, 2001, WT/DS199/3.

Article 27 affirms that patents rights shall be enjoyable *"without discrimination as to the place of invention, the field of technology and whether products are imported or locally produced"* (italics added). According to this provision, the patentee would incur patent "abuse" in the sense of Article 5.A.2 of the Paris Convention *only if* it should not provide, even by mere export, "enough" products to the country that has granted the patent. In my view, these clauses are general provisions subject to the specific exceptions contained in Articles 30 and 31 of TRIPS, and, of course, in Article 5A of the Paris Convention.

This means that when a general legal provision is in conflict with a specific legal provision, the specific provision takes precedence, in accordance with the international law principle of *lex specialis derogat legi generali*. Indeed, Article 31 of the VCLT rules of interpretation of treaties requires the application of systematic or contextual interpretation. According to this view, Article 27 of TRIPS expresses the *lex generalis* on the patentable subject-matter since it does not directly or especially address the matter of local working of the patent. On the other hand, Article 5.A.2 of the Paris Convention contains the *lex specialis* on this matter by specifically addressing the possibility of issuing a compulsory license in case of default on the part of the patentee in the local working of the patent.

Ricolfi refuses this interpretation stating that "if Article 5A still were understood as implying that import of patented goods manufactured abroad is not perfectly equivalent to local manufacture of the same goods, the municipal legislation enacted under the old Paris Convention provision would run counter to the prohibition of 'other measures' apt to restrict the free flow of trade contained in Article XI of GATT."[25] From this perspective, Article XI of GATT read in conjunction with Article 27.1 of TRIPS repeals the measure of a compulsory local working of the patent.

The systematic and teleological interpretation of the TRIPS Agreement in favour of the local working requirement. A more careful and complete interpretation of the TRIPS Agreement within its legal framework of WTO law may lead to the opposite conclusion. Ricolfi indeed advises the provider State to ground the measure of the local working requirement on the basis of the teleological interpretation of the relevant provisions.[26] In order to

[25] Ricolfi, "Is there an Antitrust Antidote?", 343–44. Article XI of the GATT 1994 addresses the elimination of quantitative restrictions introduced or maintained by countries on the importation or exportation of products. It prohibits such restrictions with the objective of encouraging countries to convert them into tariffs, a more transparent and less trade distortive instrument, www.wto.org/English/tratop_e/envir_e/envir_backgrnd_e/c7s2_e.htm. T. Cottier, "The Prospects for Intellectual Property in GATT" (1991) 28 *Common Market Law Review* 383, 408.

[26] If this is the case, then a local working requirement confined to a specific sector, such as biotechnological inventions based on local biodiversity, should be deemed TRIPS-compatible: Ricolfi, "Is there an Antitrust Antidote?", 346. On the interpretation

apply the teleological interpretation of these provisions, reference should be made to Articles 7 and 8 of TRIPS where, among the objectives of the treaty, it is stated that the "protection and enforcement of IPRs should contribute to the promotion [...] to *the mutual advantage of producers and users of technological knowledge*" (italics added). Moreover, Article 8.2 of TRIPS states that the transfer of technology needs to be referred to the concepts expressed in the CBD. The derived norms of the CBD in the Bonn Guidelines explain that this is a sector of vital importance for countries and that patents on living forms constitute a peculiar problem. The importance of such subsequent practice is that it "constitutes evidence of the understanding of the parties as to the meaning of" Article 8.2 concepts of transfer of technology. It suffices here to recall Articles 1, 8(j), 16, and 19 of the CBD and the interpretation of the legal doctrine in this field so as to conclude that if flexibility has to be used, it should be employed in favor of DCs'compliance options.[27] Article 7 also says that transfer of technology is one of the objectives of the TRIPS Agreement. Accordingly, Lowenstein and Hurtado indicate that "local working is historically recognized as the primary means for effecting this goal, [the mutual advantage of producers and users of technological knowledge and it should] not be lightly presumed that this tool for economic development was terminated by TRIPS."[28]

of the field of "vital importance" he states: "It may be conceded that the envisaged measure would entail explicitly different treatment for the specific sector, which is usually referred to as *de jure* discrimination. However, this assessment may not be conclusive, because the extent to which the prohibition of technological discrimination under Article 27(1) limits the ability of Members to target certain technological fields in dealing with the important national public policies referred to in Article 8(1) requires clarification. Indeed, the fact that some sectors are singled out as of 'vital importance' for national policies under Article 8(1) by definition implies that the corresponding measure operates selectively rather than across-the-board. Thus, I suggest that under a TRIPS general principles-based interpretation, Article 27(1) may not altogether negate Article 8(1), and vice versa. While Article 27(1) belongs to the core of patent protection and cannot be subverted by local working measures applying across-the-board, the converse also applies: if the development component expressly authorized by TRIPS is not to become altogether meaningless, then the general rule under Article 27(1) may be subject to the explicit derogation under Article 8(1) if and to the extent the important national policies indicated therein so dictate."

27 J. Reichman, "The TRIPS Agreement Comes of Age: Conflict or Cooperation with the Developing Countries?" (2000) 32 *Case Western Reserve Journal International Law* 441; J. Reichman, "From Free Riders to Fair Followers" (1996) 29 *New York University Journal of International Law and Politics* 36–39.

28 V. Lowenstein and M. F. Hurtado, "Intellectual Property and Investment Negotiations: A Stormy Marriage?", *Collection of Papers of the LLM in Intellectual Property* 592–93 (WIPO Worldwide Academy Torino, Italy 2001). Article 7 of *TRIPS*, entitled "Objectives," states that IPRs "should contribute to [...] the transfer and dissemination of technology [...] in a manner conducive to social and economic welfare." Article 8(2) of TRIPS, "Principles," notes that "appropriate measures" may be necessary to prevent patent holders from resorting to practices that "adversely affect the international transfer of technology."

Therefore, even in the hypothesis that Article 27.1 repealed the local working in all the fields of technology, Article 8.1 gives the possibility to derogate, as a matter of exception, in the specific sector of vital importance to the whole general principle of patent granting and working, even in the absence of any formal treaty reservation by a State. This is the case when, according to Article 8, the requirement of local working can be viewed as a measure necessary to protect a sector of vital importance to the socio-economic and technological development of a provider State. Ghidini stresses that this outlined interpretation/proposal does not violate Article 27.1 of TRIPS, which expresses a general principle, especially because the local working requirements would cover only, and exceptionally, the biodiversity-providing country. Meanwhile, the repeal of this principle can be applied to all other countries in which the patent holder wishes to apply.[29]

As regards the argument on the alleged repeal by Article 27.1, because patents must be granted "*without discrimination* as to the place of invention, the field of technology and whether products are imported or locally produced" (italics added), I maintain that the rebuttal is provided by the distinction between "discrimination" and "differentiation" in the EC–Canada case

> Article 27 prohibits only discrimination as to the place of invention, the field of technology, and whether products are imported or produced locally. *Article 27 does not prohibit* bona fide *exceptions to deal with problems that may exist only in certain product areas.* Moreover, to the extent the prohibition of discrimination does limit the ability to target certain products in dealing with certain of the important national policies referred to in *Articles 7 and 8.1, that fact may well constitute a deliberate limitation rather than a frustration of purpose. It is quite plausible, as the EC argued, that the TRIPS Agreement would want to require governments to apply exceptions in a non-discriminatory manner, in order to ensure that governments do not succumb to domestic pressures to limit exceptions to areas where right holders tend to be foreign producers.*[30] (italics added)

For all the reasons outlined in this section, I do not deem that this exception of the local working requirement for biotechnological patents based on GRs and TK accessed into the provider country amounts to a *discrimination* contrary to Article 27.1 but only to a *differentiation* in this *product area*, not prohibited by same.

The consolidation of the interpretation through the concept of mutual supportiveness. The provider country may defend the local working requirement

[29] Ghidini, "Equitable Sharing of Benefits".

[30] *WTO Panel Report on Canada – Patent Protection of Pharmaceutical Products*, paragraph 7.91.

stating that the teleological interpretation, the systematic interpretation, the mutual supportiveness outweigh the ban of such a measure because of the reductive interpretation of the relationship between Article A.5.2 of Paris Convention and Article 27.1 of TRIPS, through the application of the rule of solution of conflicting provisions *lex posterior derogat legi anteriori* corroborated by GATT XI. It needs to be pointed out that this rule should not be applicable in a case in which the *lex anterior* of the Paris Convention is incorporated in the *lex posterior* of the TRIPS Agreement. I argue that the aforementioned teleological interpretation is consistent with the other rules of interpretation of effectiveness – *ut res magis valeat quam pereat*, i.e. "a thing may rather have effect than be destroyed" (see Article 27.1 of the VCLT)[31] and the systematic interpretation of Article 31.3 of the VCLT about the necessity to take into account all the relevant provisions to the case at hand. The rule of effectiveness *ut res magis valeat quam pereat* is used to interpret particular provisions so to give them the fullest weight and the maximum effect consistent with the ordinary meaning of the words.[32]

The term "abuse" in Article A.5.2 of Paris Convention consists in the often mentioned process whereby a GR is extracted from a provider country and then is patented into a product that the provider-country nationals shall purchase at unaffordable prices because the GR industry is highly concentrated by reason of patents. If the provider country demonstrates the simultaneous trend of the loss of biodiversity and concentration through patent pooling this principle of interpretation leads to the valid argument that Article 27.1 of TRIPS has no repealing effect upon the "abuse" concept. Hence special measures of Article 8 of TRIPS and the exception provided by Article XX(g) of

[31] It is to be noted that the ILC did not include this principle as a rule of interpretation in the Draft Articles on the Law of Treaties because it "took the view that, in so far as [it] reflects a true general rule of interpretation, it is embodied in article 27, paragraph 1, which requires that a treaty shall be interpreted in good faith in accordance with the ordinary meaning to be given to the terms in the context of the treaty and in the light of its object and purpose," *Official Records of the UN Conference on the Law of Treaties* (1st Session Vienna, March 29–May 24, 1968 and 2nd Session 9 April 9–May 22, 1969), www.stcl.edu/library/acqmar06.htm; *Draft Articles on the Law of Treaties with Commentaries adopted by the International Law Commission at its Eighteenth Session, 39*, www.untreaty.un.org/ilc/summaries/1_1.htm.

[32] The ICJ has been cautious in the application of the this principle of interpretation in the Advisory Opinion concerning the Interpretation of Peace Treaties, where it declared: "The principle of interpretation expressed in the maxim: *ut res magis valeat quam pereat*, often referred to as the rule of effectiveness, cannot justify the Court in attributing to the provisions for the settlement of disputes in the Peace Treaties a meaning which [...] would be contrary to their letter and spirit," *Interpretation of Peace Treaties with Bulgaria, Hungary and Romania (Second Phase)*, July 18, 1950, Advisory Opinion, *International Court of Justice Reports* 229 (1950).

GATT[33] can give full effect to the CBD principles so as to justify the statutory contractual provision of the local working of the patent.

The risk of adopting this interpretation is that usually the interpreter exaggerates in giving an absolute value to the ordinary meaning, thus overlooking the relative meaning in light of the object and purpose of the treaty. But in this case there might be a harmonious coincidence stemming from both the ordinary meaning and the other rules of interpretation. Anzilotti stated that "words have no value except as an expression of the intention of the parties."[34] Accordingly, Article 27.1 of TRIPS mandates that WTO members must grant patents regardless of any discrimination as to the place of invention, the field of technology and whether products are imported or locally produced. However, the intention of the parties has never meant that this is so in absolute terms, otherwise why would they have left Article 5.2.A of Paris Convention apparently conflicting with Article 27.1 of TRIPS in the same treaty revision process? Furthermore, why would they have crafted Articles 8 and 40 that provide for these types of measures in cases of abuses as mentioned in Article 5.A.2 of the Paris Convention?

The States are the creators and the addressees of these norms of international law, and interpretation thereof means ascertaining the intentions of the parties. The shared expectation is that in a situation of "abuse" certain measures can be adopted to counter the abuse. And the conditions to trigger the measures to enable the local working requirement can also be set unilaterally by the provider country.

As to the question of whether the provider country may under this article not grant a patent, I think that this option is not on the table. It is one thing to require the local working of the patent or the issuance of a compulsory license, but another thing not to grant the patent. This option would be in stark violation to Article 27.1. I agree that Article 27.1 of TRIPS mandates the granting of a patent but it does not absolutely enter into the question of special measures of Article 8, including the local

[33] This is an exception to Article XI of GATT: "nothing in this Agreement shall be construed to prevent the adoption or enforcement by any Member of measures … (b) necessary to protect human, animal or plant life or health; … (g) relating to the conservation of exhaustible natural resources if such measures are made effective in conjunction with restrictions on domestic production or consumption". As mentioned before the WTO Appellate Body stated that the term "natural resource" in Article XX(g) is not static in its content or reference but is rather "by definition, evolutionary". On the basis of the fact that the patentability biotechnology may create an unbalance of the right and obligations at stake, the special measure of local working requirement can be accepted under the development exceptions of Article 8 of TRIPS.

[34] Anzilotti, Dissenting Opinion in *Case Concerning Diversion of Water from the Meuse*, Permanent Court of International Justice, Series A/B, No 70, 383 (1925).

working requirement as a statutory contractual obligation for certain types of patents. These include biotechnological ones based upon GRs and TK under provider country national sovereignty, that can pose a particular problem to a sector of vital importance, namely the preservation of biological resources and sustainable development.

Transfer of technology through local working of the patent can help the techniques of conservation of biological resources, which is one of the objectives of the CBD, and reduce the adverse effect or "abuse" of monopolistic rights over life forms. The provider State can indeed argue that the commercial exploitation of GRs under its own national sovereignty constitutes an "abuse" if that patent on which they are based is not locally worked, giving rise to the possibility of the State to adopt this measure.

5.2.2.1.5 Problems and solutions to the "local working requirement" contract

The "local working requirement" proposal based on Article 5.A.2 of the Paris Convention is only effective if the patent holder has the intention of applying for a patent in the provider country. This is not always the case. No abstract or sophisticated explanation can better illustrate this matter than a typical and vivid example. Imagine a type of corn preserved for several millennia in the interstices of a local farming micro-culture in Madagascar and which is naturally more resistant to current plant diseases in most western countries. Moreover, its nutritional qualities are higher than most types of corn. A Swiss bioprospecting company is orientated in its research by the knowledge of the local farmers that breed and use this type of corn. The company comes into possession of some samples of this local GR through a contract mentioning that pursuant to the granting of an eventual patent on the part of the provider State, the local working requirement shall be applied.

Once in their labs in Geneva, chemists map the corn's genome and, with a few apt strokes of genetic engineering, raise tenfold the yield of the modified species over the original one. Next, the modified plant, which is indeed novel, implies an inventive step, and is patented in Switzerland. The Board of Directors of the Swiss company strategically decide to patent the invention only in countries where biotechnological companies exist with bio-engineers skilled in the art who may reproduce the invention leading to the patented product and then commercialize it. In addition, the product is so expensive, no company will likely be interested either in reproducing it or in buying it in Madagascar. The Swiss company strategically decides not to apply for a patent in Madagascar. By so doing, it avoids (i) the registration fee that does not benefit the company since there is no risk of repetition of the invention by third parties in

Madagascar, and (ii) the expenses of transferring the technology in a country where there are no human resources or infrastructure that can welcome this project. At this point, the important provision of the MTA on the local working requirement remains vain. Not only the "local working requirement" provision remains ineffective, but, since the patented plant is so markedly more productive, the original Madagascar plant is dismissed from the international market.

How can these shortcomings be solved? Of course, the "local working requirement" could be one of the mandatory contractual obligations (of the MTA) when the bioprospecting company extracts the GR from the provider country. But that measure puts too much of a burden upon the applicant, discouraging companies from even starting the negotiations of the MTA. So it is suggested that it should not be a mandatory contractual obligation. However, the "local working requirement" cannot be enforced at the international level, unless the Swiss company applies for a patent in Madagascar.

Therefore, a smoother way to lead to the local working of the patent would be the creation of a regional patent application, whereby the regional office grants a patent valid for all the contracting parties of that region, without any further national examination. Such a regional patent would consequently lead the company to a situation where it is more likely forced to transfer its technology to the provider country that is most likely to be found among the Contracting States of the regional patent office.

5.2.2.1.6 The doctrine of fair use from infringement

A more confrontational remedy to meet the interests of the provider country, resides in the concept of "fair use from infringement."[35]

The concept of fair use from infringement borrows from copyright law and can be used as an exception to patent exclusive rights, if the material transferred from the provider country is genetically modified and then patented. An example can illustrate the thrust of this proposal: imagine a species of aloe preserved for millennia by a certain community in a certain country. If a western company patents the method to modify the gene of the aloe to create a pharmaceutical product, the very fact that the invention is based on that specific GR will immunize the community if it repeats that invention within the boundaries of the members of that group of people, i.e. not entailing manufacture or sale of propagating material, so entering into competition with the right-holder. It can be said that repetition is not considered an infringement *per se*, whereas the repetition of a

[35] M. Ricolfi, "The Interface between Intellectual Property and International Trade: the TRIPS Agreement" (2002) *Italian Intellectual Property* 29.

patented invention without proper right-holder's authorization constitutes an infringement, as a general principle of patent law.

To this exception it can be contested that it hardly fits a traditional patent system since the fair use doctrine is inherently linked to copyright exercise of rights. Especially in continental law, fair use is seldom spoken of outside copyright law.

Although there are various exceptions both in patent law and in trademark law, Article 30 of TRIPS, in my view, does not provide for a legal basis to support the concept of "fair use." As already observed in section 2.1.4 above, WTO Panels interpret Article 30 in a very restrictive way and in compliance with the literal interpretation of the ordinary meaning of the text.

Moreover, how can a local community have the financial and technological means to manufacture the invention within its boundaries without a transfer of technology by the patent holder? This skepticism leads, in my view, to the conclusion that the "local working requirement" (section 5.2.2.1.4 above) in the form of a statutory contractual provision remains the most realistic solution to manufacturing the invention in the biodiversity-provider country.

5.2.3 *Unconscionability of contract clauses*

When the MTA between the industrial party and the indigenous community does not contain any of the aforementioned provisions protecting the weaker party and it is entirely unbalanced in favor of the industrial party, it must be noted that the common law provides for the concept of unconscionability as a way to invalidate the contract or some of its clauses. This concept can be a valid argument for the provider community at the negotiation stage of the MTA in order to lead the industrial party to include a substantive number of clauses protecting the weaker party (see section 5.2.2.1 above).

I have observed how in this area of contract law, the parties do not meet on a footing of economic equality. The suggested mandatory contractual provisions would create contractual freedom through equitable benefit sharing arising from the patentability of GRs. In case of unbalanced contracts or contract clauses, the concept of unconscionability can aid the weaker party in any litigation. Indeed, cases of unconscionability are not limited to cases of clear and blatant "unfair persuasion."[36]

[36] M. Trebilcock, *The Limits of Freedom of Contract* (Harvard University Press, Cambridge, Mass. and London, 1993) 118.

I note that, at common law every contract is concluded through the exchange of consideration and is subject to the parole evidence rule: "The parole evidence rule states that an agreement or contract, signed by the parties, is conclusively presumed to represent an integration or meeting of the minds of the parties."[37] Unconscionability is a well-established exception to the parole evidence rule and can admit evidence of speech or conduct prior to or contemporaneous with the adoption written agreement, notwithstanding the rule. If the evidence shows an invalidating cause of the written agreement, such as lack of consideration, duress, mistaken illegality, or fraud, the contract can be rendered invalid in whole or in part.

The industrial party has to counter-argue and show that the State entity (in this case the provider country) is not a weaker party. If the industrial party does not succeed, the State entity can qualify as the weaker party to the point of letting the court apply the concept of unconscionability. This can comprise cases in which there is a threat to the social order because of the inequitable exploitation of the genetic resource of an indigenous community or of a provider country.

It is a well-established fact that standardized contracts of raw material transfer are prepared by industrial enterprises that have a prodigious bargaining power and position. For instance, a very small provider country or a poor provider indigenous community can be the weaker party, in need of revenues to maintain its developing economy. The provider country or the community holding the TK and GR is frequently not in a position to shop around for better terms, either because the patent holder is the author of the standard contract and has a monopoly on the invention in question that is based upon the provider country's GRs and/or related TK, or because other competitors use the same clauses. In these circumstances, the concept of unconscionability can help the weaker party to invalidate the unconscionable clauses.

The threat of rendering the contract or specific separable clauses unconscionable can also be used as an argument during the negotiation. It must be noted that certain clauses or even the whole contract that is found unconscionable can be unenforceable. Under Anglo-Saxon law, a contract is unconscionable when its inequality is so strong that it would be impossible to explain to a man of common sense without eliciting an exclamation of surprise at the inequality. The qualification of unconscionability depends upon the common law judge and the specific facts of the case:

[37] *Weaver v. American Oil Co. Supreme Court of Indiana*, 1971, 257 Ind. 458, 276 N.E.2d 144 in L. Fuller and M. Eisenberg, *Basic Contract Law* 678 (Thomson West, 2001).

An "unconscionable contract" has been defined to be such as no sensible man not under delusion, duress or in distress would make, and such as no honest and fair man would accept. There exists here an inequality so strong, gross and manifest, that it is impossible to state it to a man of common sense without producing an exclamation of the inequality of it. [...]

It is not the policy of the law to restrict business dealings or to relieve a party of his own mistakes of judgment but where one party has taken advantage of another's necessities and distress to obtain an unfair advantage over him, and the latter, owing to his condition, has encumbered himself with heavy liability or an onerous obligation for the sake of a small or inadequate present gain there will be relief granted.[38]

The question of whether a contract between an industrial party (biotechnological companies, research institutes, bioprospecting companies) and indigenous and local communities lack fair and equitable benefit sharing arising from the IPRs based upon it will certainly depend on the legal system in which the decision shall take place. Justice Frankfurter of the US Supreme Court spoke on the question of inequality of bargaining power in a dissenting opinion:

But is there any principle which is more familiar or more firmly embedded in the history of Anglo-American law than the basic doctrine that the courts will not permit themselves to be used as instruments of inequity and injustice? Does any principle in our law have more universal application than the doctrine that courts will not enforce transactions in which the relative positions of the parties are such that one has unconscionably taken advantage of the necessities of the other?[39]

There are two types of unconscionability in common law: procedural and substantive. There is procedural unconscionability when there is unequal bargaining power, absence of meaningful choice, fine-print clauses, ignorance of important facts, unfair surprise, difference in age or in bargaining power, and mistakes. Substantive unconscionability occurs when the terms are unreasonably favorable to one side – the terms are mutually unfair: one-sided and oppressive. In other words, there is an imbalance in the obligations and rights imposed by the contract. This imbalance can be an independent basis for voiding the contract or proof of procedural unconscionability. The scope of substantive unconscionability encompasses the lack of adequate benefit sharing in contracts for the transfer of raw material.

The common law tries to hold parties to their bargains, so courts are more likely to strike down a contract for procedural unconscionability

38 *Stiefler v. McCullough* (1933) 97 Ind. App. 123, 174 N.E. 823.

39 *United States v. Bethlehem Steel Corp.* (1942) 315 US 289, 326, 62 S.Ct. 581, 599, 86 L.Ed. 855, 876.

than for substantive unconscionability. It is the international customary principle of fair and equitable benefit sharing that should lead the common law judge to declare the substantive unconscionability of the clauses or of the contract because of unfair terms.

In an hypothetical situation of strong unfairness in the agreement between the industrial party and, say, the indigenous community, the common law judge may invalidate the whole contract. This is so because it would not be possible for the judge to enforce more equitable sharing of the benefits.[40] Since this situation is not beneficial for both parties, they would be encouraged to renegotiate the contract.

5.2.4 *Conclusions on the limits of access- and benefit-sharing regimes*

One of the major shortcomings of the ABS regime lies in its enforcement in foreign jurisdictions. In the case of misappropriation, the legal system of the provider country may encompass remedies such as injunctive relief that can only stop the unauthorized accessed GR at the geographical borders of the State concerned. Once the border is crossed, the restitution is more problematic, unless international cooperation by biodiversity-recipient countries is created.

An ABS legislation in a provider country that invalidates the patents granted in violation of access legislation is ineffective for the patents issued in the recipient country (Article 4*bis* Paris Convention). These remedies should rather be provided in the recipient countries against third parties who fail to comply with provider-State access legislation. Here the question arises whether the recipient countries are bound to do so by the CBD (especially Article 16.3) and the ITPGRFA or whether the normative value of the provisions therein is too soft for a direct implementation in patent law. In order to avoid the intricacies and the fragmentation of different interpretation of non-IP treaties, it is of paramount importance that relevant patent law treaties and systems along with the national patent laws be formally amended so to eliminate conflicts with non-IP treaties and meet the needs of provider countries.

Harmonization of patent law is an area in which the legal doctrine is very active and in which diplomatic negotiations are going to be involved over the next years. The protection of the provider country's interest against misappropriation of GRs and related TK can be achieved

[40] US Restatement, Contracts, paragraph 211, Principles of European Contract Law Arts. 4.109 and 4.110. Unidroit Principles of International Commercial Contracts Arts 2.19 and 2.20 in S. Burton and M. Eisenberg, *Contract Law Selected Source Materials* (Thomson West, Minn., 2003).

through the adaptation of existing IPRs allowing for two types of TK protection:[41]

(i) Defensive protection of TK. Its main goal is the preventing or revoking of patents based upon GRs or related TK that has been "misappropriated." The IP system can be accordingly adapted, taking into account legitimate claims of TK holders and proposed measures in recipient countries to render the patent system more suitable to redress these claims (see chapter 6 below).
(ii) Positive protection of TK. Its main objective is the creation of an incentive to innovation and future marketing and the licensing of TK related to GRs. Benefits would flow to the provider country from such economic and legal practices.

[41] For additional considerations cf. the establishment of a Working Group for the issue of a *sui generis* right for protection of TK, *CBD Conference of Parties Decision VI/10 on Article 8(j) and Related Provisions* (June 26, 2002), www.biodiv.org/doc/notifications/2002/ntf-2002-049-tk-en.pdf.

6 The defensive protection of traditional knowledge in international patent law

Defensive protection of GR-related TK refers to legal methods of avoiding its misappropriation by IP holders or applicants. In order to grant a defensive protection to TK in the international IP system, it is necessary to rethink the role of the legal concepts underlying IPRs in their international context.

Since TK is viewed as an external element to the classic patent system, most patent attorneys (and indeed lawyers) think that protecting it within the patent system amounts to imposing hurdles to the progress of science and to the mandate of patent law. In their opinion, patent law should exclusively deal with technological innovation, should stand as a neutral tool for rewarding investment in innovation, and should function as the main incentive for more innovation. The discussion on the challenges of the patentability of biotechnology and of the IP system as a whole presented in section 2.2 above has shown that the IP system is evolving from a purely technical field to a broader tool of economic policy encompassing elements that were once external to its nature.

This chapter looks first at the methods by which TK can be defended within the current laws of the patent system (*lex lata*) and then considers the methods that are more remote from the current patent system (*lex ferenda*).

The introduction of a certificate of origin/source is the major *de lege ferenda* proposal. Accordingly, section 6.1 explores the ways in which this objective can be realized to fully reconcile CBD obligation of PIC with TRIPS Agreement. This section discusses the ongoing doctrinal and diplomatic debates in all the relevant fora on the feasibility of this integration in international patent law.

Section 6.2 deals with the possibility of opposing the grant of a patent in case it draws upon pre-existing TK, novelty-destroying prior art. I will argue that the patent system may achieve a real "global credibility" by instituting a regime of absolute standard of novelty for plant and TK inventions and strict interpretations of the criteria for patentability. This process of making TK novelty-destroying prior art will most effectively go hand-in-hand with the development of global databases of TK to facilitate international searches of prior art.

Section 6.3 broadens the scope of the defensive protection of TK in the wider context of the legal relationship between TRIPS and the CBD. The ban on patentability of Article 27.2 of TRIPS of *ordre public* (including the environment) and morality needs to be analyzed and, in relation with the misappropriation of TK and GRs, needs to be described.

6.1 The certificate on the disclosure of origin/source

The idea of the certificate on the disclosure of origin/source is generally extremely politicized even in the most relevant fora, like the TRIPS Council. The diplomatic debate focuses on the protection of TK and GRs within the patenting process through the introduction of a mandatory international requirement of disclosure of their origin/source in the patent application. This section supports a much more careful legal and technical approach that takes into account all the other possibilities of implementation of the CBD mandate PIC obligation in international legal instruments and monitoring bodies.

This requirement would also implement the CBD-mandated PIC obligation and foster mutual supportiveness between the TRIPS Agreement regime and the CBD. The international legal status of PIC and benefit sharing and its interaction with IP law must be assessed (i) to determine the existence of any obligation pending on both the international community and on the individual States, and (ii) to interpret the obligation and the methods of compliance.

This legal analysis will lead me to distinguish between the concepts of PIC and disclosure in patent law. After assessing the normative value and the scope of the concepts of PIC and benefit sharing in section 6.1.1, I will in section 6.1.2 analyze the controversial aspects of application arising from the introduction of a certificate of disclosure of source/origin in patent law.[1]

[1] This idea appears first to have been suggested in print by F. Hendrickx V. Koester, and C. Prip, "Access to Genetic Resources: A Legal Analysis" (1993) 23 (6) *Environmental Policy and Law* 250–58. M. Gadgil and P. Devasia, "Intellectual Property Rights and Biological Resources: Specifying Geographical Origins and Prior Knowledge of Uses' (1995) 69 *Current Science* 637–39; B. Tobin, "Alternative Mechanisms for Protection of Indigenous Rights", paper presented at Symposium of Indigenous Peoples of Latin America: *Indigenous Peoples, Biodiversity and Intellectual Property* (Santa Cruz: Bolivia, September 27–30, 1994).

6.1.1 *The normative value of the concepts of prior informed consent and benefit sharing in international patent law*

This section demonstrates how disclosing the source of origin in the patent application would be critical for ensuring that the TRIPS Agreement and the CBD are implemented in a mutually supportive manner. There are various opinions concerning the objectives of the certificate of origin/source requirement as a means of ensuring benefit sharing and PIC. As such, the scope of the certificate of origin must remain very limited in order to be compatible with the current international IP system.

Two approaches are possible in this regard. The first deals with the possible integration of transparency measures that reflect the CBD-mandated objectives and obligations within the constraints of the *lex lata* (positive law) of the international patent system. Accordingly, I shall determine why the TRIPS Agreement provisions do not allow States to introduce the disclosure of origin/source as evidence for PIC and benefit sharing. The second objective is the implementation of the CBD obligations of PIC and benefit sharing through amending the TRIPS Agreement. A correct implementation requires both a political and legal analysis to propose adequate *lex ferenda* (how the law should be).

National and regional patent laws have been progressively enacted to require that the origin of the biological material stemming from plants or animals be disclosed to qualify for patent protection. These laws also mandate that disclosure be accompanied by evidence that the material was used in accordance with the relevant laws on access in the provider country.

One of the most acrimonious legal problems related to the introduction of this new requirement in domestic and international laws arises from its alleged incompatibility with Article 27 of TRIPS. This provision sets forth an exclusive number of requirements limited to novelty, inventive step, and usefulness of an invention. Additionally, Article 29 of the TRIPS Agreement arguably precludes the imposition of any further disclosure requirements.

I will examine the substantive or formal legal nature of this requirement by analyzing the scope of its implementation in international patent law. In this respect, a Swiss Proposal provides the possibility of introducing this type of disclosure requirement in the PCT and PLT and will be compared with other submitted proposals.

I will outline the scope of this patentability requirement while determining in which treaty forum this transparency measure should be introduced. It is my aim to achieve a fine-tuned balance between opposing interests of recipient and provider countries. I will then analyze the possible consequences of non-compliance and other private international law enforcement issues relating to the outlined disclosure requirement.

6.1.1.1 Treaty law on prior informed consent and its relationship to intellectual property law

The concept of PIC has been crafted by the CBD, the soft law of the Bonn Guidelines and the ITPGRFA. The analysis of the constitutive elements of this type of norm may also reveal that the basic concept has become customary PIC (i.e. without all the elements of implementation in an ABS regime or patent laws).

6.1.1.1.1 The CBD and the Bonn Guidelines

Several provisions of the CBD – particularly Articles 8(j), 15.4, 15.5, 15.7 and 16.5[2] – regulate access to GRs and TK after PIC and benefit sharing. More specifically, Article 15.7 requires Contracting Parties to take legislative, administrative, or policy measures with the aim of sharing the benefits arising from the commercial use of genetic GRs with the Contracting Party providing these resources. A systematic interpretation of this provision (see Article 31.1 of the VCLT) requires that it be read in conjunction with Article 16.5 of the CBD, which recognizes "*that patents and other intellectual property rights may have an influence on the implementation of this Convention*"; for this reason, States "*shall cooperate in this regard subject to national legislation and international law in order to ensure that such rights*" – including patents – "*are supportive of and do not run counter to its objectives*" (italics added).

Transparency measures under patent law could certainly be an effective way to implement Articles 15.7 and 16.5 of the CBD. Their introduction at the international level would provide the cooperation required from Contracting Parties of the CBD. These provisions call for a teleological interpretation. Indeed, Article 1 of the CBD states among its objectives: "*the conservation of biological diversity, the sustainable use of its components and the fair and equitable sharing of the benefits arising from the use of genetic resources*, including [...] *appropriate access to genetic resources* and [...] appropriate transfer of relevant technologies, taking into account all rights over those resources and to technologies, and by appropriate funding"[3] (italics added).

[2] In the context of access to and transfer of technology, Article 16.5 of the CBD: "The Contracting Parties, recognizing that patents and other IP rights may have an influence on the implementation of this Convention, shall co-operate in this regard subject to national legislation and international law in order to ensure that such rights are supportive of and do not run counter to its objectives."

[3] Articles 5, 6(b), 7–11, and 14 of the CBD.

A step towards a clearer method of implementing the transparency measures has been undertaken by the Sixth Meeting of the Conference of the Parties (COP) to the CBD, which took place in The Hague in May 2002. The meeting officially incorporated the principle of disclosure of origin into the soft law of the Bonn Guidelines on Access to GRs and Fair and Equitable Sharing of the Benefits Arising out of their Utilization.[4] GRs and TK are respectively considered in relation with IPRs when the Bonn Guidelines invite parties to disclose:

> the country of origin of genetic resources where the subject-matter of the application concerns or makes use of genetic resources in its development, as a possible contribution to tracking compliance with prior informed consent and mutually agreed terms on which access to those resources was granted.[5]

> [t]he origin of relevant TK, innovations, and practices of indigenous and local communities relevant for the conservation and sustainable use of biological diversity in applications for intellectual property rights, where the subject-matter of the application concerns or makes use of such knowledge in its development.[6]

Particular attention should be paid to paragraph 16(d) of the Bonn Guidelines:

> Contracting Parties with users of GRs under their jurisdiction should take appropriate legal, administrative, or policy measures, as appropriate, to support compliance with PIC of the Contracting Party providing such resources and mutually agreed terms on which access was granted. These countries *could consider, inter alia*, the following measures: (i) measures to encourage *the disclosure of the country of origin of the genetic resources and of the origin of TK, innovations and practices of indigenous and local communities in applications for intellectual property rights;* (ii) *measures aimed at preventing the use of genetic resources obtained without the prior informed consent of the Contracting Party providing such resources.*[7] (italics added)

6.1.1.1.2 The FAO ITPGRFA

Since the ITPGRFA deals with IP issues, it is important to observe whether it adds upon the CBD concept of PIC of which I am considering

[4] Bonn Guidelines. [5] *Ibid.*, Article 15.

[6] COP-7, 2004, *Decision VI/10, Article 8(j) and Related Provisions. Decision VII, Access and Benefit-Sharing as Related to Genetic Resources (Article 15), 2004*, paragraph 46 and section D, Annex, paragraph D (c).

[7] Secretariat of the CBD, *Report of the Sixth Meeting of the Conference of the Parties to the Convention on Biological Diversity*, UNEP/CBD/COP/6/20, paragraph 16(d)(ii) (2002). The Conference of the Parties of the CBD, held in February 2004, also mandated the Ad Hoc Open-Ended Working Group on Access and Benefit-Sharing (ABS Working Group) to negotiate an international regime on access and benefit-sharing and prior informed consent including the "Disclosure of origin/source/legal provenance of genetic resources and associated TK in applications for intellectual property rights" COP-7, 2004, section D, Annex, paragraph D(xiv) COP-7.

the normative value. As earlier observed, in section 3.3 above, the FAO ITPGRFA's objectives are "the conservation and sustainable use of PGRFA and the fair and equitable sharing of the benefits arising out of their use, in harmony with the CBD, for sustainable agriculture and food security" (Article 1.1). The ITPGRFA creates a system of ABS regarding PGRFA standard MTA (see Articles 10, 12.4 and 13). Though not directly mentioning the exercise of IPRs, this MTA provides for a transparency measure (that I will examine in the following sections).

With regard to transparency measures, it is important to note that Article 12.3(b) explicitly states that there is no "need to track individual accessions" in the Multilateral System established by the Treaty. This means that the ITPGRFA does not require PIC for accessing PGRFA covered by the Multilateral System (see Article 12.3(b)) because it mainly concerns a prohibition of patenting. Furthermore, the Treaty contains no provisions on access to TK relevant to PGRFA.[8]

6.1.1.2 The customary normative value of the CBD concepts of prior informed consent and benefit sharing and its interaction with patent law

The essential content of the CBD norms of PIC and benefit sharing concepts based upon Article 15.5 and 15.7 of the CBD is that a State cannot allow the commercialization of products based upon misappropriated GR or TK. Access to the GR and TK must occur under mutually agreed terms with the provider country. After having analyzed the norms of PIC and benefit-sharing concepts in treaty law, it needs to be determined whether the concept is a customary norm. This section determines with more precision the content of these concepts.

This analysis is important because the US – the most important recipient country of GRs and the one that grants most of the patents in this field – is not a party to the CBD. The determination of whether or not the PIC and the benefit-sharing norms are customary resides in the fact that, in case of positive result, they would bind the US, the country that files most of the biotech-patents and the only State not party to the CBD.

In the course of the discussion the VCLT expressly states that "a treaty does not create either obligations or rights for a third State without its consent" (Article 34). However, the norms contained in a treaty may bind non-parties if they codify existing norms of customary international law or become new customary norms. This process is expressly recognized in

[8] *Second Meeting of the Commission on Genetic Resources for Food and Agriculture acting as the Interim Committee for the FAO ITPGRFA*, Item 4 of the Draft Provisional Agenda, Report on the Outcome of the Expert Group on the Terms of the Standard Material Transfer Agreement (Rome, November 15–19, 2004).

Article 39 of the VCLT, which states, "nothing in Article 34 to 37 precludes a rule set forth in a treaty from becoming binding upon a third State as a customary rule of international law, recognized as such."[9] It has been argued that the CBD has not codified a pre-existing customary principle, mainly because the international community has started to focus on these concepts only at the negotiation of the Rio Declaration. Moreover, the huge increase in patentability of biotechnology started not so long before that time. It must therefore be determined whether this is a case of crystallization of international treaty norm.[10]

Customary norm results from a general and consistent practice of States having a sense of legal obligation. A customary norm is a rule of law "from the consistent conduct of States acting out of the belief that the law required them to act that way."[11] It follows that customary international law can be discerned by a "widespread repetition by States of similar international acts over time (State practice); acts must occur out of sense of obligation; acts must be taken by a significant number of States and not be rejected by a significant number of States."[12]

I will analyze the two classic elements of the practice and *opinio juris* schematically.

In cases in which the ICJ would have to decide upon the customary normative value of PIC and benefit-sharing concepts, it would certainly examine the CBD and the ITPGRFA, States' practice and *opinio juris*, decisions of State courts, and legal doctrine, so as to undertake the "final stage" for the acknowledgment of these concepts as customary norms.[13]

6.1.1.2.1 The *opinio juris communis* as reflected in the soft law of the Bonn Guidelines in conjunction with the CBD treaty law

If a judge had to decide upon the existence of an international customary norm that prevents States to grant patents based upon misappropriated

[9] R. Baxter, "Multilateral Treaties as Evidence of Customary International Law" (1965/66) *British Yearbook of International Law* 275–300; R. Baxter, "Treaty and Custom" (1970) 129 *Recueil des Cours de l'Académie de Droit International* 25–106.

[10] The crystallization of an emerging international customary norm has been analyzed by the *North Sea Continental Shelf Case*, paragraphs 60–82 and especially paragraph 69. In this case, the ICJ rejected the contentions of Denmark and the Netherlands and considered that the principle of equidistance, as it figured in Article 6 of the Geneva Convention of 1958, had not been proposed by the ILC as an emerging rule of customary international law. This Article could not be said to have reflected or crystallized such a rule; P. Weil, "Toward Relative Normativity in International Law" (1983) 77 *American Journal of International Law* 435.

[11] S. Rosenne, *Practice and Methods of International Law* (Oceana Publications, London, 1984) 55.

[12] *Ibid.*

[13] B. Carter, P. Trimble, and C. Bradley, *International Law* (Aspen Publishers, New York, 2003) 122.

GRs he would "*seltsam im Nebel* [...] *wandern*," to adapt one of the well-chosen literary images of Dupuy when he describes the judge who is looking for a non-written norm and its content as he wanders alone in the misty darkness.[14] The analysis of the Bonn Guidelines and their relationship with the CBD would certainly be helpful in this determination.

The Bonn Guidelines constitute soft law that clarifies and specifies the content of the norms of the CBD and applies it more specifically to the field of the exercise of IPR.[15]

In this regard, it can be asked whether the Bonn Guidelines are equivalent to the UN General Assembly Resolutions in their expression of the will of the parties to the CBD, which is universal except in the case of the USA. The Bonn Guidelines of the COP to the CBD may have the potential influence not only to interpret its treaty but also to set forth new customary principles in CBD matters. Indeed, for those who maintain that the custom is the product of international social consensus, this international forum in which the States discuss, negotiate and find a common denominator, is the privileged place for customary norm production.[16]

There are two ways in which the normative value of these Guidelines can be considered. On one level, they represent the authentic interpretation by the Contracting Parties of the relevant CBD provisions. On a second level, they create subsequent practice in the form of additional soft law.

Subsequent practice clarifies the authentic interpretation of treaty provisions. This rule of interpretation is found in Article 31.3(b) of the VCLT:

> (3) There shall be taken into account, together with the context:
>
> (b) any subsequent practice in the application of the treaty which establishes the agreement of the parties regarding its interpretation;

The Bonn Guidelines can be considered as subsequent practice because they establish "the agreement of parties" with regard to such an interpretation involving the common will of the original Contracting Parties. This process falls squarely within the general principle of international law

[14] P.-M. Dupuy, "Le droit des Nations unies et sa pratique dans la jurisprudence de la court internationale de justice", *La pratique et le droit international*, Colloque de Geneve de la Societe francaise pour le droit international (Pedone, Paris, 2004) 139.

[15] According to paragraph 1, the Bonn Guidelines, "*may serve as inputs when developing and drafting legislative*, administrative or policy measures on access and benefit-sharing with particular reference to provisions under Articles 8(j), 10(c), 15, 16 and 19 [of the CBD]; and contracts and other arrangements under mutually agreed terms for access and benefit-sharing".

[16] M. Kohen, "La pratique et la théorie des sources du droit international", *La pratique et le droit international*, Colloque de Genève de la Société française pour le droit international, (Pedone, Paris, 2004) 106.

stemming from the Roman law dictum, "*eius est interpretare legem cuius condere.*"[17] To meet that criterion, the laws and regulations of any Contracting Party must have been brought to the knowledge of and be accepted by all other Parties.[18] The Bonn Guidelines meet this requirement.

The Guidelines create new "soft law," which is non-binding by nature. However, public international law doctrine, particularly that of Simma, considers that this type of "soft law backs up hard law in a variety of ways."[19] The soft law of the Bonn Guidelines reformulates and renders more explicit the hard law (or binding norms) of the CBD provisions which were expressed in very vague and approximate terms.[20] Hence, the Bonn Guidelines, by "inviting" the Contracting Parties to implement the PIC principle in a precise manner in the IP system, specify the rather vague wording of the CBD hard-law obligations of 15.5 and 15.7 (on access to GRs), read in conjunction with 16.5 (access to and transfer of technology). This invitation is characterized by the conditional "should" which clearly indicates that paragraph 16(d) of the Bonn Guidelines is setting forth an invitation rather than an obligation.

Although the legal nature of the *instrumentum*[21] of the Bonn Guidelines is non-binding "soft law," its *negotium*[22] is precise enough in the type of

[17] G. Di Stefano, "La pratique subséquente des Etats parties à un traité" (1994) 40 *Annuaire français de droit international* 41, 45.

[18] *Kasiskili/Sedudu Island (Botswana v. Namibia)*, December 13, 1999, *International Court of Justice Reports* 1094, paragraph 74.

[19] B. Simma, "International Human Rights and General International Law: A Comparative Analysis" (1993) 4 (2) *Collected Courses of the Academy of European Law* 234.

[20] The same process has been observed in the *Declaration on Principles of International Law concerning Friendly Relations among States*, UN General Assembly Resolution 2625 (XXV) (October 24, 1970), that specifies certain vague principles of the Charter of the UN.

[21] The *instrumentum* of international law refers to the manner or form which the law takes (i.e. a convention, guideline or declaration, etc.). M. Virally, *La distinction entre textes internationaux de portée juridique et textes internationaux dépourvus de portée juridique. Rapport provisoire* (1983) 60 (1) *Annuaire de l'institut de droit international* 332–33. The term "soft law" generally refers to the content of a "soft instrument." In this sense, the term takes on many different forms and designations, including charters, declarations, codes of conduct, protocols, informal gentlemen's agreements, de facto agreements, etc. Others might add to this category conventions that have not been sufficiently ratified to enter into force, practices that are not widespread enough to be considered as customary, and finally, non-binding resolutions and recommendations adopted after a multilateral negotiation in international fora: C. Chinkin "The Challenge of Soft Law: Development and Change in International Law" (1989) 38 (4) *International and Comparative Law Quarterly* 851. The nature of the *instrumentum* has a relative effect upon the enforceability or the normative value of the provision itself. Handl submits that some formal international law instruments "such as treaties, whose ineffectiveness relegates them to the ranks of non-legal norms, are if you will, soft norms, notwithstanding their formal status." G. F. Handl, *et al.*, "A Hard Look at Soft Law" (1988) 82 *Proceedings of the American Society of International Law* 372.

[22] *Negotium* refers to the actual content of the international instrument, that is, the rules and obligations indicating behavioral expectations or "obligations of good will," such as the

obligation and it recommends to make it hard law in force on the matter. In other words, while Article 15.7 of the CBD may have been interpreted to set the obligation on the international community as a whole in a very vague manner, the Bonn Guidelines invite individual Member States to realize the obligation set forth in that Article.[23] This must be done by creating a transparency measure in their IP system.

The question arises on whether the obligation to disclose the origin of GR and TK in the patent application is compatible with the current international patent system. Depending on the manner in which this requirement is implemented in the patent system, a conflict may arise between this norm and some articles of the PLT, PCT, and also the TRIPS Agreement that create restrictions in the introduction of new requirements.

The legal conflict among these provisions is manifested by implementing all the provisions at the same time, if a country has ratified all these treaties. For instance, I have observed that a State may patent an invention based on a GR from a provider country without ensuring the grant of PIC or benefit sharing in the IP system because TRIPS does not provide it, or that a State may have an obligation to patent under Article 27 a foreign patent of a PGR that is in the Multilateral System of the ITPGRFA for which Article 12.3(d) forbids the granting of a patent. This is so even though the object and purpose of the CBD and its derivative laws are different from the purpose of IP treaties. The conflict arises because the two conflicting provisions regulate the same object and purpose: both the Bonn Guidelines and Article 16.5 of the CBD refer to IP law just like PLT, PCT, and the TRIPS Agreement.

In light of (i) the obligation of mutual supportiveness (as outlined in section 1.2.3 above) between the areas of environmental law and IP law and (ii) the general principles and objectives of Article 7 and 8 of TRIPS (as outlined in section 2.1.3 above), it is argued that it is necessary to add such a transparency requirement in patent law. In turn, this additional requirement needs an amendment of the aforementioned international patent treaties. However, there is no specific provision in the CBD and no mention in the Bonn Guidelines on how to amend international IP law. The CBD legal structure has called on the international community to

obligation to act in an appropriate manner or the obligation to consult, negotiate and co-operate. When States have to negotiate on a certain subject-matter it is generally easier that they reach a consensus on a *negotium* in soft law rather than in a hard law *instrumentum*. And this is so even if the wording of the *negotium* is so precise that some may define it as hard law. Consequently, soft law instruments are proliferating in international law.

23 This legislative method of "inviting" parties to create a system of submission of disclosure of origin of TK and GR is usually followed in various Multilateral Environment Agreement contexts.

render IP treaties compatible with its obligations, e.g. with the disclosure of origin of GRs and TK. Such an amendment process in international patent law has the additional advantage of leading the US (the most important recipient country of GRs) to unequivocally integrate into its patent system some of the universal CBD principles to which it is not bound through treaty law but that albeit might bind it as a customary norm, as the following paragraphs are intended to determine.

6.1.1.2.2 The state practice: comparative overview of legislative approaches

This sub-section verifies whether the scope of the PIC and benefit-sharing customary norms comprises the obligation of amending patent law. The *opinio juris* is certainly not clear on this point. The States' practice can better answer this question. A higher degree of commitment to *repetitio facti* for these principles can be traced through the legislative amendments that countries are undertaking to make their patent laws more compatible with the customary norm in question. In spite of the precision achieved through the wording of the Bonn Guidelines, a comparison of the recent amendments in regional and national patent laws reveals that countries and regions are currently adopting a variety of approaches when they introduce requirement of disclosure of origin and legal provenance.[24]

The practice will help in interpreting the treaty law along with the resolutions of international organizations such as the Bonn Guidelines. The judicial practice is only an auxiliary for the determination of the international law norms.[25]

The Andean Community adopted two binding Decisions which establish mandatory disclosure requirements: (i) Decision 391 (1996) entitles national IPR authorities to require applicants for patents to provide copies of access contracts – where use has been made of the region's GRs;[26] (ii) Decision 486 (2000) goes so as far as to make the granting of patents dependent upon compliance with international, Andean Community, and national law relating to ABS[27] through obligating applicants to provide copies of agreements that demonstrate the existence of PIC.[28]

[24] "Certification Systems: Product and Process Certification Including Certificate of Legal Provenance/Source/Origin", International Expert Workshop on Access to Genetic Resources and Benefit Sharing (Cuernavaca, Mexico, October 24-27, 2004).

[25] L. Caflisch, "La pratique dans le raisonnement du juge international", *La pratique et le droit international,* Colloque de Genève de la Société française pour le droit international, 128 (Pedone, Paris, 2004).

[26] *Andean Community, Decision* 391 (1996), www.grain.org/brl/?docid=581&lawid=1651; *Andean Community, Decision* 486 (2000), www.comunidadandina.org/normativa/dec/D486.htm.

[27] *Ibid., Andean Community, Decision* 486 (2000), Article 3. [28] *Ibid.*, Article 26.

Failure to provide copies of agreements for use of GRs or TK can lead to the patent being annulled.[29]

Among the Andean Community States, Peru and Costa Rica were the first States to require the disclosure of origin and evidence of legal provenance of biological material in patent applications for a grant of IPR.[30] While in Peru the patent applicant must provide evidence of PIC, in Costa Rica the patent authorities can search for such evidence in PIC as a condition for granting of patents. The penalty in both countries for lack of PIC is the suspension of the patent right.

The new Indian biodiversity law prohibits access by foreigners to Indian GRs and TK without PIC.[31] As regards the newly adopted Patent Act, the patent applicant is required to disclose either the source from which the genetic material is obtained or the geographical origin. This requirement is part of the technical details to be included in the specification.[32] The transparency measure has a double purpose. Since a large amount of biological material is collected and conserved by individuals and institutions inside and outside India, this issue raises the fundamental question of whether applicants should be required to provide information on the country of origin or merely on the source from which the resources were obtained (see section 4.2.3 above). The distinction between the place of origin or of source relates to the problem of identifying the legitimate original providers prior to the adoption of the CBD and the institution of the ITPGRFA collections. Tracing the origin of the GR would be so cumbersome that, in all likelihood, this transparency measure may not be compliant with Article 62.1 of the TRIPS Agreement.

A provision of the Indian legislation clearly states that the patent application can be opposed or revoked if the information is not given: "does not disclose or wrongly mentions the source or geographical origin of biological material used for the invention."[33] This provision requires the Indian patent law system to analyze the good faith of the information submission.

It is important to note that the Indian Patent Act does not oblige a patent applicant for inventions based on TK, to submit evidence of PIC in

[29] *Ibid.*, Article 75.

[30] *Government of Peru, Supreme Decree* 008–96-ITINCI (Implementing the Third Complementary Measure of Andean Community, Decision 391).

[31] *Indian Patent Act 2002*, www.nbaindia.org/act/act_ch2.htm. 3(1); N. S. Gopalakrishnan, "TRIPs and Protection of Traditional Knowledge of Genetic Resources: New Challenges to the Patent System" (2005) 27 (1) *European Intellectual Property Review* 11–18.

[32] *Ibid.*, s.10(4)(d)(D) requires the disclosure of "the source and geographical origin of the biological material in the specification, when used in an invention."

[33] *Ibid.*, Articles 25(j) and 64(p).

order to use the TK nor does it require the sharing of benefits derived. This sort of "enhanced certificate of source" would raise the problem of compatibility with the TRIPS Agreement.

Switzerland has rendered mandatory the declaration of the source of the genetic resource or the TK in Articles 49a,[34] 81a,[35] and 138 paragraph (Abschnitt) 1 (b)[36] of its new Patent Act. Article 81a includes a penalty (of CHF 100,000) for failure of submission, which seems hardly severe enough to discourage misappropriation, especially by large corporations. Given such a mild penalty, a patent applicant who uses illegally acquired materials is likely to lie about the source if pressed to disclose it because the application will not otherwise be processed. If the applicant is found to have purposely misstated the facts, the applicant will pay a fine up to CHF 100,000, while continuing to exercise the patent rights.

Notwithstanding the lack of a disclosure obligation in the EU Directive, Denmark and Belgium have undertaken legislative action in their patent law to comply with the CBD and the Bonn Guidelines. Not surprisingly, they have taken two different approaches. Denmark, the first EU member to adopt legislation requiring disclosure of origin in patent applications,[37] requires that patent applications shall include information on the geographical origin of the genetic material, if known, whereas Belgium has proposed an amendment to the patent law so that an invention developed on the basis of plant or animal material collected or exported in breach of the CBD would be banned from patentability because it would be contrary to *ordre public* and morality (see section 6.4 below).[38] Denmark and Norway have adopted the same attitude in a case of failure to comply with the submission requirement by applying the penal code.[39]

In its EU Biotech-Directive implementation, the Italian Parliament provided:

> that the origin of the plant or animal genetic resource, on which the invention is based, is indicated in the patent filing in such a way as to provide reference to the country of origin, in such a way as to enable the monitoring of compliance of

[34] Article 49(a). [35] Article 81(a) (new) *(II. Wrongful declaration of the source)*.

[36] Article 138 Abs.1(b).

[37] *Danish Patent Act* 412, 31/5/2000 amending the Patent Act (consolidated Patent Act 926 22/9/2000).

[38] This provision of ministerial regulation 1086 (December 11, 2000) replaced paragraph 3 of the existing ministerial Regulation on patents 374 (June 19, 1998).

[39] C. Barber, S. Johnston and B. Tobin, *User Measures: Options for Developing Measures in User Countries to Implement the Access and Benefit-Sharing Provisions of the Convention on Biological Diversity* (2nd edition, United Nations University Institute of Advanced Studies, Tokyo, 2003) 29–30, www.ias.unu.edu/binaries/UNUIAS_UserMeasures_2ndEd.pdf.

relevant importation and exportation legislation, as well as in connection to the living matter from which it has been derived.[40]

Ricolfi notes that the provision was introduced with the specific purpose to promote equity and fairness in the South/North exchanges.[41]

I have mentioned only some of the developing and industrialized countries that have interpreted in this way the normative value of the Bonn Guidelines and of the CBD.[42] However, increasing differences in patent laws create a legal uncertainty in the exercise of patent rights that needs to be thoroughly examined. The lack of uniformity in the introduction of the certificate of origin creates difficulties for local communities to know which law is breached and which remedies to pursue in the recipient country. Even in the case of the adoption of an international legal mechanism in this regard, if the concepts are too general, the barrier of the principle of territoriality will create the same difficulties. It is evident that only a strict and precise harmonization process of international patent law with regard to disclosure of source can curb this confusing process. Brazil and India submit that this harmonization should imply amending the TRIPS Agreement[43] while the EC and its Member States, and Switzerland more specifically propose to amend the PLT and PCT and, where appropriate, regional agreements such as the EPC.

6.1.1.2.3 Judicial interpretation of the CBD obligations: European Court of Justice decision on Biotech Directive 98/44

Within the analysis of States' practice reside important judicial interpretations. The only international judicial interpretation of the CBD has been undertaken by the ECJ when it had the task of interpreting, *inter alia*, the compatibility of the EU Biotechnology Directive with the CBD. The doctrinal comments that it has raised have a strong bearing on the interpretation of the normative value of the CBD

[40] Article 5(2) of the Act of the Italian Parliament n. 78 of February 22, 2006.

[41] With regard to the Italian national PGRs, it is interesting to note that Article 5(6) provides that, "if the patent application concerns its use or alteration from conservation varieties as defined by" its own legislation or "from plant or animal genetic resources which are protected by EU GIs under Reg. No. 2081 and 2082/92, a Prior Informed Consent by the Italian Ministry of Agriculture is required, except in the case the patent has diagnostic or therapeutic purposes." M. Ricolfi, "Intellectual Property and Biodiversity: A Review of Legal and Conceptual Issues and Policy Options" (2004) *Atti del Seminario*, Istituto Agronomico per l'Oltremare Firenze, 24.

[42] Submission of the EC and its Member States, *Disclosure of Origin or Source of Genetic Resources and Associated Traditional Knowledge in Patent Applications*, WIPO/GRTKF/8/11 (May 17, 2005).

[43] Communication from the US, *Article 27.3(b) Relationship Between the TRIPS Agreement and the CBD, and the Protection of Traditional Knowledge and Folklore*, IP/C/W/434.

mandated concept of PIC and benefit sharing. All EU members are also Contracting Parties of the CBD, and I hasten to stress that the ECJ pronounced its interpretation of the relevant CBD provisions before the adoption of the Bonn Guidelines. One can only wonder whether the ECJ would have been sensitive to the normative value of the Bonn Guidelines analyzed in section 6.1.1.1.1 above. I also emphasize the fact that judicial practice stands at the same level of the legal doctrine as an auxiliary tool of determining the norm of international law; these are evidences of second degree while the primary degree is held by States' practice and their *opinio juris*. This decision stands at the same level as the opinion of an author about the interpretation of the *lex lata*.[44]

During negotiations on the EU Biotechnology Directive 98/44, government and non-government forces tried in vain to curb the eventual misappropriation of GRs by patent applicants. They attempted to introduce a clause that would require full disclosure of a country of origin's consent to transfer genetic material in any application for a patent related to biological resources from DCs.[45] The outcome of these negotiations, however, left only a weak allusion to the PIC obligation in recital 27 of the preamble:

> Whereas if an invention is based on biological material of plant or animal origin or if it uses such a material, the patent application should, where appropriate, include information on the geographical origin of such material, if known; whereas this is without prejudice to the processing of patent applications or the validity of rights arising from granted patents.

The alleged incompatibility of the Directive with the CBD is one of the grounds on which the Directive has been challenged before the ECJ.[46] It has been argued that the obligation to disclose the country of origin is only conditional[47] and runs counter to Article 16.5 of the CBD. However, the ECJ has absolved the Directive from charges of non-compliance with obligations under the CBD. The core of the Court's reasoning leading to this conclusion reads as follows:

[44] Caflisch, "La pratique dans le raisonnement du juge international" 126, 128.

[45] S. Sterckx, "Some Ethically Problematic Aspects of the Proposal for a Directive on the Legal Protection of Biotechnological Inventions" (1998) 20 *European Intellectual Property Review* 123.

[46] Case C-377/98, *Kingdom of the Netherlands v. European Parliament*. For an excellent treatment of the issues subsequently decided by the European Court of Justice.

[47] ICC (2002), *Policy Statement: Should Patent Applicants Disclose the Origin of Biological Materials on which they File Patents? Should they Demonstrate Prior Informed Consent for their Use?*, prepared by the Commission on Intellectual Property, www.iccwbo.org/home/statements_rules/statements/2002/should_patent_applicants.as.

It cannot be assumed, in the absence of evidence, which is lacking in this case, that the mere protection of biotechnological inventions by patent would result, as is argued, in depriving developing countries of the ability to monitor their biological resources and to make use of their traditional knowledge, any more than it would result in promoting single-crop farming or in discouraging national and international efforts to preserve biodiversity.

Moreover, while Article 1 of the Convention on Biological Diversity states that its objective is the fair and equitable sharing of the benefits arising out of genetic resources, including appropriate access to genetic resources and appropriate transfer of relevant technologies, it specifies that this must take into account all rights over those resources and technologies. There is no provision of the Convention on Biological Diversity that the conditions for the grant of a patent for biotechnological inventions should include consideration of the interests of the country from which the genetic resource originates or should include measures for transferring technology.

Finally, as regards the possibility that the Directive might represent an obstacle in the context of the international cooperation necessary to achieve the objectives of the Convention on Biological Diversity, it should be borne in mind that, under Article 1(2) of the Directive, the Member States are required to apply it in accordance with the obligations they have undertaken as regards *inter alia* biological diversity.[48]

Furthermore, Advocate General Jacobs, in his lengthy Opinion of June 14, 2001, concluded that the reach of European patent legislation has inherent limits. In other words, it cannot cover matters like access to GRs in provider States.[49]

The Directive, being concerned with patents, does not seek to regulate matters outside the realm of industrial property. [...] it is not for patent legislation to provide for broader matters such as monitoring the source of biological material in respect of which a patent is sought. The Directive does not – nor can it – affect the ability of developing countries to establish controls over their genetic resources in order to prevent the unregulated plundering of such resources. At least a dozen countries have already taken such steps, in accordance with the Convention on Biological Diversity, and a similar number are currently developing controls.[50]

This opinion of the Advocate General explains how the regulation of access to GRs is the sole responsibility of provider states and not that of recipient country patent laws. This reasoning does not imply that no international obligation exists to control access to GRs, but rather that the current controlling measures within provider countries are largely inadequate. Hence, the EU IP legislation cannot ensure the achievement of these goals.

[48] Case C-377/98, *Kingdom of the Netherlands v. European Parliament*, 357.
[49] *Opinion of Advocate General Jacobs.* [50] *Ibid.*, paragraph 181.

The Advocate General continues:

> In my view, the arguments that the Directive is incompatible with the Convention on Biological Diversity betray a failure to appreciate the respective objectives and spheres of application of the two instruments.[51]
>
> The Directive [...] requires the Member States of the European Union to ensure that their national law provides patent protection for biotechnological inventions as there defined. To that effect it imposes a few highly specific obligations on the Member States in that narrow context. Patents conferred in accordance with the Directive will of course, as with all patents, be territorial in effect.
>
> The Convention, in contrast, is more in the nature of a framework agreement. Having set out its objectives in Article 1, the Convention on Biological Diversity proposes a series of approaches which Contracting Parties (which as at 5 June 2001 numbered 180 States worldwide) are to adopt, in many cases only "as far as possible and as appropriate." The scope of the Convention is rather wide; the suggested measures are rather varied and in most cases couched in general terms.[52]

The approach of the ECJ stems from the classic approach of the patent system and views it in clinical isolation from any pre-IP or post-IP issues, e.g. appropriation of genetic material on which a patentable invention is based. I hasten to note that the ruling of the ECJ was delivered before the adoption of the Bonn Guidelines that invite Contracting Parties to introduce the requirement of a certificate of origin into the patent system. This decision has been criticized by part of the legal doctrine and is superseded by the formation of subsequent practice through the Bonn Guidelines. It suffices here to cite Spada[53] and Ricolfi[54] who identify in the CBD provisions an obligation on Contracting Parties to introduce transparency measures in their national patent legislation without specifying how to do it. They submit that the "missing link" between the obligations of the CBD and their enforcement is the transparency measure in the patent application process in the recipient country.

6.1.1.2.4 Conclusion on the existence of a customary norm of prior informed consent in international law

My analysis has gathered a number of facts that support the existence of PIC and benefit sharing as international customary norms in international environmental law. The existence of the customary principles of PIC

[51] *Ibid.*, paragraph 177. [52] *Ibid.*, paragraph 178–79.

[53] P. Spada, "Liceità dell'invenzione brevettabile ed esorcismo dell'innovazione" (2000) 5(1) *Rivista di diritto privato* 5.

[54] M. Ricolfi, "Biotechnology, Patents and Epistemic Approaches" (2002) *Journal of Biolaw Business* 46.

and benefit sharing have crystallized from the adoption of the CBD (see section 1.1.2). PIC and benefit sharing are very general concepts. It needs to be specified that there is no *opinio juris* about how to specifically apply the concept of PIC or benefit sharing. The compliance with the norm occurs when the GR or the TK is extracted and used in the course of commerce under mutually agreed terms between the provider country and the recipient entity.

The arguments in favor of the existence of this customary norm can be summarized as follows: (i) the principles contained in the CBD is universally accepted through ratification or accession (its parties number 188): the CBD is one of the "international instruments of a universal or quasi-universal character;"[55] (ii) the repeated expression in various international fora (e.g. WIPO, UNEP, WTO, FAO) of the importance of complying with the obligations of PIC and benefit sharing indicates an *opinio juris sive necessitatis* (see also section 6.1.1.2.1. ff.);[56] (iii) the Bonn Guidelines also indicate the willingness of the CBD Contracting Parties to support this concept through the invitation to adopt specific measures of monitoring compliance with these obligations, even in the area of IP law; (iii) the incorporation of this concept in several ABS domestic laws indicates a *repetitio facti*, that is constantly evolving towards confirmation of the expressed *opinio juris*. Ultimately, it is the international public opinion that has also played an important role in raising awareness of the illegitimacy of misappropriating GRs and TK for private commercial exploitation without PIC.

This conclusion has not taken into account the doctrinal differences between objectivists and voluntarists. Because of the universal ratification of the CBD, recognizing the existence of the CBD principles of PIC and benefit sharing as international customary norms is a relatively plain task, even in a perspective of *coutume ancienne*.[57] Moreoever, the *opinio*

[55] *Barcelona Traction (Belgium v. Spain)*, February 5, 1970, *International Court of Justice Reports* 32. Weil P., "Toward Relative Normativity", 435–36.

[56] Carter, Trimble, and Bradley, *International Law*, 121, 135.

[57] The *coutume ancienne* (old custom) requires that the manifestation of *opinio juris* and *repetitio facti* by the international community of States be consistent through the repetition of a certain act during a certain lapse of time and space. It once required decades to create international customary law, because that law was developed through the uniform and consistent practice of nation States over times (it is spoken of *longa consuetudo, diuturnitas*). G. Abi-Saab, "La coutume dans tous ses états ou le dilemme du développement du droit international général dans un monde éclaté", in *Le droit international à l'heure de sa codification. Etudes en l'honneur de Roberto Ago*, (Giuffré, Milan, 1987) 53–65; G. Abi-Saab, "Cours général de droit international public" (1987) 207 *Recueil des cours de l'Académie de droit international* 173; R. Ago, "Nouvelles réflexions sur la codification en droit international" (1988) *Revue générale de droit international public* 539–76. More

juris is so clear and evident that the analysis of the practice may be superfluous.[58]

The fact that the concepts of PIC and benefit sharing are very vague or general in nature does not undermine the normative value of customary norms. Regardless of the framework nature of the Convention, the *negotium* of Article 15.4 and 15.5 is expressed with very precise and binding terms respectively "access, where granted, *shall* be on mutually agreed terms and subject to the provisions of this Article" and "access to genetic resources *shall* be subject to PIC of the Contracting Party providing such resources, unless otherwise determined by that Party" (emphasis added).

6.1.1.2.5 Conclusion on the existence of a customary norm of prior informed consent and benefit sharing including the modification of intellectual property law

Having determined the existence of the CBD concepts of PIC and benefit sharing as international customary norms, the second prong of the question is whether the scope of the customary norm includes the obligation to modify IP law.

recently, however, there is a new development of international customary law. In cases of quasi-universal participation to an international legal instrument or through the repetition of agreed language at international organization's diplomatic conferences, norms can be created more rapidly. This *coutume nouvelle* (new custom) requires a shorter period of time to become effective and is particularly relevant to the analysis of the normative value of the Bonn Guidelines. The temporal manifestation of the *repetitio facti* is then compensated for by a short but intense expression of the *opinio juris*. In other words, there is no need for any prolonged evidence of a State practice to qualify the norm expressed by the *opinio juris*. The unilateral act of States participating in an international diplomatic conference expresses their willingness to be bound by a certain norm which can be interpreted as a State practice. The theory of the *coutume nouvelle* is not totally revolutionary. It rests upon the preeminence of the subjective element of the *opinio juris* over the objective one of the *repetitio facti*. The analysis of a number of International Court of Justice (ICJ) decisions and of the historical interpretation of Article 38.1(b) of the ICJ Statute has demonstrated that the line of separation between the element of *opinio juris* and the one of *repetitio facti* is very thin, to the point that the first may include *in nuce* the latter. There is a difference between the process of formation of the customary norm and the process of observation of its existence. The objective element of the *repetitio facti* stems from the process of the judge's identification of all the evidence supporting the existence of the customary norm; State's practice is certainly one of them. Whereas the theory of the two elements is relevant to the process of identification of the norm, the *opinio juris* may be sufficient for the process of formation itself. Thus, the practice may also be considered merely as an expression of the *opinio juris* such as in the case where the process of generation of an international norm is clearly identifiable or circumscribed, like an international diplomatic conference. This means that the repeated expression of *opinio juris* (by each State forming the international community) in multilateral negotiations on a given norm of conduct can render it customary; see P. Haggenmacher, "La doctrine des deux éléments du droit coutumier dans la pratique de la Cour internationale" (1986) 90(1) *Revue générale de droit international public* 5, 118.

58 Caflisch, "La pratique dans le raisonnement du juge international", 128.

The elements identifying the existence of such scope of the customary norm need to be analyzed in their chronological development.

The ECJ judicial decision on the compatibility of the EC Biotechnology Directive on biotechnological patents with the CBD's obligations provides an additional argument against any possible customary norm mandating a modification of the patent system. No other international court or international organization has issued a decision or opinion, outside the realm of international environmental law, concerning the status of PIC and benefit sharing for the IP exploitation of the GRs and TK.

But, since then, it is noticeable that the customary principle has started to crystallize from the CBD. The Bonn Guidelines express the authentic interpretation of the CBD customary concepts of PIC and benefit sharing. This *opinio juris* is articulated in conditional terms: the implementation of the CBD PIC "*could*" occur through the introduction of a certificate of source/origin in their patent systems. Indeed, the Bonn Guidelines give a list of possibilities for implementing the customary obligation of the CBD PIC and benefit sharing. The Bonn Guidelines specify the introduction of the transparency measure in terms of possibility. The soft law *instrumentum* does not contain a very precise or hard law *negotium*. Hence, it can be considered as subsequent practice to this extent but only for the State parties, and thus not the US.

The Bonn Guidelines certainly mark an embryonic stage of the expansion of the customary norms of PIC and benefit sharing to include patent law. The soft law *instrumentum* of the universally accepted Bonn Guidelines contains a soft *negotium*: therefore, it must be concluded that at that time the international community viewed the amendment of the patent system to accommodate the concept of PIC as a possibility rather than as an obligation. However, it can be argued that the amendment of patent law in the recipient State may be the only way in which the customary norms of the CBD can be effectively complied with. But why then have the contracting parties adopted guidelines instead of a protocol? Certainly, the answer cannot limit itself to their consideration as a possibility rather than as an obligation. A protocol requires another formal process to be adopted. However, the very fact that the phrases in question were not contained in a binding instrument and that the language was put in the conditional indicates that serious doubts can be raised about the validity of formation of a norm of *coutume nouvelle* even from an instant custom perspective.

The domestic legislative practice of each State becomes crucial, after the adoption of the Bonn Guidelines. The specific implementation of the customary PIC obligation through the introduction of a certificate of origin is enjoying an increasing State practice. The practice observed in section 6.1.1.2.2 above should be associated with the universally accepted Article 16.5 which provides that IPRs be "supportive of and do not run counter to its objectives," and therefore this "practice is derived from a clear consensus among states, as exhibited both by widespread conduct and a discernible sense of obligation."[59] Notwithstanding the divergences in implementing the invitation of the Bonn Guidelines in State practice, a growing *opinio juris communis* and State practice affirm the customary principle that States have to render their patent laws compatible with the CBD PIC obligation.

State practice has a short history but is becoming general and consistent. Certainly, in a classic approach to custom, the scope of PIC and benefit-sharing customary norms that deal with the introduction of a certificate of origin/source in the patent system has not crystallized as yet. The supporters of the instant custom or *coutume nouvelle* approach would note that the language of the Bonn Guidelines specifies the way in which States should implement the PIC and benefit sharing. The use of the conditional, however, demonstrates the unwillingness of the CBD Parties to make such a norm immediately binding.

From the perspective of the TRIPS Council, the debate on this matter is not settled. The Doha Declaration in 2001 merely gives the TRIPS Council the mandate to explore "the relationship between the TRIPS Agreement and the CBD,"[60] its Members having not manifested a clear position on its normative status of international customary principle.

All these aforementioned reasons lead to the conclusion that there is much uncertainty over the existence of a customary principle of PIC and benefit sharing outside the realm of international environmental law, thus mandating a modification of the patent law to render it compliant with these concepts. In other words, there is no customary principle of PIC and benefit sharing beyond the field of international environmental law and therefore, at the moment, there is no direct customary obligation upon States to modify patent law. However, this latter obligation may one day

[59] *Competence of the International Labour Organization in regard to international regulation of the conditions of labour of persons employed in agriculture*, Advisory Opinion, No. 2 Permanent Court of International Justice, Ser. B., No. 2, 1922 (December 8, 1922). This advisory opinion of the PCIJ which held that ILO, as an international organization, had the competence to regulate internationally the "conditions of labor of persons employed in agriculture," quoted in Carter, Trimble and Bradley, *International Law*, 121, 125; *Reparation for Injuries Suffered in the Service of the United Nations* (1948–49), April 11, 1949, Advisory Opinion, *International Court of Justice Reports*, www.icj-cij.org/icjwww/idecisions.htm.

[60] Despite the previous reluctance stressed in the western proposals about the revision of this Article mentioned above, during the last Ministerial Conference of WTO in Doha unexpected language can be found in the *Doha Declaration*.

crystallize from the hard law of the CBD and the soft law of the Bonn Guidelines into customary law as the *opinio juris* and the practice of States points in the direction of, for example, introducing the certificate of origin into the patent system. However, even under the current trends of modification of the patent system, this certificate faces difficulties if it needs to be used as evidence of control of the compliance with PIC and benefit sharing (see more specifically sections 6.1.3.6.8 and 6.1.3.6.9 below).

Hence, the practice (*repetitio facti* or *diuturnitas*) to modify the patent system is not widespread but still limited to States (mainly DCs) that are very attentive to this matter. "There is no precise formula to indicate how widespread the practice must be,"[61] however the fact that more and more countries are amending their patent laws in this area, leads one to conclude that a customary norm of adapting patent law to the CBD is in its early stage of crystallization. Before becoming a customary norm, this concept certainly also needs to acquire more precision. It is suggested that the embryonic customary norm has passed from the stage of *conduite* to one of *pratique* and is in the process of reaching one of *norme*, thus headed towards the stage of *coutume* (custom).[62]

The very fact that States find that the best way to implement the concepts of benefit sharing and PIC is to amend their patent law in a mutually supportive manner strengthens, *a fortiori*, the determination in favor of the existence of PIC and benefit sharing as norms of general international law. The practice of international organizations will certainly go in this direction as they are helped by non-State actors in the development of the connection between the PIC and benefit sharing and its means of implementation in patent law.[63]

6.1.1.2.5.1 The US as persistent objector to the customary principle of prior informed consent In the previous paragraphs it has been determined that certain CBD provisions related to PIC and benefit sharing are customary norms. They also bind the US (the only State that has not ratified or acceded to the CBD). One should briefly remind oneself that the content of the CBD norms of PIC and benefit-sharing concepts based upon Article 15.5 and 15.7 of the CBD is that a State cannot allow the commercialization of products based upon misappropriated GR or TK

[61] Carter, Trimble and Bradley, *International Law*, 125.
[62] Dupuy, 'Le droit des Nations unies', 141 quoting the works of Abi Saab and then Cahin.
[63] L. Boisson de Chazournes, "Qu'est-ce que la pratique en droit international?", La pratique et le droit international, Colloque de Genève de la Société française pour le droit international (Pedone, Paris, 2004) 43–46.

and that access to the GR and TK must occur under mutually agreed terms with the provider country.

The considerations on the normative value of these concepts are important given the current expansion of the scope of application of the CBD from the realm of international environmental law to that of IP law with its obligation of declaring the source or the evidence of PIC and benefit sharing during the patent application.

The question arises here as to whether the US can claim the status of persistent objector (i) to prevent the customary norms in the course of crystallization entering into force, and/or (ii) in case the theory were accepted as valid by an international court, to be immunized from the opposability of such customary norms. The very existence of the concept of persistent objector is disputed. Without delving into the debate over the existence or not of the concept of persistent objector in international law, I will assume that – were it a valid concept of defense that relies on opposition to a customary norm – these should be some of the guidelines on how to determine whether the US can qualify for this status.[64]

[64] For a discussion on the theory of persistent objector in particular Abi-Saab, "Cours général de droit international public" (1987) 207 *Recueil des cours de l'Académie de droit international*, 180–82. *North Atlantic Coast Fisheries Arbitration before the Permanent Court of Arbitration at the Hague under the Provisions of the General Treaty of Arbitration of April 4, 1908, and the Special Agreement of January 27, 1909, between the United States of America and Great Britain*, US 61st Congress, 3rd Session, Senate, Document No. 870 (Washington: Government Printing Office, 1912–13, Vol. 1). The debate over the validity in international law of the theory of persistent objector goes hand in hand with the way customary norms are formed. In favor of the validity of the concept are the voluntarist publicists, *inter alia*, T. Stein, "The Approach of the Different Drummer: The Principle of the Persistent Objector in International Law" (1985) 26 *Harvard Law Journal* 463; G. Fitzmaurice, "The Law and Procedure of the International Court of Justice 1951–54: General Principles and Sources of Law" (1953) *British Yearbook of International Law* 21–26; L. Henkin, "International Law: Politics, Values and Function: General Course on Public International Law" (1989) 216 *Le recueil de cours de l'Académie de droit international* 53–58; Weil, *Toward Relative Normativity*, 413–42; M. Akehurst, "Custom as a Source of International Law" (1974–75) *British Yearbook of International Law* 24–26, 23–27; H. Waldock, "General Course on Public International Law" (1962) 106 *Le recueil de cours de l'Académie de droit international* 49–50; H. W. A. Thirlway, *International Customary Law and Codification* (A. W. Sijthoff, Leiden, 1972) 78; K. Wolfke, *Custom in Present International Law* (Martinus Nijhoff, Dordrecht, 1993) 66–67; P. Cahier, "Changements et continuité du droit international, Cours général de droit international public" (1985) 195 *Le recueil de cours de l'Académie de droit international* 231–37; M. Sorensen, "Principes de droit international public: Cours general" (1960) 101 *Le recueil de cours de l'Académie de droit international* 44; I. C. MacGibbon, "Some Observations on the Part of Protest in International Law" (1953) *British Yearbook of International Law* 318; M. Danilenko, *Law-Making in the International Community* (Martinus Nijhoff Publ., Dordrecht, 1993) 109–13; A. Steinfeld, "Nuclear Objections: The Persistent Objector and the Legality of the Use of Nuclear Weapons" (1996) 62(4) *Brooklyn Law Review* 1635–86 (deals with the question whether State possessing nuclear weapons can claim the status of persistent objector); L. Loschin, "The Persistent Objector and Customary Human Rights Law: A

To determine whether the US has achieved persistent objector status, one must consider the manifestation of its *opinio juris* by the US and its practice in opposition to the formation period from its inception. This can manifest itself through the unilateral conduct of the US vis-à-vis the formation of the customary norm in question. This unilateral conduct can be formal or informal. Formal conduct is constituted by unilateral acts that explicitly manifest the intention of the State. Informal conduct mainly stems from acts or abstentions that, as a consequence, have produced rights or obligations.[65] This analysis limits itself to the most indicative unilateral acts and the acquiescence towards the customary norms in question.

It is not possible to exhaustively analyze all the unilateral acts that allegedly support the status of persistent objector of the US: reasons for a certain vote at an international conference, a reservation or declaration upon becoming a party to an international legal instrument, protest notes, diplomatic communication, any proceeding in the course of a negotiation, domestic laws, domestic administrative and judicial decisions, declarations of members of the government with language and the intention of the State, acts or declarations by representatives of legal advisers to State governments, bilateral treaties, and press releases and official statements.

Particular attention should be paid to the concept of acquiescence. It follows from the principles of good faith and equity that "[...] acquiescence

Proposed Analytical Framework" (1996) 2 *U.C. Davis Journal of International Law & Policy* 148; M. H. Mendelson, "Formation of Customary International Law" (1998) 272 *Le recueil de cours de l'Académie de droit international* 228–33; International Law Association, *Statement of Principles Applicable to the Formation of General Customary International Law* (written by M. H. Mendelson) *Report of the Sixty-Ninth Conference* (London, 2000) 738–74; O. Elias, "Some Remarks on the Persistent Objector Rule in Customary International Law" (1991) *Denning Law Journal* 37–51; G. Pentassuglia, *La rilevanza dell'obiezione persitente nel diritto internazionale* (Laterza, Bari, 1996) 255; B. J. McClane, "How Late in the Emergence of a Norm of Customary International Law May a Persistent Objector Object?" (1989) 13 *ILSA Journal of International Law* 6; Danilenko, *Law-Making in the International Community*, 111. Against the validity of the concept are objectivist publicists, *inter alia*, J. Charney, "The Persistent Objector Rule and the Development of Customary International Law" (1985) 56 *British Yearbook of International Law* 1–24; P.-M. Dupuy, "A propos de l'opposabilité de la coutume générale", in *Le droit international au service de la paix, de la justice et du développement. Mélanges offerts à Michel Virally* (Paris, 1991) 257–72; B. Conforti, "Cours général de droit international public" (1988) 212 *Le recueil de cours de l'Académie de droit international* 74–77; Abi-Saab, *Cours général de droit international public*, 180–82; M. Bos, "The Identification of Custom in International Law" (1982) 25 *German Yearbook of International Law* 45–50; G. P. Buzzini, *Le droit international général au travers et au-delà de la coutume: Essai de conceptualization d'une réalité aux contours fluctuantes* (Graduate Institute of International Studies, Geneva, 2007); L. Condorelli, "La Coutume" in M. Bedjaoui (ed.), *International Law: Achievement and Prospects* (UNESCO, Paris, 1991) 205.

[65] Kohen, "La pratique et la théorie des sources du droit international", 108.

is equivalent to tacit recognition manifested by unilateral conduct which the other party may interpret as consent."[66] The case law of the ICJ indicates that in order to determine the tacit consensus of a State vis-à-vis a certain norm, the State's attitude must be uniform and constant and the practice must be accompanied by the intention to comply with it.[67] With regard to this concept, it should be noted that the US participated in the negotiation process and the adoption of the CBD and signed the Convention on June 4, 1993. It currently participates as an observer at the CBD COP where the progressive intensification of discussions, along with those in other international fora (such as WIPO, UNCTAD, UNESCO, human rights bodies, etc.) is leading to the creation of "soft law" on ways to implement the PIC obligation. Thus, it is possible to observe the US perception of the crystallization process of the customary norms to be that a GR needs to be accessed and commercially exploited under PIC and benefit sharing with the provider State's authority.

The US voted in favor of the Doha Declaration of the WTO Ministerial Conference in 2001, giving the TRIPS Council the mandate to explore "the relationship between the TRIPS Agreement and the Convention on Biological Diversity."[68] In its 2004 submission to the TRIPS Council, the US has restated its support for the objectives of the CBD:

> Based on recent discussions in the TRIPS Council aimed at fulfilling the Doha Ministerial mandate, and written contributions submitted in that context, Members appear to share several broad policy objectives. These objectives include: (i) ensuring authorized access to GRs, i.e. with PIC; (ii) achieving equitable sharing of the benefits arising from the use of TK and GRs; and (iii) preventing the issuance of erroneously issued patents. *The US supports these objectives* and has consistently encouraged and supported the equitable sharing of benefits arising from the utilization of TK and practices of indigenous and local communities.[69] (italics added)

In a maximalist objectivist view, it can be argued that this declaration of support stated in the TRIPS Council easily disqualifies the US as a persistent objector to the customary norms in question. However, an extreme voluntarist approach can raise doubts as to whether this

[66] *Case Concerning Delimitation of the Maritime Boundary in the Gulf of Maine Area (Canada v. United States of America)* October 12, 1984, *International Court of Justice Reports.*

[67] The acquiescence has been mainly found in territorial dispute cases: *Case Concerning the Temple of Preah Vihear(Cambodia v. Thailand)* June 15, 1962, *International Court of Justice Reports*, 30; *North Sea Continental Shelf Case*, 25 paragraph 28, 26 paragraph 30; *Certain Phosphate Lands in Nauru (Nauru v. Australia)* September 13, 1993, *International Court of Justice Reports* 247 paragraph 13 (1992), Kohen, "La pratique et la théorie", 109.

[68] *Doha Declaration.* [69] IP/C/W/434, paragraph 5.

statement demonstrates the *opinio juris* of the US to feel obliged to comply with the CBD objectives that, in turn, express the customary principles at stake. Although the US statement demonstrates that it supports the aforementioned objectives, there is no indication that the US views PIC and benefit sharing as a legal obligation. Supporting the objectives of a treaty like the CBD does not mean that the US feels bound by an obligation. It can be counter-argued that the US did not state its opposition either to the objectives or the obligations of the CBD.

Principles enjoying universal acceptance like PIC and benefit sharing imply an automatic presumption of a customary norm that binds the international community as a whole. In order not to be bound by a crystallized customary norm, a State must timeously, unambiguously, and persistently object to the norm with cogent reasons. The US did not express any opposition to the norms. This acquiescence or the lack of manifestation of opposition on the part of the US against customary norms, should be enough to disqualify it as a persistent objector.[70]

It can thus be argued that the support of these customary principles in an official statement submitted to a multilateral negotiation setting of an international organization amounts to a unilateral act manifesting an *opinio juris* to be bound by this universally accepted customary norm. Yet, the statements of the US do not directly express the *opinio juris* to be legally bound by the stated principles of the CBD. In a case in which the US would be called to defend its position when accused of alleged violation of the customary norm crystallized from the CBD of permanent sovereignty over the GRs and the prohibition of exploiting foreign GRs without PIC and benefit sharing, it could not effectively maintain its persistent objector status because of the clear absence of any other express rejection of such principles as normative binding obligations: the US did not oppose the creation and formulation of the CBD customary principles in question. On the contrary, it participated in the negotiation and in the whole process of formulating and adopting the CBD principles and continues to express its opinions within the TRIPS Council for the revision of Article 27 of TRIPS and in the WIPO IGC on IPGRTKF. This is true especially when one compares this US attitude towards the general principles of the CBD and how the US deals with the impact of norms on the patent law system. In the latter case, it has continued to manifest its robust opposition to any application of the CBD principles to patent law.

[70] K. Skubiszewski, "Les actes unilatéraux des Etats", in Mohammed Bedjaoui (ed.), *Droit international: Bilan et perspectives* (Pedone, Paris, 1991) 231 (this author studies the unilateral acts of States in relation with the determination of their value for the opposability of a customary norm of international law).

The US acquiescence in this case means decay of the persistent objector status, because this concept, if admitted in international law, is based on a positive action on the part of the State to clearly express its objection. The decay of the status of persistent objector triggers immediate and automatic opposability of the customary norm that binds the entire international community of States. The signature alone does not bind the US to the CBD principles but it commits the US not to do anything that prevents the entry into force and the implementation by the ratifying States. The subsequent unilateral acts that render null and void the effects of the signature (Article 18 of the VCLT) cannot be used as an indication of not being bound by certain customary principles contained therein. Customary principles are *ex ante* detached from a particular international legal instrument.

I have demonstrated that PIC and benefit sharing have crystallized into customary international law (see section 6.1.1.2 above). It is beyond the scope of this study to determine whether the customary concepts of PIC and benefit sharing were formed before the adoption of the CBD and then were merely codified therein. It can be noted that the customary concepts of PIC and benefit sharing for commercial exploitation of GRs and TK find the roots of their processes of crystallization as customary norms in the *travaux préparatoires* of the CBD. These principles have indeed been newly created in the context of the negotiation of the CBD. Even in presence of a process of codification and not crystallization, the more recent behavior of the US disqualifies it as a persistent objector.

Another question can be raised as to whether the lack of objecting *opinio juris* is sufficient to determine the disqualification of persistent objector. It may be asked whether this clear manifestation of the subjective element of the *opinio juris* inherently includes the objective element of the *diuturnitas, repetitio facti* or State practice. From an "instant custom" and objectivist perspective, it can be argued that a clear manifestation of *opinio juris* (in this case, lack of objection to the PIC and benefit sharing) outweighs a possible contrary practice. No official organ of the US has ever officially declared that these CBD objectives/principles were not binding upon it. This acquiescence towards the CBD objectives and concepts[71] combined with the unilateral acts of the statements like the ones above may easily lead one to conclude that the US has the perception that it is bound by

[71] B. Cheng, "Custom: The Future of General State Practice in a Divided World", in R. Macdonald and M. Douglas (eds.), *The Structure and Process of International Law: Essays in Legal Philosophy Doctrine and Theory* (Kluwer, The Hague, 1983) 513, 532, 549. B. Cheng, "United Nations Resolutions on Outer Space: 'Instant' International Customary Law, Indiana", *Journal of International Law* (1965) 5, 23–48.

these rules. The contrary practice of disregarding these principles of PIC and benefit sharing does not diminish this norm's value as custom. The view that the US *opinio juris* expresses the objective element of the practice renders less relevant the disregard (or the violation) of such customary norms by the US courts and administrative authorities (including the USPTO) granting patents that are based on misappropriated GRs and TK. The restrictive interpretation of the words of the unilateral act in light of the intention of the State is sufficient[72] to establish the perception that the US is bound by the customary norm.

In sum, there is a customary international obligation binding on the entire international community that States have to employ their best efforts to curb the phenomenon of unauthorized appropriation of GRs and TK under the sovereignty within each State in order to commercially exploit them. In light of the aforementioned elements, there is, *prima facie*, sufficient evidence to indicate that the US could not qualify as persistent objector towards the customary obligation to avoid commercially exploiting foreign GRs and TK without PIC and benefit sharing. It remains to be seen whether the scope of the customary norm of this commercial exploitation includes the obligation to modify this patent system, i.e. whether the US, as other countries, needs to modify its national patent system in a way that is compliant with these customary principles.

The obligation of PIC and benefit sharing has two sides: *obligation de faire* and *de ne pas faire* (i.e. obligation to do something or not to do something, due diligence).[73] The scope of the international customary norm of PIC or benefit sharing does not entail any particular positive action on the part of the State, except the exercise of its due diligence in preventing violations of the obligation. In other words, the fact that the scope of the obligation does not encompass a duty of action does not immunize the State from being responsible for allowing the violation of the obligation to occur without undertaking any measure to prevent it.

6.1.1.2.5.2 The US as persistent objector to the implementation of prior informed consent as a customary norm in the patent system In reality, two questions arise in this context. The first question is whether the scope of the customary obligation of PIC and benefit sharing comprises the

[72] Kohen, "La pratique et la théorie," 109.

[73] *Cour Permanente d'Arbitrage dans l'affaire des Pêcheries de l'Atlantique Nord du 7 Septembre 1910* (*US. v. Great Britain*), RSA, Vol. XI and *Corfu Channel (UK v. Albania)* December 17, 1948, *International Court of Justice Reports* 124 (1947–48). This type of obligation triggers a *préjudice médiat* or *immmédiat* especially in the human rights field; this explanation stems from the course of Public International Law of Prof. Condorelli at the Law School of the University of Geneva, Switzerland (1999).

prohibition to grant patents for biotech-inventions based on misappropriated GRs and TK. This entails the modification of the domestic patent system so as to incorporate a control to ensure that these two norms are complied with by the patent applicant. The second question regards the possibility that the US would qualify as a persistent objector vis-à-vis the possible development of this customary norm.

The first question has been addressed in section 6.1.1.2.5 above. It has been determined that the norm is not customary as yet and concluded that this obligation lies *in statu nascendi*. In the context of the mutable and dynamic international law-making process, it may become a new customary norm.

In order to verify the qualification of the US as a persistent objector towards this customary norm in its stages of formation, its unilateral acts need to be observed. The US has stated that the CBD provisions do "not require countries to modify their patent laws in any way."[74] This statement is a clear indication that the US refuses to grant the customary norm of PIC and benefit sharing an immediate scope, including the modification of IP.

The US has specifically opposed a patent system that monitors these obligations. It states that courts and other authorities of those jurisdictions providing the GRs or TK would be more appropriately situated to examine these matters.[75] The US has articulated its position in three concepts. Firstly, the US maintains that a "completely separate, transparent mechanism needs to be established outside the patent system that implements these criteria regardless of any disclosure [of source] made in a patent application."[76] Secondly, the US believes that a new patent disclosure requirement would indeed fail to address benefit sharing resulting from commercialization that occurs outside the patent system.[77] Thirdly, the US argues that the most effective means of achieving the shared objectives of obtaining appropriate ABS is through development of national laws outside the patent system which can more directly and effectively regulate conduct relevant to these issues.[78] The US expresses views stemming from the classic patent approach that opposes the integration of a disclosure requirement.

The US continues to argue that:

> patent law was not designed to regulate or enforce misconduct issues, such as misappropriation of TK or genetic resources, but to promote the progress of the

[74] *Communication of the United States of America, Article 27.3(b), Relationship between the TRIPS Agreement and the CBD, and the Protection of Traditional Knowledge and Folklore*, IP/C/W/449 (June 10, 2005), www.wto.org/english/tratop_e/trips_e/art27_3b_e.htm.

[75] *Ibid.*, 4. [76] *Ibid.*, 3. [77] *Ibid.*, 4. [78] *Ibid.*, 6.

useful arts. Patent rights permit an inventor to exclude others from certain acts [Article 28 of the TRIPS Agreement], but do not permit an inventor to use the invention without restriction. Restrictions can and are placed on the use of certain inventions to ensure safety and efficacy (e.g. health regulations governing pharmaceutical products), to protect the environment (e.g. regulations governing emissions from automotive engines) or to protect domestic or national security (e.g. regulating firearms). These restrictions, notably, are enforced outside the patent system by separate regulatory mechanisms.[79]

More specifically, advocates of the US defending its position as a persistent objector may use this statement to demonstrate that Article 16.5 – that puts States under an obligation to cooperate to ensure that IPRs are "supportive of and do not run counter to its [CBD's] objectives" (Article 16.5) – is not binding on the US. The aforementioned statement immunizes it from opposability in case of crystallization of an eventual customary norm mandating that States must render their patent systems supportive of the CBD's objectives by modifying its patent law so to ascertain that TK is not patented or misappropriated and that GRs under the sovereignty of foreign States are not patented without PIC and benefit sharing.

In my view, in case the concept of persistent objector is valid in international law, it is clear that the US qualifies as such in the period of formation of this potential customary norm. This does not mean that the opposition of the US to the *de lege ferenda* proposals on the disclosure of origin are absolutely grounded on the right reasons. It can be said, for instance, that stating that patent law does not deal with similar issues is not correct: patent law comprises the concepts of evident abuse of application and the regime of inventor/employer that are similar concepts of abuse for misappropriation of the essential resource on which the invention is based. The findings of the present study lead me to maintain that the manner in which essential elements of patentability, like GR and TK, are acquired is related to patent law. But the way in which the patent system should be modified is not clear. Again this obligation rests more on the international community than upon each individual State.

Regardless of the reasonableness of the arguments of the US, this important unilateral act of objection clarifies its status of persistent objector at the early stage of crystallization of the norm that mandates modification of patent law in accordance with the principles of PIC benefit sharing. No conflicting unilateral act to the aforementioned statement can be found. The manifestation of the *opinio juris* is coherent and constant on this point. The fact that an applicant receives a monopoly right on the

[79] *Ibid.*, 7.

basis of a GR and TK accessed in a provider country without PIC does not engage the State responsibility of the US that has neither ratified the CBD nor adopted the Bonn Guidelines. These latter are non-binding soft law but are contributing to the formation of certain customary norms at a very early stage of crystallization.

In a very objectivist perspective, however, the US will have to face the theory of "graduated normativity" in international legal contexts. Accordingly, some commentators have begun to talk about "soft responsibility" in the sense of a responsibility that ranks lower than "classic responsibility" of the State in terms of its normative contents. Once this process of crystallization is complete, any State, including the US, by acting neither within the patent system nor outside of it, may be found responsible for commercialization of patented products based on GRs acquired through blatant acts of misappropriation.[80]

For all these reasons, if the theory of persistent objector is not a valid concept of international law and if the US is found constantly in breach of the aforementioned concepts of PIC and benefit sharing that have become customary norms, it becomes a "persistent violator"[81] instead of a "persistent objector."

6.1.1.2.5.3 The decay of the status of persistent objector before the peremptory norm of state sovereignty over its genetic resources If a State is admitted to be a persistent objector vis-à-vis a customary rule of *jus dispositivum*, what happens to this status if the same customary norm becomes a *jus cogens* norm? The answer is simple: the status of persistent objector decays. I have found no author contesting the same. It is argued that the CBD mandated concept of national sovereignty over GRs is not only a treaty or customary norm but also a *jus cogens*, an inderogable norm. The legal discourse should discuss whether the norm would include a prohibition against commercially exploiting GRs in the IP system without PIC by the sovereign provider country or whether they can be accessed without compliance with the provider State rules (which may include benefit sharing and PIC). Since TRIPS Agreement does not include these concepts a potential conflict arises to be solved under the relevant rules of international law.

[80] "A Hard Look at Soft Law" (1988) 82 *American Society of International Law and Procedure* 371–77 (remarks by Michael Reisman); P. M. Dupuy, "Soft Law and the International Law of the Environment" (1991) 12 *Michigan Journal of International Law* 420, 422–25, 428–31.

[81] Kohen, "La pratique et la théorie", 90.

The conflict between the CBD provisions and TRIPS Agreement acquires another dimension. While Article 27 of TRIPS Agreement mandates that States grant patents over GRs based innovation, the CBD affirms the sovereignty of States. Some scholars maintain that the concept of State sovereignty upon biodiversity is a *jus cogens* norm.[82] This falls within the general scope of the *jus cogens* norm of the protection of the environment. This means that, in case of conflict between the two norms, the *jus cogens* norm of national sovereignty over its own GRs trumps the treaty provision of TRIPS Agreement in case the GR on which the innovation is based has been exploited in a way that breaches the peremptory norm. This is so for the acquisition of GRs in a manner that is not compliant with the domestic laws of the provider country.

On the one hand, it can be stated that the two concepts are not on the same level: the principle of permanent sovereignty does not exclude, as in any other field, its limitation and qualification by other international agreements. States may express their sovereign powers by making provision in a treaty that allows, in this case, that private parties may own a monopoly right for an innovation based upon GRs under their sovereignty.

On the other hand, the concept of State sovereignty upon its biodiversity is a *jus cogens* norm. Consequently, in case of conflict between the norm of the CBD which expresses a *jus cogens* norm and a treaty norm of TRIPS Agreement, Article 53 of VCLT clearly mandates that a *jus cogens* norm trumps a treaty norm since "a peremptory norm of general international law is a norm accepted and recognized by the international community of States as a whole as a norm from which no derogation is permitted and which can be modified only by a subsequent norm of general international law having the same character."

The *jus cogens* norm of permanent sovereignty over the State's GRs would allow private individuals to obtain IPRs for innovations based upon GRs under the sovereignty of another State (as Article 27 of TRIPS Agreement mandates), but the access to the GRs must be in compliance with the laws of the provider State. I do not think that the scope of this *jus cogens* norm comprises a ban against patenting innovations based upon foreign GRs if they are acquired in a way that does not respect the concept of national sovereignty of the provider State over their GRs. However, in case of illegal acquisition of GRs, the State that grants patents on

[82] S. R. Chowdhury, "Permanent Sovereignty over Natural Resources", in K. Hossain and S. R. Chowdury, *Permanent Sovereignty over Natural Resources in International Law* (Frances Pinter Publ. London, 1984) ix; K. Baslar, *The Concept of the Common Heritage of Mankind in International Law* (Martinus Nijhoff, The Hague, 1998) 136.

innovations based upon misappropriated GRs may become responsible by omission for a violation of a *jus cogens* norm, to which it cannot raise a defense based on persistent objector status.

In sum, in case the theory of persistent objector is valid, it would allow a State to be a persistent objector to a customary rule of *jus dispositivum* but the status diminishes in the face of a *jus cogens* norm.[83] It would be very hard to maintain that the scope of this *jus cogens* norm comprises the ban of patenting innovations based upon foreign GRs if they are legally acquired. But if the way in which GRs are appropriated, commercialized and originate a patent (on which the innovation is based) is made in violation of the laws of access to it, then the question does arise as to whether the State that grants such patents becomes responsible by omission for a violation of a *jus cogens* norm, to which it cannot raise persistent objector status as a defense. The concept of omission in this regard is explained in the following section.

6.1.1.2.6 State responsibility for breach of the customary principle of prior informed consent and benefit sharing in the patent system

The question arises as to which level of responsibility a State incurs in case it breaches PIC and benefit-sharing customary norms. This subsection observes how the regime of State responsibility can be triggered in case of breaches of these obligations.

I begin by discussing the rules that should be applied in case of breach of a customary norm. The ultimate responsibility of the State with regard to the CBD should be viewed in light of the principles set out by the ILC. The ILC has distinguished between (i) internationally wrongful acts – that give rise to State responsibility – and (ii) activities not contrary to international law – that give rise to liability for injurious consequences.[84]

[83] For instance, the objection raised by South Africa and Rhodesia in the *South West Africa, ICJ Pleadings, Oral Arguments, Documents*, Vol. X, 9–11 (1960) with respect to the *apartheid* regimes of South Africa and Rhodesia did not immunize them from the opposability of a *jus cogens* norm and other human rights standards although they had persistently objected to these peremptory and underogable *jus cogens* customary norms; Charney, "The Persistent Objector Rule", 13–14; Stein "The Approach of the Different Drammer", 462.

[84] Some scholars dispute the validity of this distinction; P. Allott, "State Responsibility and the Unmaking of International Law" (1988) 29 *Harvard International Law Journal* 1; A. Boyle, "State Responsibility and International Liability for Injurious Consequences of Acts Prohibited by International Law: A Necessary Distinction?" (1990) 39 *International and Comparative Law Quarterly* 1; D. Magraw, "Transboundary Harm: The International Law Commission's Study of International Liability" (1986) 80 *American Journal of International Law* 305; J. Combacau and D. Alland, "Primary and 'Secondary' Rules in

The alleged responsibility of a State falls within the scope of the above point (i) and needs to be assessed under Article 2 of the Draft Articles on Responsibility of States for Internationally Wrongful Acts, which reads: "there is an internationally wrongful act of a State when conduct consisting of an action or omission: (a) is attributable to the State under international law, and (b) constitutes a breach of an international obligation of the State."[85]

For instance, the EU may be found allegedly breaching the CBD and its derivative law because of its Biotech-Directive (discussed in section 6.1.1.2.3 above). In this case, the alleged breach by the EU consists in failing to ensure compliance with the CBD PIC-mandated obligation in its Biotech-Directive. It follows that dereliction of such duty falls under the category of the ILC's primary rules. The secondary rules trigger the responsibility and the legal consequences of failure to abide by the primary rules.[86]

The ILC has developed three tests to determine State responsibility:

Was there an obligation of international law?

Was the duty breached?

Can responsibility be attributed to a State for the violation of international law?

I answer these questions through the analysis of the position, e.g. of the EU Members that need to comply both with the Biotech-Directive and the CBD. The same reasoning is *mutatis mutandis* applicable to all the other CBD parties in a similar situation.

I note two different interpretative approaches regarding the existence of an obligation to create a system ensuring that the obligation of PIC and benefit sharing has been complied with between the patent applicant and the provider country counterpart. On the one hand, Spada[87] and Ricolfi[88]

the Law of State Responsibility: Categorizing International Obligations" (1985) 16 *Netherlands Yearbook of International Law* 81; G. Handl, "Liability as an Obligation Established by a Primary Rule of International Law" (1985) 16 *Netherlands Yearbook of International Law* 49; M. Spinedi and B. Simma (eds.), *United Nations Codification of State Responsibility* (Oceana, New York, London, 1987).

85 *Draft Articles on Responsibility of States for Internationally Wrongful Acts*, at its 2683rd meeting held on May 31, 2001, and at its 2701st meeting held on August 3, 2001 Report of the Drafting Committee, A/CN.4/L.602/Rev, www.un.org/law/ilc/texts/State_responsibility/responsibility_articles(e).pdf#pagemode=bookmarks. P.-M. Dupuy, "The International Law of State Responsibility: Revolution or Evolution?" (1989) 11 *Michigan Journal of International Law* 105; P.-M. Dupuy, "Reviewing the Difficulties of Codification: On Ago's Classification and Obligations of Means and Obligations of Result in Relation to State Responsibility" (1999) 10 *European Journal of International Law* 371.

86 *Contra* see Allott, "State Responsibility and the Unmaking of International Law"; V. Lowe, "Precluding Wrongfulness or Responsibility: A Plea for Excuses" (1999) 10 *European Journal of International Law* 405 (1999).

87 Spada, "Liceità dell'invenzione brevettabile", 18.

88 Ricolfi, "Biotechnology, Patents and Epistemic Approaches", 46.

identify in the CBD provisions an obligation on Contracting Parties to introduce transparency measures in their national patent legislation. However, they do not specify exactly how this should be done. There are good reasons to agree with Ricolfi and Wells' criticism vis-à-vis the ECJ Advocate General's position (see section 6.1.1.2.3 above) stating that the recipient countries' patent laws should not be solicited in order to avoid the unauthorized appropriation of genetic information.[89] Indeed, Ricolfi and Wells' approach maintains that Articles 15.7 and 16.5 of the CBD set forth a general obligation on the legislative action that is needed at the *national level* so that the exercise of IPRs does not run counter to the objectives of the CBD.

There is no precise obligation that this legislative action should be done by introducing an additional requirement in patent law requiring the evidence of PIC from the provider country or by disclosure of the origin/ source of GR and TK. This obligation is largely couched in generalities, broad objectives, caveats, and other qualifiers carrying little specificity in terms of State obligation. The provisions of Articles 15.5 and 16.5 do not necessarily mean that the *negotium* of these provisions are hard law since they are contained in an *instrumentum* of hard law.

These obligations are imprecise and subject to the discretion of the parties. The provisions are of a simple exhortative character. However, they are more precise than the treaty provisions that contain terms like "should" instead of "shall," as in other provisions of the CBD. Girsberger, on the other hand, maintains that this CBD-mandated obligation rests on the shoulders of the international community as a whole rather than on the individual Contracting Parties.[90] I support Girsberger's view for two reasons: first, because a literal interpretation of Article 16.5 of the CBD provides that States "shall cooperate" in this respect instead of mandating a *specific* obligation upon individual Contracting Parties; second, because the introduction of an additional patentability requirement, like the disclosure of origin, allegedly conflicts with other international patent treaties like PCT, PLT and the TRIPS Agreement to which all EU countries are parties.

[89] A. J. Wells, "Patenting Life Forms: An Ecological Perspective" (1994) 3 *European Intellectual Property Review* 117; Ricolfi, "Biotechnology, Patents and Epistemic Approaches"; R. Baxter, "International Law in 'Her Infinite Variety'" (1980) 29 *International and Comparative Law Quarterly* 550.

[90] M. Girsberger, "Transparency Measures under Patent Law regarding Genetic Resources and Traditional Knowledge – Disclosure of Source and Evidence of Prior Informed Consent and Benefit-Sharing", (2004) 7(4) *The Journal of World Intellectual Property* 465–67.

Given that there is a general obligation on the international community of States rather than a specific one on individual States, the lack of substantive provisions on PIC in the EU Directive is not in direct breach of the duty described in Articles 15.7 and 16.5 of the CBD. The CBD parties that do not introduce transparency measures are not in direct breach of international law; *a fortiori* the US is not bound by this obligation.

Thus, Articles 16.5 and 15.5 set forth an obligation of *cooperation* among States so that patent rights are not contrary to the objectives of the CBD. This obligation of cooperation does not require that individual States legislate in the field of patent law.

Since there is no international obligation upon States to introduce particular measures in the patent system, the EU Directive does not engage the responsibility of the EU by omission under Article 2(b) of the Draft Articles on Responsibility of States for Internationally Wrongful Acts. Nevertheless, (b) can be relevant in determining the responsibility of States that are still under the obligations of the CBD. The EU, taking the minimum level of commitment towards its CBD obligations, makes it more possible that some patent applications breach the objectives and the substantive provisions of the CBD. Consequently, the EU Member countries have an obligation of due diligence to control access over GRs and TK after PIC and benefit sharing for the purpose of IP exploitation. This due diligence obligation does not imply a certain specific legislative measure to be undertaken on the part of the State. However, if the State fails to exercise the control required it breaches the due diligence obligation. Hence, some legislative measures are essential to comply with this type of obligation.

This field of analysis is similar to the principle of international law that states that "acts by non-state entities, such as a citizen or official for whose acts a State is not responsible, do not give rise to State responsibility."[91] Complying with or violating the PIC (or benefit-sharing) obligation is mainly carried out by private entities when they patent an invention based on GR or TK that has been unlawfully accessed and appropriated from a provider country. However, States must comply with the obligation of due diligence to prevent such breaches of the CBD occurring. The concept of due diligence has been applied in the context of the international protection of human rights (minimum standard of protection)[92] but it can be

[91] A. Pellet, "Can a State Commit a Crime? Definitely Yes", (1999) 10 *European Journal of International Law* 425.

[92] *Nationality Decrees Issued in Tunis and Morocco* Advisory Opinion, October 4, 1922, Permanent Court of International Justice 3; more recently, the General

applied to assess the conduct of a State vis-à-vis any international treaty or customary obligation. The mechanism that triggers responsibility for breaching this principle has been particularly established by the ICJ in the *Case Concerning US and Consular Staff in Tehran USA v. Iran*. The Court first sought to "determine to what degree the actions in question are legally imputable to the State."[93] It went on to say:

> the policy thus announced by the Ayatollah Khomeini, of maintaining the occupation of the embassy and the detention of its inmates as hostages for the purpose of exerting pressure on the US government was complied with by other Iranian authorities and endorsed by them repeatedly in statements made in various contexts. The result of that policy was fundamentally to transform the legal nature of the system created by occupation of the embassy and the detention of its diplomatic and consular staff as hostages. The approval given to these facts by the Ayatollah Khomeini and other organs of the Iranian state, and *the decision to perpetuate them translated continuing occupation of the embassy and detention of the hostages into acts of that State*. The militants, authors of the invasion and jailors of the hostages, *had now become agents of the Iranian state* for whose acts the state itself was internationally responsible.[94] (italics added)

The situation described by the ICJ is certainly different from the one of alleged cases of misappropriation of GRs. However, the principle expressed in this ICJ case can *mutatis mutandis* be applied to our situation. The State, by granting patents based upon misappropriated GRs, gives an approval towards the acts of misappropriation of private companies breaching the national sovereignty of the provider States' ABS regimes. The total inaction of the State in failing to exercise minimal control (e.g. through the application of the CBD principle PIC) can transform the acts of the private companies into the acts of the State. It is through the application of the principle of due diligence that a State (in this case the EU) may prevent misappropriation of GR and TK in the course of trade. Granting patents without ensuring the respect of PIC and benefit sharing may trigger the State responsibility for breaching the substantive provisions of the CBD. This omission amounts to a State's official approbation of the unlawful conduct of its private citizens.[95] This fact fundamentally transforms the acts of the private entities into wrongful acts attributable to the State. This theoretical construction has not explained the particularity of the wrongful act. As observed earlier, the wrongful act of biopiracy or

Recommendation 19 of the *UN Committee on the Elimination of All Forms of Discrimination against Women* states: "Under general international law and specific human rights covenants, States may also be responsible for private acts if they fail to act with due diligence to prevent violations of rights or to investigate and punish acts of violence, and for providing compensation" UN Doc. E/CN.4/1999/68, paragraph 25.

93 *United States Diplomatic and Consular Staff in Tehran* (1979) *ICJ Reports* 136, paragraph 56.

94 *Ibid.*, paragraph 74. 95 *Ibid.*, paragraphs 62, 72–73, 76–79.

misappropriation of GR or related TK has not been enshrined in an internationally agreed-upon definition.

There are different gradations of seriousness of wrongful acts. The bioprospecting activity of a US pharmaceutical company in the Amazon forest to examine the TK of indigenous people to find active compounds for research in the field of pharmaceutical product development is potentially different from the acquisition of a special banana in Cuba that is thereafter analyzed in the lab and accidentally found to have special characteristics of flavor that can be extracted and introduced in the banana trees in the south of Spain. In both cases, patents are obtained and the final product is very successful. No PIC or benefit sharing has been negotiated with the country of origin for the GR and TK. At what point does the recipient State become responsible for letting these allegedly wrongful acts of misappropriation go unchecked? How do the primary rules in the CBD principles and secondary rules of State responsibility apply to these acts? And what is the catalyzing factor that triggers the imputability to the State? At what point can the patenting inventions without controlling compliance with PIC and benefit-sharing concepts be considered a serious and blatant act of unjust enrichment? At which stage of the commercial exploitation is the breach of the CBD principles materialized: at the start of the technical research, at the moment of commercialization, or at the moment of the realizing of a certain amount of benefit? Can the States defend themselves by stating that their nationals are using the information and not stealing their GRs? It can be argued that this activity does not take away from or deplete the biological resources of any State. They have just used the information contained in their GRs and have invented around it.

In light of these questions, it is not surprising that the US or other countries are extremely uncertain about the precise nature of the CBD obligations at stake. Only a treaty addressing all the legal questions arising from these customary principles with a potentially expanding scope into the field of IP can clarify the relationship between the acts of the State and those of private companies.

6.1.1.2.7 The impact of the customary norm of prior informed consent and benefit sharing on patent law upon national legislation

Confusion can easily arise between the concept of PIC – that has been qualified as a customary principle – and the introduction of a certificate of origin of the GR and TK. Though acting in the same area, the two concepts are distinct in the sense that the normative value of one does not influence the other. It is true, however, that the disclosure of origin in patent law is one of the possible ways of implementing the PIC concept. The existence of this international obligation can be challenged because it is grounded only in the soft law of the Bonn Guidelines.

It cannot be said that all the States have this perception with regard to the disclosure of origin/source in patent law because this norm has been adopted within the CBD, i.e. outside the setting of a treaty having the object and purpose of IP. CBD Contracting Parties have been cautiously using a legally non-binding instrument like the Bonn Guidelines to implement this precise obligation, while, at the same time, many of the same countries are participating in the reformation process of the PCT and the drafting of the SPLT. This indicates that the very existence of the international treaty norm on the introduction of a disclosure requirement in patent law is not only debatable in the legal doctrine, it is also in conflict with customary norms.

The fact that only a few States have undertaken a legislative action to introduce this requirement and that this matter is currently discussed at the PCT and PLT manifests a lack of necessary widespread and consistent practice. For all these reasons, it would be unwarranted to consider this norm of the introduction of the certificate of origin as a customary norm (see section 6.1.1.2.5 above).

In sum, the concepts of PIC and benefit sharing have crystallized in customary norms of international law. However, these norms do not entail any particular obligation upon any State to amend their national patent system. CBD Contracting Parties are (i) empowered to make use of their legislative discretion in shaping and implementing legislation in accordance with the best possible expression of the vague provisions of the CBD, and (ii) after the adoption of the Bonn Guidelines, they are under the soft law obligation to adopt legislative measures in their patent system to disclose the origin/source of the GR and TK in patent applications. Inaction on the part of CBD Contracting Parties to integrate transparency measures in the patent system engages a soft responsibility, whereas the omission of due diligence in assuring, through any other measure, that PIC and benefit-sharing obligations are complied with amounts to an internationally wrongful act. As to the question of whether the EU Biotech-Directive prevents EU Member States from introducing transparency measures, the answer is negative since countries can comply with the cumulative obligations of the European Directive and of the CBD legislative *acquis*.

This network of hard law and soft law obligates the international community to solve the problem between the CBD and the rest of the patent treaties.

6.1.2 *Implementation of the disclosure of source in international patent law*

Disclosure is at the core of patent law.[96] The prime task of a patent examiner is to receive all the information required to disclose the claimed invention: a description of the best mode to carry it out, any known technology and prior art, the identity of the true inventor, and the legal basis for entitlement. It is interesting to note that PIC is used in the case of joint inventors and employee's inventions requiring evidence of their mutual consent for the grant of a patent.[97]

I have observed how the TRIPS Agreement, PCT, and PLT constrain the ability of each individual State to modify its patent laws by introducing transparency measures. In this regard, a distinction should be drawn between the obligations laid upon the international community and those resting on each individual State. If, on the one hand, an obligation is laid upon each individual State to insure that the patent system is compliant with the obligations of the CBD and its derivative laws, a stronger obligation is, *a fortiori*, created on the international community as a whole.

The conflict between the derivative laws of the CBD and the body of the aforementioned international patent treaties imposes an obligation upon the international community as a whole to amend the international treaties. This amendment process is the only way in which reconciliation among the CBD and international patent law can be realized. It is quite a task to identify the exact treaty on which to act within the dense network of interlocking treaties on this subject-matter. Identifying the exact treaty along with the negotiating strategy amounts to a surgical operation. It must be performed with care and precision so that the whole body of legally protected interests in international patent treaties remains intact. This section analyzes the various paths that the international community can follow to perform this surgical intervention on the international patent system so as to include the submission of a certificate of origin.

6.1.2.1 Policy objectives

Disclosing the source of GR should achieve three policy objectives at the international level. The first consists in transparency, since disclosure of the source would allow the authorities of the provider country to trace the patent application on the basis of the GR that they provided.[98] Second, disclosure

[96] *WIPO Technical Study on Patent Disclosure Requirements Related to Genetic Resources and Traditional Knowledge* (World Intellectual Property, November 15, 2004) 20.

[97] C. Heath and S. Weidlich, "Intellectual Property: Suitable for Protecting Traditional Medicine" (2003) 1 *Intellectual Property Quarterly* 82, quoted in K. Verma, "Protecting Traditional Knowledge" (2004) 6 *The Journal of World Intellectual Property* 788.

[98] Bonn Guidelines, 2002, paragraph 16.

would contribute to the provider State's ability to track the compliance with PIC and the mutually agreed-upon terms on which access to those resources has been granted under the provider country's rules.[99] Third, disclosure can help examiners assess the authentic novelty and inventive step involved in the patent application as well as simplifying the search in databases on TK – that, as it shall be observed in section 6.2.3 below, are increasingly being established at the local, regional, and national levels.[100] The overall objective is to stall the process of rewarding "bad patents" based on knowledge or information misappropriated from another provider country.

6.1.3 *From basic problems to a possible solution*

Any policy-making innovation in the patent law system needs precise terminology. This need for precision requires the clarification of terms like "genetic resources," "biological resources" or "biological material," and "TK" or "knowledge, innovations and practices of indigenous and local communities," "country of source," "country of origin," etc. For this purpose, the definitions that shall be used are the ones already given in section 4.2 above.

Precise terminology is just one step towards clarity in an area of international law, which is fraught with difficult questions. Legal literature has posed the following questions about introducing this transparency measure in biotechnology patent applications. What is the relationship between the invention and the TK that would trigger the disclosure of origin requirement? What information needs to be disclosed in the patent application? Is the "source" the "country of origin" or the "geographic origin"? When does a disclosure have to be made, that is, what mechanism "triggers" the requirement? Are there any exceptions to the requirement? In what international instrument should the requirement be introduced? Should it be optional or mandatory for States to implement the requirement at the national level? If the patent applicant does not fulfill the requirement, should there be any sanctions and, if so, what kind?

Who are the partners to negotiate with – a central public body, a local authority, or even a private association of citizens? Who is to represent a local community? What is the extent of his or her powers? If no authority

99 *Ibid.* paragraph 16(d).

100 *Article 27.3(b), the Relationship Between the TRIPS Agreement and the CBD, and the Protection of Traditional Knowledge*, Communication from Switzerland, IP/C/W/400/Rev. 1 paragraph 9 (June 18, 2003), www.docsonline.wto.org/PDFDocuments/t/IP/C/W40ORl; B. Tobin, "Certificates of Origin: A Role for IPR Regimes in Securing Prior Informed Consent", in *Access To Genetic Resources. Strategies for Sharing Benefits* (ACTS Press, Nairobi, 1997).

exists and there is no mechanism for granting certification of origin in the provider country, should the requirement for a certification be waived? What can restrain a company from claiming that a resource was obtained from such a country when it was actually collected illegally from another country with ABS regulations? In the case of patenting a plant variety, what should be done when genetic material may come from numerous sources, some of which may no longer be identifiable because of lack of documentation and the length of time between the source's acquisition and its use in breeding programs? Would the requirement apply when the invention consists of a synthesized substance isolated or derived from active compounds of an accessed resource? And if this is the case, should any derived product be subjected to this requirement? Should this requirement also apply to the GRs accessed from the FAO ITPGRFA Multilateral System? What are the consequences of non-compliance with the requirement?[101] These are some of the essential questions that the law-makers will have to address.

As regards the obligation of cooperation between provider and recipient countries, this issue is even more complex. It appears that a PIC requirement within the recipient State patent law would not have any effect if the provider State does not create the adequate legislative framework of negotiated access with mutual agreement on benefits. Indeed, sections 5.1 and 5.2 above have demonstrated that this legislative task – to be performed by DCs – is fraught with several difficulties. Nevertheless, the inverse is also true: whatever effort a provider State may make in enforcing all the indispensable legislative measures and making them workable in actual practice, it is likewise essential that there is an appropriate and simultaneous cooperation by recipient States.

This is so because of the ubiquitous nature of IPRs, particularly when applied to the international exchange of GRs. Since IPRs are really a form of information, they can be hidden and stored away and made finally untraceable until the borders of the provider country are crossed and an act of "biopiracy" is committed. What can the provider State do in this case? Of course, as the US suggests in its submission, it can declare the transfer invalid or stipulate for sanctions ranging from tort liability to administrative or criminal sanctions, with all the problems of cross-border effectiveness. While these cases of biopiracy are rarely discovered at airports or at the borders, discoveries of biopiracy acts could more frequently occur in the course of transparent patent grant proceedings in recipient

[101] Ricolfi, "Biotechnology, Patents and Epistemic Approaches", 77–90; M. Girsberger, "Transparency". Communication from the US, Article 27.3(b) IP/C/W/434 and IP/C/W/449, paragraphs 12–13, www.wto.org/english/tratop_e/TRIPS_e/art27_3b_e.htm.

States if a certificate of origin was submitted.[102] Thus, it is essential for international treaties to be amended to include the disclosure of origin/source. Given all these legal uncertainties, it is understandable that some recipient States, like the US and Japan, may not be willing to engage in setting up another patentability requirement at the international level as long as all the pertinent legal questions have not been properly addressed at the international level. Broadly speaking, it is likely that the deposit of a certificate of origin/source accompanying the patent application may eventually be subjected (i) to the establishment of ABS regimes in provider countries; (ii) to the increase in understanding by industrialized recipient countries of the definitions of key terms, such as biopiracy, TK, and its registration systems in databases; and (iii) to the establishment of efficient authorities that grant the "certificate of origin" and that should also act according to internationally agreed-upon standards.

6.1.3.1 The debate over the compatibility of the mandatory requirement of disclosure of source/origin under current international patent law

The debate over the compatibility of the mandatory requirement of disclosure of source/origin is an appendix to the wider confrontation between two main approaches to patent law: the classic and the radical approach.[103]

On the one side, the "classic" approach – that inspires the majority of industrialized countries' patent systems – views the patent system as neutral and mainly isolated from pre- or post-IP matters. In other words, the patent system has the exclusive mandate to reward technological innovation regardless of what happens before or after the achievement of the invention. Since the proposed patentability requirement of disclosure of origin is inspired by a CBD obligation, it is no wonder that the US maintains that this problem of disclosure lies outside international patent law. The EU and the US submit that the best solution to the problem of benefit sharing and biopiracy should be found in the international fora dealing with the problem of access to GRs and not in those dealing with IP *stricto sensu.*

[102] This argument is vociferously expressed in the documents submitted by the Group of Countries of Latin America and the Caribbean, in *WIPO Intergovernmental Committee on Intellectual Property and Genetic Resources, Traditional Knowledge and Folklore, First Session* (April 30 to May 3, 2001), www.wipo.int/eng/meetings/2001/igc/document.htm; *Traditional Knowledge and the Need to Give It Adequate Intellectual Property Protection, WIPO Committee on the Relationship between Intellectual Property, Genetic Resources and Traditional Knowledge*, WIPO/GRTKF/IC/I/5 (March 16, 2001).

[103] Ricolfi, "Biotechnology, Patents and Epistemic Approaches".

On the other side, the "radical" approach sees the international patent system as unequivocally embedded in the realm of the *droit commun* (general law) at the national level and argues that it needs to be made compatible with all the principles of equity that form a part of international law. In this view, when any international, regional, or national authority grants a patent, society is in reality not only rewarding the efforts of the technical innovator, but it is also justifying the resulting legal monopoly, i.e. an exception to free competition in order to create an incentive to innovate. At the same time, patent law tries to strike a balance between the monopoly of the right-holder, the competitors, and the users of the invention. Since the patent system is viewed as a fundamental tool for increasing the level of technological innovation in a given society, it cannot be seen as socially neutral. A political argument regarding the present negotiations at WIPO and WTO is appropriate at this juncture.

As earlier observed, this *impasse* creates additional frustration in some DCs that view the IP system as an imposition by the industrialized countries, i.e. those who hold the biotech-patents and have the potential to continue to patent in this field of technology. In my view, maintaining that the international patent system is socially neutral and does not need any adjustment is actually detrimental to the functioning of the international patent system itself. Indeed, the patent system needs the support and cooperation of the DCs to affirm its legitimacy. For these reasons, many DCs insist that the method for the international community to verify whether GRs have been lawfully appropriated and the CBD duty of negotiating truly implemented is through the patent applicant providing some proof prior to the grant of a patent on a biodiversity-based invention. While a solution to this complex problem could consist of creating a system of certification of the GR and TK inside the patent system, attention shall be concentrated on the arguments in favor and against introducing an additional requirement as part of the current international patent law. In this approach, my proposal is a "middle ground" between the opposing views among States as well as among legal scholars.

6.1.3.1.1 Arguments against compatibility with the TRIPS Agreement

The debate over introducing a certificate of origin/source in an international treaty stems from its alleged incompatibility with Articles 27.1, 29, 30, and 62 of the TRIPS Agreement that, if read conjunctively, would not allow the introduction of additional conditions of patentability other than the ones expressly and exclusively provided for by

Article 27.1.[104] Pires de Carvalho has expounded the legal arguments related to the incompatibility of a mandatory additional requirement of the disclosure of origin/source with the TRIPS Agreement.[105]

Since some countries have created a mandatory requirement to disclose the country of origin/source, the question of the compatibility of this measure with the TRIPS Agreement may arise within the dispute settlement mechanism of WTO. The following arguments may support a WTO member State wishing to resort to the WTO dispute settlement mechanism to challenge the aforementioned national patent laws that have introduced a new patentability requirement that disregards the following TRIPS Agreement obligations:

(i) *Incompatibility with the* numerus clausus *of patentability requirements of Article 27.1.* In accordance with the interpretation of Article 27.1 jointly read with Article 32[106] of the TRIPS Agreement, Member States cannot turn non-disclosure of origin of the genetic material into a violation of the law of patents that results in rejection or cancellation of the patent. Article 27 provides an exclusive list of substantive requirements, i.e. of novelty, inventive step or non-obviousness, for an invention to be patented. The requirement to disclose the source/origin would amount to adding a new requirement not provided for in this Article.

(ii) *Incompatibility with Article 29 of TRIPS Agreement.* The disclosure requirement is not compatible with Article 29 of TRIPS which only obligates the applicant to disclose the invention in a clear and complete manner so that a person skilled in the art can practice the invention. This Article does not mention the necessity of disclosure of the origin/source of the GR (see also Articles 9–11).

(iii) *Incompatibility with Article 62 of the TRIPS Agreement.* The disclosure requirement would not be compatible with Article 62 which establishes the conditions for the acquisition of IPRs. According to Article 62.1 of the TRIPS Agreement: "Members may require, as a condition of the acquisition or maintenance of intellectual property rights [...] compliance with reasonable procedures and formalities."[107]

[104] *Matters concerning Intellectual Property and Genetic Resources, Traditional Knowledge and Folklore – an Overview* WIPO/GRTKF/IC/1/3 (April 30–May 3, 2001).

[105] N. De Carvalho, "Requiring Disclosure of the Origin of Genetic Resources and Prior Informed Consent in Patent Applications Without Infringing the TRIPS Agreement: the Problem and the Solution" (2000) 2 *Washington University Journal of Law and Policy* 371–401, http://ls.I stl.edu/Journal/2/p271carvalho.pdf.2000.

[106] TRIPS, Article 32 reads: "An opportunity for judicial review of any decision to revoke or forfeit a patent shall be available."

[107] TRIPS.

A mandatory obligation to disclose all information known to the applicant about the country providing the biological material and the associated TK would amount to adding an "unreasonable" requirement, under Article 62.1. This reasoning acknowledges the fundamental uncertainties related to acquiring such a certificate. Only certain *routine* maintenance requirements can be admitted, e.g. the right to payment of certain fees for renewing patent rights. A disclosure requirement would place heavy administrative burdens on patent offices and applicants and thus would be "unreasonable" under the terms of Article 62.1.[108]

(iv) *Incompatibility with the prohibition of discrimination.* This disclosure requirement would be contrary to Article 27.1 which states that "patents shall be available and patent rights enjoyable without discrimination as to [...] the field of technology." Adding a new requirement in the field of biotechnology would discriminate against this field of technology vis-à-vis other fields.

De Carvalho concludes his essay by describing the only manner in which such a requirement can be implemented without breaching any provision of the TRIPS Agreement, that is at the enforcement stage rather than at the application stage. This view is inspired by the equitable common law principle of "clean hands," which prevents a judgment for a petitioner who is morally culpable in this matter. Judicial authorities may use this doctrine to prevent the enforcement of an exclusive right when it has been obtained in a fraudulent or abusive way until the right-holder cleans his hands by correcting the wrongful act.

6.1.3.1.2 Arguments in favor of compatibility of the disclosure requirement with the TRIPS Agreement

On the other hand, cogent and convincing arguments have been raised in favor of the compatibility of this requirement with the aforementioned TRIPS provisions. The following arguments can be raised in this regard:[109]

[108] De Carvalho, "Requiring Disclosure of the Origin" 386–88, www.ls.I stl.edu/Journal/2/p271carvalho.pdf.

[109] D. Vivas, *Análisis de la relación del ADPIC y la implementación nacional de la CDB: El caso del sistema de acceso y de divulgación del origen de los recursos genéticos en Venezuela*, www.iprsonline.org; C. Correa, *Establishing a Disclosure of Origin Obligation in the TRIPS Agreement* (Quaker UN Office, Geneva, August 2003), www.geneva.quno.inficl/pdf/disclosure %200P %2012.pdf or www.iprsonline.org; D. Fritz, *Patente auf der Grundlage biologischer Ressourcen aus Entwicklungsländern, Mitteilungen der deutschen Patentanwälte* (2003) 94. Jg., Heft 8/9, 349–72; G. Dutfield, "Sharing the Benefits of Biodiversity – Is there a Role for the Patent System?" (2002) 5(6) *World Journal of Intellectual Property* 899–931. M. Hassemer, "Genetic Resources", in S. Von Lewinski (ed.), *Indigenous Heritage and Intellectual Property* (Kluwer, The Hague, 2004) 211. On this issue, there is no need to quote all the NGOs in favor of compatibility.

(i) A literal interpretation of Articles 1, 7, and 8 of the TRIPS Agreement makes it clear that WTO members are free to establish a method of applying the provisions in their own system. Particular emphasis should be placed on Article 8 that goes the furthest, allowing countries to adopt appropriate measures in order to circumvent IPR-holder abuses.[110] Proponents claim that patenting GR and TK without PIC of the provider country is an abuse of patent law. Yet this policy freedom does not have to directly contravene the other provisions of the TRIPS Agreement. The interpretation of the reasonableness criterion set forth by TRIPS Article 62.1[111] plays a vital role in this regard (see section 6.1.3.6.7 below).

(ii) As regards the exclusivity of the requirements of Article 27.1, a distinction needs to be drawn between the requirements for *patentability of the invention* as such, and those for *entitlement of the applicant to practice the invention*. Article 27.1 addresses the first and does not regulate the second. The technical patentability of any disclosed invention does not necessarily mean that any applicant is entitled to a patent. Requiring a certificate of source would only apply to entitling the patent applicant to become the right-holder of the patent.[112]

(iii) Article 27.1 and 29 do not contain *per se* a *numerus clausus* (closed number) of requirements. They only prohibit Members from introducing a disclosure requirement as a substantive requirement. Payments of fees and presentation of documents related to corporate capability are, for instance, formal requirements that many national systems have implemented, although these requirements are not specifically provided for by the TRIPS Agreement.

(iv) Similarly, Article 32 does not establish an exclusive list of causes for revocation; otherwise several States would be entirely in breach of this provision when they revoke patents for the following reasons not mentioned in TRIPS Agreement: "(i) non payment of fees, taxes or annuities, (ii) the grant of the patent to a person who was not entitled to it, (iii) the extension of the patent's subject-matter beyond the subject-matter in the application as filed; and (iv) the failure of the applicant to disclose the invention clearly enough and

[110] TRIPS, Article 8. [111] TRIPS, Article 62.1.

[112] F. Dodler, *Improving the Legal Position of Stakeholders of Bioresources in the Statutory Law of Developed Industrial Countries*, Professor, Faculty of Law, University of Basel, Switzerland, counseling the Berne Declaration, www.evb.ch/cm_data/Dolder_Heymanns_E.pdf.

completely enough for it to be performed by a person skilled in the art."[113]

(v) Regarding the incompatibility with Article 27.1 which prohibits discrimination between fields of technology, this matter has been solved in the WTO Panel decision in *Canada v. EU*. The Panel made clear that whereas the *discrimination* of one field over another is not permitted, a *differentiation* may be admitted if it sets forth justified exceptions to solve a particular problem in a specific sector of technology.[114] Different rules can be applied for particular product areas provided that they are adopted for bona fide objectives.[115] Even though this decision did not deal with biotechnological patents, and in spite of the fact that the decision of the panel does not indicate an authoritative legal interpretation of the WTO provisions and did not define what a bona fide objective is, it at least sets forth the general principle. The qualification of benefit sharing and the balance of rights as bona fide objectives are clearly found in the soft law declarations and guidelines in the WIPO IGC on IPGRTKF and in the CBD subsequent practice. It is therefore to be concluded that the teleological interpretation of this provision in light of Article 7 leads to its relaxation (as indicated above in (ii)).

(vi) Article 62.1[116] prohibits Members from burdening patent applicants with formal requirements which are not "reasonable" within the meaning of the provision, but Article 62.1 allows States to require compliance with "reasonable" procedures and formalities. In light of the teleological interpretation outlined in the above point (v) and the application of the rules of interpretation of treaties of Article 31.3(c) of the VCLT, the soft law and subsequent practice of the CBD, it can be concluded that the disputed measure requiring disclosure is undoubtedly reasonable. "In general the term 'reasonable' may be interpreted as letting Members impose formalities that are adequate to their purpose,"[117] but, on the other hand, not overly

113 D. Vivas, "Requiring the Disclosure of the Origin of Genetic Resources and Traditional Knowledge: The Current Debate and Possible Legal Alternative", in C. Bellman, G. Dutfield, and R. Meléndez-Ortiz (eds.), *Trading in Knowledge* (Earthscan, London, 2003) 202.

114 *Ibid.*, 202.

115 *WTO Panel Report on Canada – Patent Protection of Pharmaceutical Products*, paragraph 7.91.

116 Article 62 of *TRIPS*, reads as follows: "1. Members may require, as a condition of the acquisition or maintenance of the intellectual property rights provided for under Sections 2 through 6 of Part II, compliance with reasonable procedures and formalities. Such procedures and formalities shall be consistent with the provisions of this Agreement."

117 *Resource Book on TRIPS and Development*, 622.

restrictive on the applicant and not conflicting with the substantive provisions of TRIPS. It must be remembered that the TRIPS Agreement sets forth substantive rules and contains only minimum standards. A disclosure certificate can, therefore, be introduced as a formal requirement and be reasonable according to the Member States' domestic policy objectives (see Article 8 of TRIPS Agreement).

(vii) Article 62.4 states that "procedures concerning the acquisition or maintenance of IPRs and, where a Member's law provides for such procedures, administrative revocation and *inter partes* procedures such as opposition, revocation, and cancellation, shall be governed by the general principles set out in sections 2 and 3 of Article 41." This provision must be interpreted in light of Article 41.2 that states that these procedures must be "fair and equitable." Given all the reasons for a reconciliation between the TRIPS Agreement and the principles of the CBD (see particularly Articles 1 and 15.7), a formal requirement to disclose the origin is "fair and equitable."

(viii) The concept of a disclosure requirement is included in Article 29 of TRIPS Agreement that provides for a firm requirement for disclosure as a specific obligation on the patent systems of WTO members. It more precisely requires that the invention be disclosed "in a manner sufficiently clear and complete for the invention to be carried out by a person skilled in the art." States may also require that the applicant indicate the best mode for carrying out the invention known to the inventor.[118] Of course, Article 29 does not mandate any particular type of disclosure of origin of the GR at the basis of a patent application. It does, however, demonstrate that disclosure of all the relevant information is not a concept far from patent law.

(ix) If including a certificate of origin is considered too burdensome for the classical patent system, it is possible to make the submission of this document a simple and integrated part of the examination, just like the renewal fees paid regularly by applicants.

(x) Ultimately, the introduction of this transparency measure can also be seen as a subset of the concept of *ordre public* of Article 27.2. The introduction of a certificate of origin can be also justified on the legal basis of the concept of *ordre public* of Article 27.2. This transparency measure can be viewed as necessary, lest the banning effects of *ordre public* exception be triggered. The link between the defensive

[118] Vivas, "Requiring the Disclosure of the Origin", 202.

protection of TK and the concept of *ordre public* in patent law is set out in section 6.3.

In light of the overwhelming legal arguments in favor of the compatibility of this new requirement with the TRIPS Agreement, it must be affirmed that such requirements are valid if they do not create unnecessary burdens on the applicant. Indeed, careful attention must be paid to Article 62.1 that describes eventual requirements as "reasonable procedures and formalities." The "reasonableness" criterion is fundamental in evaluating the characteristics of the certificate of origin requirement. The introduction of a formal or substantive requirement depends on whether this requirement is reasonable or not in terms of Article 62.1. The reasonableness criterion also applies to the consequences for not complying with this disclosure requirement: these consequences must also be reasonable in relationship to the objective the requirement seeks to achieve.

In light of the above findings, there is no absolute answer as to compatibility of the disclosure requirement with the TRIPS Agreement. It very much depends on the characteristics of this requirement. For instance, the following paragraphs will observe how implementing this disclosure requirement in national laws as a substantive requirement would be incompatible with the TRIPS Agreement. Introducing the disclosure of source/origin as a substantive requirement would require the amendment of TRIPS Agreement. A second option consists in implementing it as a formal requirement. Before entering into the technicalities of how to introduce this requirement at the international and national level, it is necessary to explain the difference between a formal and a substantive patent requirement.

6.1.3.2 The difference between a formal and a substantive patent requirement

The *summa divisio* between formal and substantive requirements for patent applications stems from the PCT[119] and the PLT,[120] which are the main technical treaties on international patent law. These treaties operate on the distinction between formal or substantive patent elements, i.e. whether the elements refer to the form or the contents of an application.

A formal requirement refers to the need to disclose information, to the need to submit certain types of documents or to a required physical format. A substantive requirement refers to the nature of the invention and to the underlying standards of patentability (see, e.g. Article 27 of the

[119] *Patent Cooperation Treaty of June 19, 1970*, as amended and modified, www.wipo.org/eng/main.htm (the US became a Party on January 24, 1978).

[120] www.wipo.int/treaties/en/ip/plt.

TRIPS Agreement). This distinction can be illustrated by considering the consequences of non-compliance. Failure to comply with a substantive requirement, like novelty, prevents the applicant from receiving the patent. Conversely, failure to meet certain formality requirements may only jeopardize the exercise of the rights of a patent.[121] Besides this theoretical distinction, however, the line separating the two qualifications of substantial and formal requirements is not always clear.

The complexity of the distinction between a substantive and a formal requirement can be illustrated by an example of the "description requirement." The interaction between the PCT and the PLT on this matter is revealing. Article 5.1.a.iii of the PLT identifies as a formal requirement "a part which on the face of it appears to be a description." This is one of the elements that forms part of an application sufficient to establish a filing date. While the PLT sets forth this requirement as a formal one, it is more difficult to assess its legal nature as formulated in Article 3.2 of the PCT, where an "international application shall contain [...] a description" to establish a filing date. The description "shall disclose the invention in a manner sufficiently clear and complete for the invention to be carried out by a person skilled in the art" (Article 5).

This same hybrid requirement appears clearly substantive in Article 29 of TRIPS which requires that an applicant for a patent shall disclose the invention in a manner sufficiently clear and complete for the invention to be carried out by a person skilled in the art. This Article requires that the applicant indicate "the best mode for carrying out the invention known to the inventor," leaving this in effect as an optional additional requirement for a patent application.[122] However, disclosure is not part of Article 27.1, which sets forth the substantive requirements. The consequences of non-compliance are the invalidation of the patent, which makes this requirement formal as well as substantive.

Rule 5.1 of the PCT Regulation provides that the description should "set forth at least the best mode contemplated by the applicant for carrying out the invention claimed: this shall be done in terms of examples, where appropriate, and with reference to the drawings, if any: where the national law of the designated State does not require the description of the best mode but is satisfied with the description of any mode (whether it is the

[121] *SCM Corp. v. Radio Corp. of America*. When a patent affects the public interest, the court has to discourage that type of conduct by not enforcing "a patent obtained under these circumstances"; *Corning Glass Works v. Anchor Hocking Glass Corp.*, quoted in De Carvalho, *Requiring Disclosure of the Origin*, 397–98.

[122] *Resource Book on TRIPS and Development*, 448–54.

best contemplated or not), failure to describe the best mode contemplated shall have no effect in that State."

This example of the description of the invention in PLT as a formal requirement and in the PCT as a substantive requirement demonstrates that if one scratches the surface of the dogmatic distinction between a formal and substantive requirement, one finds out that the difference is not absolute. Another good example is the requirement of the "enabling disclosure," that operates in the UK, US, Germany and other countries especially in relation to biotechnological patents. It requires that the invention be disclosed in a way that "any information that is obtained as a result of an analysis undertaken by a person skilled in the art must be obtained without undue burden or without the need to exercise any additional inventive effort."[123]

Although the line separating the conceptual definitions of formal and substantive requirement is very thin, it is suggested that the disclosure of origin requirement should be crafted in accordance with the classic elements of formal criteria, in a way in which it can more easily comply with the "reasonable" criterion set out in Article 62 of the TRIPS Agreement. In this way there would be no undue burden imposed on the patent application examiner in the tracing of the sources of single compounds, e.g. of new drugs from pharmaceutical companies.

6.1.3.3 State of the law and proposals de lege ferenda *of implementation of disclosure of origin*

Designing transparency measures like the disclosure of origin under patent law with regard to GRs and TK requires the legal analysis of various provisions of several international agreements, e.g. the TRIPS Agreement, the PCT and PLT, the CBD, and ITPGRFA. The reasoning of the previous sections has demonstrated that without amending the TRIPS Agreement, the international community has to amend other treaties. This section discusses the legal implications of the most important proposals submitted to the WTO, WIPO, and the CBD.

6.1.3.4 The TRIPS Agreement

Amending certain TRIPS provisions by introducing a substantive requirement or any other system of cooperation for the disclosure of the origin/source of the GR, would solve this whole legal problem and debate.

[123] L. Bently and B. Sherman, *Intellectual Property Law* (Oxford University Press, 2001) 420–22. *Biogen and Medeva plc* (1997) R. P. C 1, in F. Abbott, T. Cottier and F. Gurry (eds.), *The Intellectual Property System: Commentary and Materials* (Kluwer, The Hague, 1999) 42–45.

I have noted a shift in the proposals for modification of Article 27.3(b) from advocating the ban on patents on life forms to advocating a certificate of origin requirement as evidence of PIC and benefit sharing. For example, section 19 of the Doha Ministerial Declaration of the 4th WTO Ministerial Conference Doha, Qatar (November 2001),[124] instructed the TRIPS Council to address issues related to GR and TK in its review of Articles 27.3(b) and 71.1 of the TRIPS Agreement. The negotiations[125] on this matter have been overshadowed by discussions on the issue of the TRIPS Agreement and public health.[126] No particular result was achieved in the 5th WTO Ministerial Conference held in September 2003 in Cancún.

Within the TRIPS Council, the like-minded megadiverse countries[127] and the African countries propose incorporating the principles of Article 15 of the CBD (on the methods of ABS) into the TRIPS Agreement with the aim of banning patents that run counter to this Article. The boldest position has been held by the group of like-minded megadiverse countries

[124] Paragraph 19 of the *Doha Declaration*, instructs the TRIPS Council, "in pursuing its work program including under the review of Article 27.3(b), the review of the implementation of the TRIPS Agreement under Article 71.1 and the work foreseen pursuant to paragraph 12 of this declaration, to examine, *inter alia*, the relationship between the TRIPS Agreement and the CBD, the protection of Traditional Knowledge and Folklore, and other relevant new developments raised by Members pursuant to Article 71.1."

[125] *The Relationship between the TRIPS Agreement and the Convention on Biological Diversity and the Protection of Traditional Knowledge, submission by Bolivia, Brazil, Cuba, the Dominican Republic, Ecuador, India, Peru, Thailand, and Venezuela*, IP/C/W/403 (June 24, 2003), www.wto.org/english/tratop_e/TRIPS_e/art27_3b_e.htm paragraph 1. *Taking Forward the Review of Article 27.3(b) of the TRIPS Agreement, Joint Communication from the African Group*, IP/C/W/404 5–6 (June 26, 2003), www.docsonline.wto.org/PDFDocuments/t/IP/C/W404.doc.; *EC and Member States: Review of Article 27.3(b) of the TRIPS Agreement, and the Relationship Between the TRIPS Agreement and the CBD and the Protection of Traditional Knowledge and Folklore: "A Concept Paper" Communication from the European Communities and Their Member States*, IP/C/W/383 paragraphs 49–58 (October 17, 2002), www.docsonline.wto.org/DDFDocuments/t/ip/c/w383.doc. The US instead defends the contractual approach, asserting that it may effectively ensure an equitable sharing of benefits between the GR holders of the country of origin and the researchers who apply for a patent. The US supports its position by noting that the system would be similar to the regime for access to genetic materials in US national parks and could be adapted to the legal systems and government structures of other countries; *Access to Genetic Resources Regime of the United States National Parks, Communication from the United States*, IP/C/W/393 (January 28, 2003), www.docsonline.wto.org/PDFDocuments/t/IP/C/W393.doc.

[126] J. Curci and M. Vittori, "Improving Access to Life-Saving Patented Drugs – Between Compulsory Licensing and Differential Pricing" (2004) 7 *The Journal of World Intellectual Property* 747.

[127] "The megadiverse countries are a group of countries in which less than the (10 percent) of the global surface has more than the (70 percent) of the biodiversity. Most of these countries are located in the tropics. In 2002, an organization '*Like-Minded Megadiverse Countries*' was formed to recognize these countries as biodiversity hotspots"; www.en.wikipedia.org/wiki/Megadiverse_Countries.

(composed of Bolivia, Brazil, Cuba, the Dominican Republic, Ecuador, India, Peru, Thailand and Venezuela). This proposes inserting a provision in the TRIPS Agreement which would require patent applications for inventions using biological resources and TK to (i) disclose the source and the country of origin of such resources and knowledge (see section 4.2.3 above); (ii) provide "evidence of PIC through approval of authorities under the relevant national regime;"[128] and (iii) provide "evidence of fair and equitable benefit-sharing under the relevant national regime." The fulfillment of these requirements would be a condition for acquiring patent rights.

On the other hand, the African Group proposes to add a new section 3 to Article 29 of the TRIPS Agreement, which reads as follows: "3. Members shall require an applicant for a patent to disclose the country and area of origin of any biological resources and traditional knowledge used or involved in the invention, and to provide confirmation of compliance with all access regulations in the country of origin."[129]

Although the EC Member States have expressed their willingness to discuss in the TRIPS Council a system for disclosing and sharing information about the geographical origin of biological material in patent applications,[130] they have opposed the idea that failing to comply with this requirement should stand in the way of granting a patent or that it would have an effect on the validity of the patent once it was granted (as opposed to the proposals of the megadiverse and the African countries). The eventual new requirement in the TRIPS Agreement will not be a *de facto* or *de jure* additional formal or substantive patentability criterion.[131] The EC proposes that the requirement should, however, be introduced in PCT and PLT. This position is very similar to Switzerland's proposed amendment "to the PCT to enable Contracting Parties to require patent applicants [...] to declare the source of genetic resources and/or TK, if an invention is based on or uses such resources or knowledge."[132] Switzerland's proposal stands on the middle ground within the acrimonious debate at the WIPO IGC on IPGRTKF, between the US desire to keep the status quo[133] and the position of most DCs under the leadership

[128] *The Relationship between the TRIPS Agreement.*

[129] These requirements would formalize what in the view of the African Group should be expected of all such patent applications, IP/C/W/404, 707.

[130] *Review of Article 27.3(b) of the TRIPS Agreement, and the Relationship Between the TRIPS Agreement and the Convention on Biological Diversity and the Protection of Traditional Knowledge and Folklore*: "A Concept Paper," Communication from the European Communities and Their Member States, IP/C/W/383 (October 17, 2002), www.docsonline.wto.org/PDFDocuments/t/ip/c/w383. doc, paragraph 56.

[131] *Ibid.*, paragraph 55. [132] IP/C/W/400/Rev. 1. [133] IP/C/W/434.

of Brazil and India that are pushing for a global mandatory requirement of disclosure in the TRIPS Agreement of PIC and benefit sharing.

6.1.3.5 The Patent Cooperation Treaty and Patent Law Treaty connection

In light of the strongly opposed views expressed in the TRIPS Council with regard to the amendment of Article 27, within the articulated international patent treaty framework, the PCT and PLT are the only places in which this requirement may be introduced within a reasonably short time.

The megadiverse countries and the African Group (and recently pushed by Brazil and India) want to amend the TRIPS Agreement by introducing a certificate of origin as a substantive requirement. Such an amendment of TRIPS would not render necessary the amendment of PCT in accordance with Article 27(5) that reads:

> Nothing in this Treaty and the Regulations *is intended to be construed as prescribing anything that would limit the freedom of each Contracting State to prescribe such substantive conditions of patentability as it desires*. In particular, any provision in this Treaty and the Regulations concerning the definition of prior art is exclusively for the purposes of the international procedure and, consequently, any Contracting State is free to apply, when determining the patentability of an invention claimed in an international application, the criteria of its national law in respect of prior art and other conditions of patentability not constituting requirements as to the form and contents of applications. (italics added)

The same is indicated in Article 2(2) of PLT:

> Nothing in this Treaty or the Regulations is intended to be construed as prescribing anything *that would limit the freedom of a Contracting Party to prescribe such requirements of the applicable substantive law relating to patents as it desires*. (italics added)

However, it appears that this option is either unrealistic or impossible as the US and the EU have raised political opposition to it because of their interpretation of Article 27.1 of TRIPS Agreement.

Therefore, pursuant to Switzerland's proposal, the international community should follow the option that consists in amending the PCT to introduce this requirement as a formal one. This procedure has the advantage of timeliness since the PCT is in the process of being revised.[134] If the PCT were amended, it would also be necessary to amend PLT as well. Amending the PCT is the most expeditious political choice. It is in

[134] Switzerland introduced this proposal at the 4th Session of the PCT Reform Working Group held in May 2003, www.wipo.int/pct/reform/en/index.html; *Working Group on Reform of the Patent Cooperation Treaty*, PCT/R/WG/5/13 (November 17–21, 2003).

the interest of provider countries and in harmony with already existing international law provisions.

In addition, there is another way that the PCT and PLT will need to be amended. There is a general political consensus that this certification requirement can only be implemented as a formal requirement. Therefore, amending the introduction of a formal requirement in the TRIPS Agreement would not be sufficient since the Contracting Parties of the two Agreements would not be able to introduce the disclosure requirement in their national legislations unless the PCT and PLT were also amended.[135] Crafting a new formal requirement will be subject to the interpretation of Rule 4.1[136] of the PCT Regulations as well as Article 6(1) of the PLT:

> Except where otherwise provided for by this Treaty, no *Contracting Party shall require compliance with any requirement relating to the form or contents of an application different from or additional* to: (i) the requirements relating to form or contents which are provided for in respect of international applications under the [PCT]; (ii) the requirements relating to form or contents compliance with which, under the [PCT], may be required by the Office of, or acting for, any State party to that Treaty once the processing or examination of an international application, as referred to in Article 23 or 40 of the said Treaty, has started; (iii) any further requirements prescribed in the Regulations [of the PLT]. (italics added)

The literal and historical interpretation[137] of this Article prevents PLT Contracting Parties from introducing any additional transparency measure as a formal requirement. This interpretation implies that PCT or PLT Contracting Parties currently introducing new formal requirements are in breach of this rule. Amendment of the PLT/PCT is thus necessary for the sake of harmonization with CBD and with the eventual emerging customary norm in this regard.

The PLT establishes procedures and contains no provisions on substantive patent law with respect to national and regional patent applications or to international applications under the PCT after the applications enter the national phase. The PLT does not establish a completely uniform procedure for all Contracting Parties, but provides assurance for applications and owners, for example, that an application that complies with the minimum requirements permitted under the Treaty and

[135] Girsberger, "Transparency", 478 and Carvalho, "Requiring Disclosure of the Origin", 377–78.

[136] This rule contains the mandatory and optional elements forming the patent application.

[137] Girsberger, "Transparency", 464.

Regulations will comply with the formal requirements applied by any Contracting Party.[138]

It is important to note that this treaty introduces additional formal requirements and harmonizes national patent laws regarding the acquisition and maintenance of patents.[139]

Even if the certificate of origin is made a formal requirement for patentability, it can still have a significant impact on the grant of a patent, as provided by the PLT.

6.1.3.5.1 Proposal to amend the Patent Law Treaty

The diplomatic debate concerning the introduction of this requirement was jump-started by Columbia's proposed amendment of Article 6(1)(iii) of the PLT:

> every document shall specify the registration number of the contract affording access to genetic resources and a copy thereof where the goods or services for which protection is sought have been manufactured or developed from genetic resources, or products thereof, of which one of the Member countries is the country of origin.[140]

This proposed article provides for protection of the country's biological and genetic heritage by mandating that patent offices require applicants to undergo an administrative procedure that would include a sworn declaration as to the GRs and related knowledge, innovations, and practices of indigenous peoples and local communities that were used, directly or indirectly, in the R&D of the IPR application (including samples helpful for the research but that did not form the basis of the final product). The proposed article would also require evidence of PIC of the country of origin and/or of the indigenous or local community. Such an international certification system would standardize how these conditions would be fulfilled. This standardization would lead to a national system in the provider State that would issue certificates only after ascertaining that all obligations concerning access to GRs had been fulfilled, such as PIC, equitable benefit sharing, and perhaps other conditions imposing limitations on the use of the genetic material or knowledge. A patent would then be granted upon inclusion of such certificates. Without the certificate, a

[138] Paragraph 2.01 *Patent Law Treaty and Regulations under the Patent Law Treaty done at Geneva and Explanatory Notes on the Patent Law Treaty and the Regulations under the Patent Law Treaty* (June 1, 2000), www.wipo.int/treaties/en/ip/plt/.

[139] www.wipo.int/patent/law/eii/plt.html.

[140] *Protection of Biological and Genetic Resources, Proposal by the Delegation of Colombia,* WIPO Doc.Scv/3/10 paragraph 2 (September 8, 1999), www.wipo.int/scp/en/documents/session_3/pdf/scp3_l0.pd6.

patent would automatically be rejected. In this proposed system, a patent would only be granted if the acquisition of GRs and TK were made legally, and every patent application document would specify the registration number of the contract affording access to GRs from the specified country of origin. This proposal of including an indication of origin of the genetic material in the patent application was also discussed in 2000 at the WIPO General Assembly.[141] The US objected to this proposal on the grounds that this requirement was a modification to substantive law and not only to procedural law. In compensation for the failure to introduce such a proposal in the PLT, WIPO General Assembly found a consensus to establish the WIPO IGC on IPGRTKF that held its first meeting from April 30 to May 5, 2001.

The ongoing discussions at the WIPO IGC IPGRTKF have been accompanied by scientific studies that have greatly influenced the position of various States. For instance, authors like Fernandez, Tobin, Bellot, Langford, Davis, and Young have made proposals for implementing these ABS provisions in national laws and at the global level.[142] Any policy maker should take into account these works when shaping an ABS regime, including the certificate of origin. Their full analysis falls outside the scope of the present study since they concentrate on the technical aspects of implementation. This book focuses on the international legal aspects of its introduction into the network of patent law treaties.

In this context, the debate strategically moved to the PCT reform negotiations where Switzerland proposed amending the PCT with reference also to the PLT in a manner acceptable to the opposing States' views.

6.1.3.5.2 Proposal to amend the Patent Cooperation Treaty

The PCT provides a centralized procedure for international patent applications. The PCT and its Regulations contain an exclusive number of requirements relating to the form or contents of such international patent applications.[143] Article 27.1 of the PCT states that:

[141] The Group of Countries of Latin America and the Caribbean submitted two documents to the WIPO General Assembly called *Traditional Knowledge and the Need to Give it Adequate Intellectual Property Protection*, WO/GA/26/9 (September 14, 2000) and *Matters Concerning Intellectual Property and Genetic Resources, Traditional Knowledge and Folklore* WO/GA/26/6 (September 25–October 3, 2000) .

[142] *Certification Systems: Product and Process Certification Including Certificate of Legal Provenance/Source/Origin*, International Expert, 265–84.

[143] In contrast, according to Article 27(5), the PCT does not "limit the freedom of each Contracting State to prescribe such substantive conditions of patentability as it desires."

> [n]o national law shall require compliance with requirements relating to the form or contents of the international application different from or additional to those which are provided for in this treaty and the regulations.

Particular transparency measures, contained in Rule 51*bis*(i) and Rule 4 of the PCT Regulations, can be invoked to introduce into national law the requirement to disclose the source/origin of GRs and TK. A brief analysis of these provisions is necessary.

Article 51*bis*(i) elaborates on Article 27 and specifies that "the national law applicable by the designated Office may [...] require the applicant to furnish, in particular: (i) any document relating to the identity of the inventor, (ii) any document relating to the applicant's entitlement to apply for or be granted a patent," as well as details, where applicable, in relation to priority documentation, oath or declaration of inventorship, and evidence concerning exceptions to lack of novelty. The other transparency measures are listed exhaustively in sub-paragraphs (a) to (c) of Rule 4.1 of the PCT Regulations. Rule 4.1(c)(iii) is worth mentioning since it states that the request may contain "declarations as provided in Rule 4.17," which, in turn, allows patent offices to require applicants to include the declarations relating to the optional national requirements referred to in Rule 51*bis*(a) from (i) – (v). These transparency measures may be required at the time an international patent application is filed or later during the national phase. Rule 51*bis*, entitled, "Certain National Requirements Allowed under Article 27," lists in paragraph 1 (a) to (f) certain optional requirements regarding the submission of documents as evidence.

The aforementioned exclusive measures relate "to the form or contents of the international application" (Article 27(1)). These exclusive measures mean that the current provisions of the PCT would prevent the national legislature from introducing such measures. Hence, the amendment of the PCT and its Regulations is necessary in order to enable the national legislature to introduce the certificate of origin requirement.

Under the PCT Regulations, a national patent system may expand the number of documents required to demonstrate an applicant's entitlement to a patent. When Article 51*bis* mentions the submission of "any document relating to the applicant's entitlement to apply" for a patent, this documentation may include issues such as whether the applicant is party to an agreement, e.g. an MTA concerning inputs to the inventive process that affect the applicant's legal entitlement to apply for or to hold a patent. The WIPO technical study has found that

> a national office of the Patent Cooperation Treaty may require under their national law to provide a declaration concerning their entitlement to apply for and be

granted a patent (in the case of the majority of designated States): this can be complied with already upon filing or at a later stage during the international phase (by providing the appropriate declaration), or upon or after entry into the national phase before the designated offices concerned.[144]

Although the PCT system has a specific mechanism for disclosing requirements in the form of deposit of biological materials and nucleotide or amino acid sequence listings,[145] neither Rule 4 nor Rule 51*bis*(1) of the PCT Regulations provides for a specific declaration concerning the source of GR/TK as a separate element of the form or content of an international application.

Article 27.5 of the PCT provides that nothing in the PCT or its Regulations "is intended to be construed *as prescribing anything that would limit the freedom of each contracting State to prescribe such substantive conditions of patentability* as it desires" (italics added) and that "national law may require that the applicant furnish evidence in respect of any substantive condition of patentability prescribed by such law."

This requirement of the disclosure of origin is a substantive requirement because it affects the applicant's entitlement to apply for and be granted a patent. The entitlement is very important in terms of the ultimate ownership and exercise of the patent. Nevertheless, such a measure is, on the one hand, prohibited under Article 27 of the TRIPS Agreement, and, on the other hand, politically incompatible with the positions of the industrialized countries. If this disclosure of the source is designed as a requirement "relating to the form or contents of the international application," i.e. a formal requirement, then the PCT needs to be amended, otherwise national patent systems cannot mandate the disclosure of origin of the GR to their applicants.

For all these reasons, a mandatory form of the submission of a certificate of origin in national patent laws necessitates an amendment of the PCT and its Regulations. This amendment will, in turn, harmonize and clarify patent laws. The political argument that not all DCs are party to the PCT does not have any relevance to the introduction of this requirement. DCs, being mainly provider countries, will have a natural incentive to create such requirements in their patent system.

6.1.3.6 A more comprehensive proposal

During the process of revising the PCT and its Regulations, various proposals have been submitted on the disclosure of the origin/source of

[144] *WIPO Technical Study on Patent Disclosure Requirements Related to Genetic Resources and Traditional Knowledge* 59 (WIPO, 2003).

[145] Rule 13*bis*(1), www.wipo.int/pct/en/texts/rules/r13bis.htm#_13bis_1.

GRs and TK in patent applications. The most comprehensive proposals will be analyzed and compared.

The Swiss proposal will play a pivotal role in this comparative analysis, since it spurred a large diplomatic debate (that has not reached a conclusion yet) while at the same time catalyzing various States' opposing views.[146] The EU proposal is largely inspired by the Swiss proposal which, in turn, achieves a fine balance among States' opposing views with the aim of making the international patent system more cooperative. Switzerland proposes to amend the PCT Regulations to explicitly enable national patent legislation to require declaration of the source of GRs and TK in patent applications, upon or after entry of the international application into the national phase of the PCT procedure, if an invention is directly based on such resource or knowledge. The proposed amendment to the PCT would also apply to WIPO's PLT. Accordingly, the Contracting Parties of the PLT would be able to require in their national patent laws that patent applicants declare the source of GRs and/or TK in national patent applications.

6.1.3.6.1 Use of terminology

The Swiss and the EU proposals employ very precise and practical definitions such as "genetic resources," "biological resources" or "biological material," "TK" or "knowledge, innovations and practices of indigenous and local communities," "source," the "country of origin" or the "geographic origin." The use of these terms ensures consistency with the CBD, the Bonn Guidelines on access to GRs

[146] Switzerland introduced this proposal at the 4th Session of the PCT Reform Working Group held in May 2003. *Proposals by Switzerland Regarding the Declaration*, Annex, Part Two, Section i. These Proposals were again discussed at the 5th Session of this Working Group held in November 2003, WIPO PCT, Article 27.3(b), *The Relationship Between the TRIPS Agreement and the Convention on Biological Diversity, and the Protection of Traditional Knowledge*, Communication from Switzerland, IP/C/W/400/Rev. 1, paragraphs 92–96 (June 18, 2003), www.docsonline.wto.org/PDFDocuments/t/IP/C/W40ORl.doc; and at the WIPO PCT, *Proposals by Switzerland Regarding the Declaration of the Source of Genetic Resources and Traditional Knowledge in Patent Applications*, paragraphs 124–24 and were on the Agenda of its 6th Session, held in May 2004, www.wipo.int/pct/reform/en/index.html, which Switzerland submitted to the PCT Reform Working Group held in May 2003. It is to be noted that this matter has been discussed also by the SCP and the Substantive Patent Law Treaty at its 9th Session held in May 2003; the SCP discussed the issue of the disclosure of the origin in patent applications in the context of Article 5.1 of the draft Substantive Patent Law Treaty. For this reason, it has urged that the WIPO IGC on IPGRTKF accelerate its work and present a progress report to the session of the General Assembly WIPO GA; Report, WO/GA/30/8 (October 1, 2003), www.wipo.int/documents/en/document/govbody/wo_gb_ga/pdf/wo_ga.30_8.pdf, 6. The Swiss proposal has been supported by Norway and a number of DCs but rejected by the US, while Japan supported the proposal but took the position it should be discussed further in the WIPO IGC on IPGRTKF; the EU believed the proposal should be examined further.

and fair and equitable sharing of the benefits arising out of their ûtilization, and the ITPGRFA.

6.1.3.6.2 The concept of the "source" of genetic resources and traditional knowledge

When an applicant submits a patent application, it may be difficult to establish the origin of a given resource on which his invention is based. Sometimes there are many countries of origin for a single GR since its ecological distribution is not limited to a single country. With the aim of making the submission of the certificate of origin the least cumbersome possible, Switzerland and the EU propose that the certificate should indicate the "source" rather than "origin."[147] Alternatively, Dutfield[148] maintains that the term "legal provenance" may be the most appropriate term to use since the source country may of course not have acquired the resource legally. This terminology, however, would change the scope and the objective of the requirement, since legal provenance would not only disclose the country of origin of GRs, but would also prove that the GR and TK had been accessed in compliance with the ABS legislation of the country of source. These authors envision that this certificate would be granted by a national authority as evidence of PIC. Girsberger indicated why the certificate of origin/source should not serve to provide legal evidence of PIC or benefit sharing (sections 6.1.1.2.6 and 6.1.1.2.7 above).[149] However, the EU does not include this concept in its proposal.

The term "source" should be understood in the broadest sense possible, since according to the CBD and the ITPGRFA, a multitude of entities may be involved in ABS. Depending on the GR or TK in question, one can distinguish primary sources, including Contracting Parties providing GRs (see Articles 15, 16, 19 of the CBD), the Multilateral System of FAO's International Treaty (see Articles 10–13 FAO International Treaty), indigenous and local communities (see Article 8 (j)), and secondary sources, including *ex situ* collections and scientific literature. Accordingly, there are a variety of possible primary and secondary sources. Patent applicants must declare the primary source to fulfill the requirement if they have information about this primary source at hand. A secondary source may only be declared if patent applicants have no information at hand about the primary source. For example, if the

[147] Submission of the EC and its Member States, 2.

[148] G. Dutfield, "Disclosure of Origin: Time for a Reality Check?", *Dialogue on Disclosure Requirements: Incorporating the CBD Principles in the TRIPS Agreement on the Road to Hong Kong* 2 (WTO Public Symposium, Geneva, April 21, 2005).

[149] Girsberger, 'Transparency', 484–85.

patent applicant received the GR from a botanical garden, but does not know the Contracting Party providing the genetic resource, the botanical garden must be disclosed as the source.

Other proposals, by using the terms "source" or "origin" indiscriminately, may lead to an ambiguous implementation. The requirements may also become burdensome, especially if the patent holder is then obliged to disclose the origin of the material in all cases.

6.1.3.6.3 Scope and trigger mechanism

Switzerland and the EU maintain that the relationship between the invention and the GR and related TK constitutes one of the most difficult questions regarding the certificate of origin proposals. If the GR and related TK is distantly or only tangentially related to the invention, then the disclosure of origin should not be required. Thus, I suggest that in such amendment the expressions "based on," "used in" and "derived from" should be points of reference for the triggering mechanism that would require a certificate of origin.

Switzerland has proposed a new PCT Rule 51*bis*1(g)(i) which sets out the trigger mechanism and the scope of the obligation to declare the source. It "makes clear (1) that the invention must make immediate use of the genetic resource, that is, depend on the specific properties of this resource, and (2) that the inventor must have had physical access to this resource, that is, its possession or at least contact which is sufficient to identify the properties of the genetic resource relevant for the invention." With regard to TK, the proposed new Rule 51*bis*l(g)(ii) makes clear that the inventor must know that the invention is directly based on such knowledge, that is, the inventor must "consciously derive the invention from this knowledge."[150]

Because the Swiss proposal is a first pioneering experiment for the amendment, it is inherently vague on a number of points. Further discussions in more complex detail in the PCT and PLT fora would be needed, as there is no single way to create a disclosure requirement on the origin of GRs.

First, I hypothesize cases in which the application should be accompanied by a requirement to submit a certificate of origin. The simplest case would be one in which the person skilled in the art needs the GR to carry out the patentable invention, and the GR is not readily available to that person. This case implies that the applicant may be obliged to disclose the source of the GR so that third parties can carry out the invention. In cases

[150] *Additional Contents of the Swiss Proposals Regarding the Declaration of the Source of Genetic Resources and Traditional Knowledge in Patent Applications*, PCT/R/WG/6/11 (April 21, 2004), www.wipo.int/pct/en/meetings/reform-wg/pdf/pct_r_wg_6_11.6.

where the GR is necessary but is readily available to third parties skilled in the relevant art, the disclosure requirements entail that the GR be fully described. In another case, the TK may constitute an inventive contribution to the invention; in this case, the applicant should be required to disclose the TK holder as a joint inventor.

The primary role of the certificate of origin in these cases is verification with disclosure requirements. The certificate can also facilitate assessment of the novelty and non-obviousness of the invention by alerting the patent examiner to the fact that the invention is based on TK and thus that the TK is important prior art that should be included in the examination.

In formulating the trigger mechanism, the guidelines sketched by the WIPO IGC on IPGRTKF should also be taken into account

(i) If access to a genetic resource is required to enable a person skilled in the art to carry out the invention (or to carry out the best known mode where applicable), and it is not readily available (including through depositary authorities), then there may be an obligation to disclose its source, because it may otherwise be impossible for third parties to carry out the invention.

(ii) If, however, the genetic resource is readily available to third parties who are skilled in the relevant art, then established disclosure requirements may not necessarily create an obligation to identify the specific source (the nature of the genetic resource must however be fully described).

(iii) If, on the other hand, the genetic resource is so remote from the claimed inventive concept, as not to be needed in carrying out the invention, then it may not be relevant to the enablement or best-mode test (where applicable) for disclosure; in this case it would be necessary to clarify how the claimed invention could be determined to be based on or derived from the genetic resource.

(iv) If TK (known to the applicant) is so close to the claimed invention that it is in fact intrinsic to it under the legal doctrine that determines "inventive contribution" in the jurisdiction concerned, then it may be necessary either to declare the provider of the TK as a joint inventor (or indeed as the sole inventor, where the TK in itself provides the inventive concept of the claimed invention), or to amend the claimed invention to exclude the TK element (in which case it is likely to be highly relevant prior art, and thus may need to be disclosed in any case).

(v) If TK (known to the applicant) is so remote from the claimed inventive concept that it is neither relevant to the assessment of validity or determination of inventorship, then it may be necessary to clarify how the claimed invention could be determined to be based on or derived from the TK.[151]

As these are guidelines for the present, their normative value is merely hortatory, lacking any binding force. These guidelines are the softest version of non-legal "soft law,"[152] without any binding effect whatsoever. They, however, possess the status of *lex ferenda* that should be incorporated by domestic laws and international institutions.

The WIPO IGC on IPGRTKF points out that the following issues related to the requirement to certify source or origin are still unresolved

(i) would the requirement also apply when the invention, for which the application is filed, concerns synthesized substances that were isolated or derived from active compounds of an accessed genetic resource and, if so, what is an agreed definition of "derived?";
(ii) whether and how the requirement would apply for genetic resources accessed from multilateral systems for facilitated access to genetic resources, which may be established in the agricultural sector; and
(iii) what the consequences of non-compliance with the requirement would be, ranging from a fine to invalidation or revocation of the patent.[153]

In spite of the Swiss and EU efforts to bring legal certainty to this complex matter, the trigger mechanism remains vague. A precise taxonomy of all the possible trigger mechanisms lies outside the scope of this research. It is, however, essential that they are negotiated for the sake of legal certainty. Whereas, traditionally, legal certainty in patent law has been often used to place the IP system in total clinical isolation from the rest of international law, including the CBD obligations, in this case a serious implementation needs a clear list of circumstances in which the patent application needs to be accompanied by a certificate of source.

6.1.3.6.4 Exceptions to the disclosure requirement

According to the Swiss proposal, only patent applicants who have information about the source of GRs used in their inventions are required to submit a certification of source or origin; all other patent applicants are required to "declare that the source is unknown to them":

[151] *Initial Report on the Technical Study on Disclosure Requirements Related to Genetic Resources and Traditional Knowledge* WIPO/GRTKF/IC/4/11, 27–28 (December 9 to 17, 2002); Tobin, *Certificates of Origin.*

[152] P. Drahos, "Indigenous Knowledge and the Duties of Intellectual Property Owners" (1997) 11 *Intellectual Property Journal* 179.

[153] *Initial Report on the Technical Study on Disclosure Requirements Related to Genetic Resources and Traditional Knowledge*, 11.

Consequently, if an invention fulfills the conditions of the new Rule 51*bis*.1(g), the proposed wording would explicitly enable national legislation to require patent applicants to either declare the source of the genetic resource or knowledge, innovations and practices, or to declare that this source is unknown to them.[154]

While EC countries also seek to exempt the patent applicant who does not know the geographic origin of the GR,[155] the proposals of the mega-diverse countries and the African Group do not provide for any exceptions to the disclosure requirement. However, Girsberger provides the following example that illustrates the necessity of an exception: a patent applicant has an invention based on a plant obtained from a gene bank under the ITPGRFA that was collected decades ago; the source of the plant is not known and it can be traced to various countries. Under these circumstances, it would be unduly burdensome to require the applicant to trace the path of the material to certify its origin. For this reason, in order to be compliant with the "reasonableness" criterion of Article 62.1 of TRIPS, an exemption is needed. Simply removing applicable sanctions without providing an exception to the requirement would not be enough to lighten the burden on the applicants. An express exception or waiver would eliminate any presumption of non-compliance with the requirement. An exemption seems to be a reasonable solution that would be in compliance with Article 62.1 of the TRIPS Agreement (see *infra* section 6.1.3.6.8).

6.1.3.6.5 Establishing a list of government agencies competent to receive information on certificate of source

Switzerland proposes that the transparency measure be further strengthened by establishing a list of government agencies competent to receive information about patent applications containing a declaration of source of GRs and/or TK. For easy reference, this list could be accessible on the internet. Patent offices that receive applications containing such a declaration could inform the competent government agency that the respective State is declared as the source. This information could be provided in a standardized letter sent to the competent government agency. Switzerland invites WIPO, in close collaboration with the CBD, to further consider the possible establishment of such a list of competent government agencies. I hasten to add that a clearing house mechanism[156]

[154] *Proposals by Switzerland Regarding the Declaration of the Source of Genetic Resources and Traditional Knowledge in Patent Applications*, Annex, paragraph 21.

[155] *EC Submission, Disclosure of Origin or Source of Genetic Resources and Associated Traditional Knowledge in Patent Applications* WIPO/GRTF/IC/8/11, 4 (Geneva 6–10 June, 2005).

[156] S. Biber-Klemm and J. Curci, "Clearing House Mechanisms", in T. Cottier and S. Biber-Klemm (eds.), *Rights to Plant Genetic Resources and Traditional Knowledge* (CABI, 2006) 269 ff.

could serve the purposes that Switzerland allocated to these so-called "government agencies." The EU has suggested that the CBD CHM serve as the "central body to which the patent offices should send the information available from the declaration on disclosure."[157]

6.1.3.6.6 The nature of the disclosure requirement

Switzerland proposes amending the PCT by introducing an "enabling clause" that makes certificates of origin a requirement that States can optionally impose. Girsberger, on the other hand, prefers an optional requirement in the PCT since implementing a mandatory requirement at the PCT level would not create a consensus any time soon.[158] However, an optional requirement would not have any further effect than the soft law obligation expressed by the CBD and the Bonn Guidelines. Since my position is that there is an obligation under customary international law related to the provisions in the CBD and the Bonn Guidelines for implementing this disclosure of origin, it would be contradictory for me to support the Girsberger view. It is incumbent on the international community to reconcile the CBD norms and the TRIPS Agreement.

The EU proposal would directly address this objective by introducing this requirement as a "binding disclosure requirement," as stated in the *incipit* of its proposal.[159] Without a binding disclosure requirement in PCT/PLT, States' law-makers would hardly feel bound by such unclear and ambiguous obligations contained in this new patentability requirement. Simply spelling out in the PCT the manner in which such a requirement could be introduced is not enough. Failure to introduce a mandatory disclosure requirement would frustrate important policy objectives stated in section 6.1.2.1 above. Furthermore, the CBD body of norms and the TRIPS Agreement would remain in conflict. This conflict is something that the international community should avoid in order to create a well-balanced international patent system that is functional and in compliance with the objectives of Articles 7 and 8 of the TRIPS Agreement. A mandatory requirement would create legal certainty within a harmonized PCT system on a global scale that, in turn, would increase the trust of provider countries in the patent system.

This call for harmonizing the CBD and TRIPS is indeed the position of the group of like-minded mega diverse countries expressed in the TRIPS Council, which states that the TRIPS Agreement should be amended to include this new certification requirement.[160] The EC agrees with this

[157] *EC Submission*, WIPO/GRTF/IC/8/11, 5. [158] Girsberger, "Transparency", 476–7.
[159] *EC Submission*, WIPO/GRTF/IC/8/11, 1.
[160] *Taking Forward the Review of Article 27.3(b) of the TRIPS Agreement.*

proposal, provided that this requirement is only applied when the invention has in fact been based on GR and TK.

In order to achieve their objectives within the shortest period of time, DCs that are parties to those treaties should focus their effort on amending the PCT and PLT by introducing a mandatory and formal disclosure requirement along the lines suggested by Switzerland and the EU. It is currently impossible for me to predict the scenario under which the US may shift from its current adamant opposition to any amendment to the TRIPS Agreement to a position supporting the EU proposal. Once the WIPO IGC on IPGRTKF has established a way in which such a system can work, the real diplomatic discussion should take place in the PCT and PLT to proceed for an amendment. The TRIPS Council debate between Brazil/India and the US (and the accompanying reply and rebuttal) can also go on for a further TRIPS Agreement amendment.[161]

At this point, it is likely that the international community will have to accept the disclosure of source requirement through amending the PCT and PLT. However, it must be noted that few DCs are parties to the PCT and PLT. Thus, only after amending the PCT and PLT will the international community be ready for a modification of Article 27.3(b) of TRIPS. This matter of negotiation needs to be viewed in the context of the Doha Round that was fraught with major differences on farm subsidies and tariffs, where countries were unwilling to reduce barriers to farm goods and industrial products. It seems unlikely that Brazil and India, although supported by a wide number of countries, will convince the US to yield on this matter. The outcome of this debate depends on the solution of other WTO matters. Moreover, from an academic point of view, it seems premature to push for a global and all-encompassing requirement of disclosure of origin as evidence of PIC and benefit sharing. Until CBD parties have more clearly implemented other CBD provisions by adopting a clear ABS regime with appropriate and well-trained national authorities that fully understand the relationship between IPRs and their GRs and TK, there is no point in pushing for disclosure requirements.

6.1.3.6.7 Towards the implementation of the disclosure requirement in compliance with the "reasonableness" criterion

The implementation of a certificate of origin requirement in the international IP system navigates between the provider-friendly provisions of

[161] The submissions of Brazil, India and the US constitute an intense debate within the concert of submissions of other States, www.docsonline.wto.org/PDFDocuments/t/IP/C/W404.doc.

the CBD and the TRIPS Agreement constraints. This dilemma has led the CBD Conference of Parties (COP) in Decision VI/24 to invite the WIPO to prepare a technical study on possible transparency measures with the purpose of disclosing: (a) GRs used in the development of the claimed inventions; (b) the country of origin of GRs used in the claimed inventions; (c) associated TK, innovations, and practices used in the development of the claimed inventions; (d) the source of TK, innovations, and practices used; and (e) evidence of PIC.[162] This report demonstrates the difficulty of coordinating the work of various international organizations on the same subject. This type of all-encompassing disclosure of source outlined in the WIPO report overlooks, in my view, the provisions of the TRIPS Agreement. It indeed contains major constraints vis-à-vis the wide scope of disclosure of source that the report actually sets forth. This observation brings us back to a well-known problem: the unity of the international IP system that is threatened by a plurality of international organizations with competing authority to give authentic interpretations of IP treaties.

By the same token, the discussions and deliberations within the COP of the CBD have inspired the group of megadiverse countries to propose within the WTO TRIPS Agreement Council an amendment to this treaty to oblige Members States to require that applicants for a patent relating to biological material or to TK submit a disclosure of the origin of the GR and related TK as evidence of PIC and of fair and equitable sharing of benefits.[163] This proposal can hardly be accepted by the industrialized countries because of the constraints imposed by the TRIPS Agreement and inherent to patent law.

6.1.3.6.8 Disclosure of origin as evidence of prior informed consent

The megadiverse and the African countries want the disclosure of origin to serve as evidence of PIC. Paragraphs 24–40 of the Bonn Guidelines implementing Article 15.5 of the CBD encourages Contracting Parties to access foreign GRs on a system of PIC. As regards TK, Article 8(j) contains only an obligation of involvement and approval on the part of the indigenous and local community without expressly requiring their PIC.

Like Gopalakrishnan, many authors and NGOs support an all-encompassing certificate of origin, containing evidence of PIC for access

[162] *The Draft Technical Study on Disclosure Requirements Related to Genetic Resources and Traditional Knowledge* (Draft Study) WIPO/GRTKF/IC/5/10, 71 (November 1–5, 2004).

[163] *The Relationship between the TRIPS Agreement*, paragraph 1. Gopalakrishnan, "TRIPS and Protection", 11–18.

to the GR and TK. This all-encompassing certificate seems to be a practical and simple way to verify that national legislation on access to GRs and TK is complied with.[164] A more detailed analysis, however, reveals that, if the submission of the certification of origin has to serve as evidence for a PIC requirement, this additional requirement would transform the very nature of the formal requirement by making it incompatible with the "reasonableness" criterion of Article 62.1. The reasons can be outlined as follows:

(i) *Lack of ABS legislation.* At present, only a limited number of States, as Correa demonstrates, have drafted national ABS legislation and the respective administrative procedures aimed at granting a documented PIC to the bioprospector.[165] As a consequence, if submission of a certificate of origin as evidence of PIC becomes mandatory, patent applicants may not be in a position to provide the required evidence and would in many instances be excluded from patent protection. Fulfilling this obligation would require the recipient State patent authorities to appropriately interpret the provider country's law. In this case, in order to determine whether the certificate of origin is valid, the recipient country patent authority would be obliged to analyze: (a) whether the obligation of PIC is provided for by the national legislation; (b) what national authority can give this evidence; and c) whether the evidence has been correctly supplied. This task clearly goes far beyond the competence and the reasonable capacity of the recipient country patent authorities.

(ii) *Necessity of competences beyond the capacity of patent examiners.* In order to determine whether the PIC has been complied with, the patent authorities would also have to acquire special interpretative expertise in specific international treaties like the CBD and ITPGRFA. This additional requirement would clearly burden patent authorities with substantial administrative work and would also pose legal and practical problems.[166]

(iii) *No obligation under international law.* Requiring the patent applicant to submit evidence of PIC in order to acquire patent rights goes beyond the treaty obligations of Articles 15.5 and 8(j) of the CBD and of the ITPGRFA. Article 8(j), in particular, only requires the

164 *Ibid.*, 18 (for instance, this author states that "Article 8(j) of the CBD obligates that prior informed consent of the community must be obtained before utilizing their [indigenous and local communities] knowledge system)."

165 C. Correa, *Establishing a Disclosure of Origin*, 9; *EC and Member States*, paragraph 54.

166 The whole argument of the US revolves around the notion that patent examiners (and a system to monitor the examiners) are not prepared to deal with a global disclosure requirement; IP/C/W/443, paragraph 22; IP/C/W/449, paragraph 18.

approval and the involvement of the TK stakeholders without specifically requiring PIC. The ITPGRFA does not require PIC with regard to PGRFA and related TK. Facilitated access to PGRFA through the Multilateral System does not need PIC procedures on a case-by-case basis.[167]

Regarding the inclusion of evidence of PIC in the certificate of origin, the doubts of Ricolfi,[168] the opinion of Advocate General Jacobs of the ECJ,[169] and the position of Girsberger and Switzerland are to be taken equally into account:[170] the task of controlling the GRs and related TK in accordance with the ABS legislation, including PIC, should be primarily performed by the provider country legislative authority. This evidence of PIC is typically a pre-IP matter that should not disrupt the patent system.

While the submission of a certificate of origin remains entirely within the bounds of a formal "reasonable" requirement, burdening the patent applicant and patent office with evidence of PIC would tip it over these internationally acceptable boundaries. With the introduction of a simple certificate of origin, the IP system would be fully compatible with the CBD obligations. The rest of the work mandated by the CBD falls outside the scope of IP. The submission provides government authorities of the provider countries willing to verify compliance with the CBD obligations with easy access to the information on the disclosure of source, even in a foreign patent application, and allows the authorities to immediately assess whether their ABS legislation has been complied with or not. If not, they can undertake all the legal remedies in collaboration with the recipient country judicial authorities, to limit the enforcement of the patent. All these uncertainties make it clear that, if the certificate of origin requirement is accompanied by required evidence of PIC, it becomes incompatible with Article 62.1 of the TRIPS Agreement.

6.1.3.6.9 Disclosure of origin as evidence of benefit sharing

The megadiverse countries have proposed that the certificate of origin also be accompanied by evidence of benefit sharing. The obligation of benefit sharing in this context is inspired by the goals of the CBD and the Bonn Guidelines, and even the ITPGRFA. Article 15.7 of the CBD states that:

167 Correa, *Establishing a Disclosure of Origin*, 9.

168 Ricolfi, "Biotechnology, Patents and Epistemic Approaches", 85.

169 *Opinion of the Advocate General Jacobs.*

170 M. Girsberger, *Biodiversity and the Concept of Farmers' Rights in International Law* (Peter Lang, Berne, 1999) 176–77.

> each Contracting Party shall take legislative, administrative or policy measures, as appropriate [...] with the aim of sharing in a fair and equitable way the results of research and development and the benefits arising from the commercial and other utilization of GRs with the Contracting Party providing such resources.

Paragraph 48 of the Bonn Guidelines says the following under the title of "distribution of benefits":

> benefits should be shared fairly and equitably with all those who have been identified as having contributed to the resource management, scientific and/or commercial process. The latter may include governmental, non-governmental or academic institutions and indigenous and local communities.

Article 8(j) of the CBD mandates benefit sharing arising from both the use of the GR and the related knowledge.

A more detailed analysis, however, reveals that if the submission of the certification of origin were to serve as evidence for a benefit-sharing obligation, this additional requirement would transform the very nature of the formal requirement by making it incompatible with the "reasonableness" criterion of Article 62.1 of TRIPS (for additional procedures and formalities needed to acquire or maintain IPRs). The reasons why this obligation would be incompatible can be outlined as follows:

i) *No obligation in international law.* The relevant CBD Articles and subsequent practice do not mandate that the benefit sharing be specifically evidenced by the submission of a certificate of origin in the patent application. They do not mandate a special relationship between the obligation to share the benefit and patent law.

   Moreover, the ITPGRFA involves a variety of entities in benefit sharing. As already observed in sections 3.3.3.1 ff., Article 13.2(d)(ii) institutes an appropriate mechanism for benefit sharing, including a trust fund to be established by the Governing Body of the Treaty.[171] The benefits arising from the use of ITPGRFA that are shared under the Multilateral System should, according to Article 13.3, flow primarily to farmers in all countries. Article 9.2(b) of the Treaty considers "the right to equitably participate in sharing benefits arising from the utilization of [PGRFA]" as a measure to protect and promote farmers' rights. Also in this case, no particular provision expressly connects this treaty with IP treaties.

ii) *The requirement would not be within the reasonable competence of patent authorities.* In this regard, the benefit-sharing procedures arising

[171] According to Article 19.3(f) of the ITPGRFA, this mechanism is foreseen to receive and utilize the "financial resources that will accrue to it for purposes of implementing this Treaty."

from the commercial exploitation of GR and TK fall largely outside the scope of patentability *per se*. A patent grants an exclusive right, and the financial benefits do not arise at the moment of the patenting. In any case, there would be a practical problem for the patent authorities to verify compliance with benefit-sharing agreements. In most cases, no benefits are shared at the moment of the patent application since they flow from the commercialization rendered possible by the granting of the patent. It would be an anachronism to require such evidence at the moment of application. If the patent authorities were required to verify the validity of the evidence of benefit sharing, they would indeed have to exercise authority that clearly goes beyond their competence and capacity. Since both the CBD and ITPGRFA have been developing articulated systems of benefit sharing, it would be unreasonable to require the patent authorities to acquire in-depth legal expertise to determine which treaty and related national legislation applies, especially if we think that these treaties do not belong to the international IP system. The task of verifying whether the benefits arising from the exploitation of TK are shared should occur after the patent has been granted and should be the task of national authorities who have the necessary expertise in the applicable laws and have jurisdiction over the parties.

iii) *Confidentiality of benefit sharing*. As Girsberger notes, the public disclosure of contractual arrangements at the moment of the patent application would undermine the contractual autonomy of the parties.[172] Parties can agree under Article 15.7 of the CBD to submit their agreement to the principles of confidentiality. Disclosing such arrangements would breach this principle.

This analysis leads me to conclude that ensuring compliance with benefit-sharing provisions arising from commercialization of innovations based on GR cannot be reasonably carried out by the recipient country patent authorities. They have neither the competence nor the expertise to perform a verification at the time of the patent examination and granting. Thus, introducing this requirement would be incompatible with Article 61.2 of the TRIPS Agreement. In my view, the legal and practical problems related to disclosure of origin requirements as a proof of benefit sharing are not in compliance with Article 62.1 of the TRIPS Agreement.

[172] Girsberger, "Transparency", 485.

The best alternative to achieve this objective is to amend the TRIPS Agreement by introducing a new substantive requirement for patentability in Article 27 or Article 29.

6.1.3.6.10 Consequences of non-compliance

The consequences of non-compliance would determine whether the requirement is formal or substantive. This is another important aspect that deserves particular attention. In case it needs to be introduced in the PLT, Switzerland and the group of megadiverse countries are in agreement in proposing that the sanctions to be applied are those that already exist within the international patent treaties.[173] In particular, Switzerland proposes to apply Article 10 in conjunction with Articles 6(1) and 6(8) of the PLT.[174]

Article 6(8) admits sanctions in these terms:

a) Where one or more of the requirements applied by the Contracting Party under paragraphs (1) to (6) are not complied with within the time limit prescribed in the Regulations, the Contracting Party may, subject to sub-paragraph (b) and Articles 5 and 10, apply such sanction as is provided for in its law.

(b) Where any requirement applied by the Contracting Party under paragraph (1), (5) or (6) in respect of a priority claim is not complied with within the time limit prescribed in the Regulations, the priority claim may, subject to Article 13, be deemed non-existent. Subject to Article 5.7(b), no other sanctions may be applied.

In this regard Article 10 of PLT explains the consequences of non-compliance with a formal requirement (see Article 10 of the PLT):

> Non-compliance with one or more of the formal requirements referred to in Article 6(1) […] with respect to an application may not be a ground for revocation or invalidation of a patent, either totally or in part, except where the non-compliance with the formal requirement occurred as a result of a fraudulent intention.

Industrialized countries have expressed their reluctance to apply sanctions that invalidate a patent in case of non-compliance. They argue that such invalidation would transform the requirement into a substantive one. Indeed, the EC clearly states

[173] *Proposals by Switzerland Regarding the Declaration of the Source of Genetic Resources and Traditional Knowledge in Patent Applications*, paragraph 1.

[174] *Draft Technical Study on Disclosure Requirements Related to Genetic Resources and Traditional Knowledge*, WO/GA/30/7 Add. 1, paragraph 165 (August 15, 2003), www.wipo.int/documents/en/document/govbody/wo_gb_ga/pdf/wo_ga_30-7addl.pdf.

Disclosure requirement should not act, *de facto* or *de jure*, as an additional formal or substantial patentability criterion. Failure to disclose, or the submission of false information should not stand in the way of the grant of the patent and should have no effect on the validity of the patent, once it is granted. Legal consequences to the non-respect of the requirement should lie outside the ambit of patent law, such as for example in civil law (claim for compensation) or in administrative law (fee for refusal to submit information to the authorities or for submitting wrong information). Patent law should not be used to sanction non-respect of domestic access and benefit-sharing requirements through the rejection of the patent application or the invalidation of the patent.[175]

Notwithstanding all the cogent reasons why this requirement could be implemented as a formal one, the position of the EC opposes the introduction of *any* additional requirement; *a fortiori* it opposes the sanction of patent invalidation for non-compliance. Some have inferred from this position that these types of sanctions would make benefit sharing a *de facto* substantive requirement. I disagree with this inference, since the aforementioned sanctions of the PLT are for failure to comply with formal requirements and not substantive ones. At a stage when the international community is not ready for an amendment in TRIPs Agreement, I maintain that the implementation of this requirement as a formal one through amending PCT and PLT is the most adapted to achieve the goals and objectives. The sanctions provided for failure to fulfill PLT formal requirements should be extended to the failure to comply with this additional proposed requirement; in addition, domestic law may add any applicable further criminal sanctions.

6.1.3.6.11 Private international law issues related to the certificate of disclosure of source

A thorough analysis of the certificate of origin would be incomplete if it did not include its aspects related to private international law. For instance, one of the most important objectives for the introduction of transparency measures is to provide access to information concerning patents based on GRs and TK preserved by indigenous peoples, by DC universities and research institutions, or by public interest civil society groups in provider countries.

In this connection, an area that deserves much attention (but that can only be sketched in this paragraph), is "access to justice" against infringement of legitimate interests of provider countries in a foreign recipient country jurisdiction. The many relevant technical legal issues include

[175] *Communication by the European Communities and their Member States to the TRIPS Council*, 14.

enforcement of foreign judgments, standing to sue in foreign jurisdictions, evidentiary standards and the burden of proof,

> access to information regarding existence of rights, the breach of rights, the existence of judicial or administrative processes offering relief, alternative dispute resolution mechanisms, as well as [...] opportunities for obtaining legal representation, and [...] means for covering the costs of actions.[176]

In order to facilitate a real access to justice, serious possible impediments must be addressed, including the cost and complexity of bringing an action against major corporations or research institutions in a foreign jurisdiction.

Much research is needed on private international law surrounding the implementation of a certificate of origin. A new field of research explores the legal remedies in foreign jurisdictions initiated either by the local and indigenous communities or by the provider country itself. Within the scope of this field of research also falls the applicability or the development of tailor-made alternative dispute resolution procedures like arbitration, the application of the Convention on the Recognition and Enforcement of Foreign Arbitral Awards[177] and the Convention on the Settlement of Investment Disputes between States and Nationals of Other States.[178]

The main private international law problem here arises from the interaction between the territorial nature of patent rights and the enforcement of foreign judgments that are contrary to the exercise of such rights. In other words, if the court of one country invalidates a patent because it is contrary to that country's ABS legislation, how would that invalidation affect the patentee's rights in other jurisdictions that have granted such a patent?

The enforcement of foreign judgments varies among jurisdictions. A general distinction between common law jurisdictions and civil law jurisdictions can be outlined as follows: in common law countries, before determining whether a foreign judgment may be enforced, there are a number of procedural aspects that the judges will examine on a case-by-case basis. Civil law countries tend to determine cases in which the rules of reciprocity will apply. It is difficult to pursue a thorough study of the probability of a foreign patent being enforced in a common law jurisdiction, given the diversity of procedural statutes and case law relevant to each jurisdiction. The *travaux préparatoires* of the Hague Convention on the Recognition and Enforcement of Foreign Judgments – which has been

[176] Barber, Johnston and Tobin, "User Measures"; Tobin, "Certificates of Origin".
[177] Done at New York (June 10, 1958) 330 *UNTS* 38 (1959).
[178] Done in Washington (March 18, 1965) 575 *UNTS* 159 (No. 8359); *ILM* (1965) 532.

under discussion since 1996 and involves more than 45 countries – may reveal much about the differences of approach in civil law and common law countries. This Convention seeks to establish a regime whereby a provider country may secure the enforcement of its judgments in the courts of a recipient country in both torts and contracts, if both countries are parties to the Convention.

In sum, there are already various ways to bring such cases before national courts under tort and contract law. A technical study (that falls outside the scope of the present study) would be needed to explore the legal issues underpinning these actions, such as rules on enforcement of foreign judgments, standing before the courts, evidentiary standards, burden of proof, knowledge of rights and of the possibility of obtaining relief, legal representation, language, availability of visas, and costs.

6.1.3.7 Conclusion

The analysis has been necessarily limited to the IP aspects of this important issue of transparency through disclosure of the source of TK in patent applications. This disclosure is just one of the potential measures that may be considered to facilitate the harmonization between the CBD objectives and the exercise of IPRs. As a matter of fact, most of the CBD objectives are realized through actions outside the IP system, because many products on the market are based on GRs that are accessed and bioprospected without having been patented.

Moreover, my analysis has consciously overlooked the social, anthropological and economic aspects of PIC. There is little or no research on the feasibility, practicality and cost of introducing disclosure requirements, whether mandatory or voluntary, into international patent law. The possibility of successful adoption of this transparency measure as a binding obligation will also depend on a full assessment of the costs that this measure would entail for the IP system. Dutfield has demonstrated how international IP law has always been driven by industrialized countries (as explained in section 1.1.3 above). Furthermore, industry lobbyists have always blocked provisions that would burden bioprospecting activities with new bureaucratic procedures and would raise the cost of business transactions, especially if the introduction of provisions such as the certificate of origin become mandatory. There are both practical and inherent limitations in the IP system that prevent the integration of the CBD objectives. My analysis has focused solely on finding ways in which the international legal system can be best harmonized by working with the relevant treaty provisions.

My solution has been largely but not totally inspired by the Swiss proposal. The elements of the Swiss proposal could be introduced in a timely manner and would not require extensive changes to the provisions of the relevant international agreements of the PCT and PLT. A major difference between my proposal and the Swiss one is that my proposal considers this requirement as mandatory.

The proposals outlined in this section would thus enable the Contracting Parties of relevant international agreements, including the TRIPS Agreement, the PCT, the PLT, the CBD, and the FAO ITPGRFA to fulfill their respective obligations. This harmonization applies in particular to Articles 27.1 and 62.1 of the TRIPS Agreement, as well as to Articles 8(j), 15.4, 15.5, 15.7, and 16.5 of the CBD. These proposals aim to provide the means of ensuring that the international agreements on IP and the CBD be implemented in a mutually supportive manner. Furthermore, the Swiss proposal would enable the Contracting Parties of the CBD to implement the provisions of the Bonn Guidelines, in particular paragraph 16(d), as well several of the decisions adopted by the CBD Conferences of Parties Nos. 6 and 7. Finally, requiring the declaration of the source of GR used in inventions would alert the patent examiner that any TK associated with that GR would be relevant prior art. The examiner would then be able to find the TK through a search of databases on TK established at the local and regional levels.

Although the requirement of the disclosure of source cannot *per se* constitute evidence of PIC and benefit sharing, it would still allow for verification of whether or not PIC of the country providing the GRs has been obtained and whether provisions have been made for fair and equitable benefit sharing.

From a policy-making perspective, I do not maintain that this amendment to the PCT and PLT should render unnecessary any further amendment of Article 27 of the TRIPS Agreement in a way that favors the interests of bioculturally diverse DCs. On the contrary, this reform of the PCT and PLT can be a first experiment in conciliation between the CBD and patent law that can create a common understanding for a review of Article 27 of the TRIPS Agreement, making its integration with the CBD explicit, clear, and harmonious.

6.2 Traditional knowledge as prior art

The second method through which the defensive protection of TK can be realized in the IP system is in relation to the concept of "novelty-destroying prior art"[179] during or after the prosecution of a patent

[179] I specify that novelty-destroying prior art refers to prior art (the state of the art prior to the patent application, that is, knowledge and practices already existing) which shows that a patent application does not present new matter (that it existed in the prior art) and thus does not meet the statutory requirements for patent protection.

application. This section identifies the circumstances in which TK can be considered novelty-destroying prior art during or after a patent application. It is a requirement in all patent regimes that an innovation must be novel and involve an inventive step (or be non-obvious) to merit protection under the law,[180] and the concept of prior art is linked to the procedures instituted by patent offices to assess whether the claims in the patent application meet the novelty and inventive step (or non-obviousness)[181] requirements for patentability. There are three progressive elements of innovation that need to be addressed in examining the inventive step requirement: (i) the raw material of the GR, (ii) the TK and other forms of informal knowledge related to the raw material, and (iii) the industrial knowledge applied to (i) and (ii). Distinguishing among the three is the task of the patent examiner in order to grant a quality patent that effectively balances the competing goals of the patent system (in light of the objectives of Article 7 of the TRIPS Agreement).

In order to identify instances in which TK can be viewed as novelty-destroying prior art in a patent application opposition, this section first examines novelty and prior art concepts in both US and European patent law. The concepts of novelty, inventive step, and prior art will be treated separately. It is important to understand how they relate to each other, and so this section will briefly explain how a claim of novelty and inventive step is analyzed in light of the prior art.

The complexities surrounding searches to find prior art relevant to a patent application are due to the relative difficulty in gaining access to knowledge in the public domain and to the sheer enormity of the public domain. My comparative analysis will explore how the territoriality of patents affects prior art legal principles and the search process. The differences between the law and practice of the EPO and USPTO will be examined. Within this interface between novelty and prior art, special attention will be given to the geographic limitations on prior art in the US patent system where unwritten or unpublished knowledge from a foreign country does not count as proof of prior art, whereas in the EPO there is no geographic disparity. Generally speaking, the concept of geographic

180 Article 27.1 of the TRIPS; Article 15 of the PCT; and Rule 33 of the related Regulations, www.wipo.int/edocs/notdocs/en/pct/treaty_pct_130.html.

181 The inventive step requirement of the European Patent Convention and some European national systems essentially corresponds to the non-obviousness requirement in US law. The footnote to Article 27.1 of *TRIPS* states that "[f]or the purposes of this Article, the term 'inventive step' [...] may be deemed by a Member to be synonymous with the term 'non-obvious' [...]". Thus, whatever differences of nuance between the US and EU systems may exist with regard to the particular application of the general concept, these differences are not taken into account in the TRIPS Agreement.

disparity implies that even if the patent office is not aware of the written source in a foreign country and grants the patent, this lack of knowledge does not bar subsequent opposition.

The rules of international treaties and customary norms governing this matter will clarify some of the improvements that are proposed for the regional systems in order to take into account TK as prior art. This section finally discusses the international prior art search mechanism currently in place, and also explores the ways in which effective TK database systems can facilitate international prior art searches.

6.2.1 *The interrelation between novelty, inventive step, or non-obviousness and prior art*

Information which is in the public domain cannot be subject to patent claims. In order for a patent to be awarded on an invention, that invention must be new and involve an inventive step. Since there is no internationally agreed-upon definition of what an invention is, its constitutive elements, i.e. prior art and novelty, are the subject of endless scholarly and judicial discussion.[182] While the concept of novelty as a prerequisite for obtaining a patent is simple to comprehend, complexities arise in the relationship between novelty and prior art.[183] In order to assess the novelty and inventive step of an invention, the examiner must compare it with the state of the art prior to the invention. If every element of the claimed invention has been used together before, then the invention is not novel and will be rejected. However, an application is usually not so straightforward, because so much innovation consists of adding a new element to the prior art or combining elements existing in the prior art in a new way. The problem then is deciding to what degree an invention incorporating prior art involves an inventive step or is non-obvious. Thus, the novelty and inventive step requirements must always be judged against existing knowledge in the public domain.

In order to understand how prior art may destroy novelty or inventive step, the present study will analyze these concepts as described in the TRIPS Agreement as well as in US and European patent law. These two

[182] "[T]he TRIPS Agreement contains no definition of invention and therefore leaves member countries relatively free to draw the line between patentable 'discoveries' and actual inventions in the biological field [...] the lack of consensus concerning biological patents thus allows countries considerable leeway in fashioning their policy options"; *The TRIPS Agreement and Developing Countries* 34 (UNCTAD, New York, 1996).

[183] Novelty, at least in US law, is straightforward. It is rarely the focus of arguments. If the invention exists in the prior art (in a single reference), then there is no novelty; otherwise, there is.

regional systems have significant similarities and differences in their approaches to the question.

Novelty and inventive step in Article 27.1 of the TRIPS Agreement draw upon the general principles common to the main (US and EU) traditional regional patent systems. The general interpretation of this TRIPS-enshrined concept is similar to the specification made in Article 15.2 of PCT that sets out procedures for international prior art searches. For the most part, the concept of prior art in Article 27.1 of the TRIPS Agreement follows the implementation that has been enshrined in Rule 33.1(a) of the PCT Regulation. It states that prior art is everything that has been made available to the public anywhere in the world by means of written disclosure, and which can be of assistance in determining that a claimed invention is novel or non-obvious.[184] The impact of TK as prior art depends on its relationship to the patentability requirements of non-obviousness, inventive step, and novelty. Inventive step and non-obviousness consist of the same general patentability requirement present in most patent systems. Novelty reflects a lower threshold than non-obviousness. So if a patent application matches the requirement of non-obviousness/inventive step, it also satisfies the requirement of novelty. Prior art is used to reject patent applications for lack of novelty and inventive step. Thus, if TK were available as prior art, it could be used to reject patents on uses of biodiversity that are identical to those of indigenous TK holders (for lack of novelty) and for uses that incorporate minor advances (for lack of inventive step).

It is not possible in a few pages to explain all the possibe ways in which specific TK could or would destroy novelty, non-obviousness, or inventive step of a biotechnological patent. This topic falls outside the scope of the present study. Instead, I will concentrate on the way in which TK can become prior art in the patentability examination process.

[184] *Resource Book on TRIPS and Development*, 359. See also the PCT Regulations Rule 33.1(a). The current situation is portrayed by "(a) In certain countries, prior art is constituted by everything that has been made available to the public anywhere in the world by any means before the filing or priority date of the application. On the other hand, in other countries, non-written disclosures, such as oral disclosures, or use outside their jurisdiction, do not form part of the prior art, and thus do not constitute a bar to patentability. (b) While certain patent systems require a concrete disclosure for complying with the standard of 'availability to the public,' others provide that the theoretical possibility of having access to the information is sufficient." *Further Development of International Patent Law* SCP/4/2 (November 6–10, 2000), www.wipo.int/meetings/fr/doc_details.jsp?doc_id=1499.

6.2.1.1 Novelty and prior art in the US

In US law, it is essential that an invention be novel and non-obvious[185] in order to qualify for a patent. Section 35 of US Code paragraph 102 (a) describes the novelty requirement and specifies the prior art that is available to negate both novelty and non-obviousness: "A person shall be entitled to patent unless the invention was known or used by others in this country, or patented or described in a printed publication in this or a foreign country, before the invention thereof by the applicant for patent."

The novelty requirement serves to protect information in the public domain; it also motivates inventors to seek a patent as soon as they are able, as it is not possible to apply retroactively for an invention that has already been disclosed.[186] US law imposes geographic limitations on what is available as prior art to challenge patent applications. As earlier observed, paragraph 102 (a) states "[p]rinted publication[s]" are available from around the world, but evidence that an invention was "known or used" may only be presented from the US. This limitation is particularly relevant to TK, which is often not present in a printed publication, but which does fall under the "known or used" category. Ironically, the first US patent act did not contain any geographical limitation,[187] and, as Mgbeoji points out, early US jurisprudence had defined the US novelty requirement as pertaining "in relation to every part of the world."[188] The geographical limitation was added in 1836.[189]

Using the same prior art that is available to destroy novelty, an examiner also assesses whether a person with ordinary skill in this particular field would consider the innovation obvious. The test is "what the combined

185 Non-obviousness is the US equivalent of involving an inventive step. The non-obviousness requirement is contained in 35 US Code paragraph 103(a): "A patent may not be obtained though the invention is not identically disclosed or described [...] if the differences between the subject-matter sought to be patented and the prior art are such that the subject-matter as a whole would have been obvious at the time the invention was made to a person having ordinary skill in the art to which said subject-matter pertains [...]"

186 N. Roht-Arriaza, "Of Seeds and Shamans: The Appropriation of the Scientific and Technical Knowledge of Indigenous and Local Communities" (1996) 17 *Michigan Journal of International Law* 919, 937.

187 Section 1 of the 1790 US Patent Act allowed patents for inventions "not before known or used," without any reference to where they were known or used, *An Act to Promote the Progress of Useful Arts*, 1 Stat. 109–12 (April 10, 1790).

188 I. Mgbeoji, "Rethinking the Role of International Law in Relation to the Appropriation of Traditional Knowledge of the Uses of Plants" (dissertation submitted for the Degree of Doctor in the Science of Law, Dalhousie University Halifax, November 2001, copy on file with the author), 197 footnote 806, quoting *Dawson v. Follen*, 7 F. Cas. 216 (C. C. D. Pa. 1808).

189 M. Bagley, "Patently Unconstitutional: The Geographical Limitation on Prior Art in a Small World", (2003) 87 *Minnesota Law Review*.

teachings of the references [in the prior art] would have suggested to those of ordinary skill in the art."[190] Thus, both prior art and all information readily obtained from that prior art are not patentable innovations. To repeat, foreign prior art must be in written form in order to be taken into account by the USPTO: "A person shall be entitled to a patent unless: The invention was patented or described in a *printed publication in this or a foreign country or in public use or on sale in this country*, more than one year prior to the date of the application for patent in the United States" (italics added).

In case of direct misappropriation of TK, a patent application can be challenged under Section 35 of US Code paragraph 102(f) that reads: "A person shall be entitled to a patent unless he did not himself invent the subject-matter sought to be patented." Thus, where a researcher is claiming a patent on the subject of TK, then that TK can be brought forward as evidence that the researcher has not invented the subject of the patent application, regardless of whether the TK is found in a printed publication. However, this requirement presents a very narrow opportunity for the use of TK. If the researcher has altered the substance or process from the way it is used by the traditional community, then the researcher has contributed innovation and is thus an inventor, and the TK cannot be presented to address questions of obviousness. Also, the information must be presented during the examination of the patent. If it is not discovered that the TK has been misappropriated until after the patent is granted, then the TK cannot be used for a re-examination of the patent, which can only occur on the basis of printed publications.

From all these considerations, it is apparent that unpublished TK needs a more effective place as prior art in the US system.

6.2.1.2 Novelty, inventive step, and prior art in Europe

Prior art is broader in EU law than US law, in that it does not have any geographical limitations. So where US law would grant a domestic patent on an undocumented foreign practice, European law would not. Furthermore, European law makes specific reference to the distinction between a discovery and novelty in EPC Article 52(2)(a): "(2) The following in particular shall not be regarded as inventions within the meaning of paragraph 1: (a) discoveries, scientific theories, and mathematical methods."

A mere discovery of the property of an existing plant or other naturally occurring process is not sufficient in and of itself to qualify as a novelty unless the plant or process has been substantially altered or modified through the substantive investment of human ingenuity.

[190] Bagley, "Patently Unconstitutional", 694, quoting *Cable Elec. Prods, Inc. v. Genmark, Inc.*, 770F21015, 1025 (Fed. Cir. 1995).

Article 54(2) of the EPC defines prior art (the state of the art) as "*everything made available to the public by means of a written or oral description*, by use, or in any other way, before the filing of the European Patent application" (italics added).

This definition emphasizes that, according to the EPC, all known practices, whether present in a written document or not, are prior art for the purposes of patent law categorization, regardless of their geographical origin. As I have shown in the section on US law above, this view is not universally accepted. EPC jurisprudence elaborates this definition by specifying that the general public need not be aware of the existence of the information for it to qualify as prior art, but rather the information need only be available to any person at any time prior to the application.[191] It is to be noted that the "publication" requirement under US law is rather broad; it includes anything that is findable by someone that is searching for it.

Prior art plays an important role in the incentive to develop new inventions and innovations. Disregarding prior art in a certain field of technology in the patent examination process is detrimental to fostering the appropriate innovation that justifies the patent system. Prior art in any form merits recognition and acknowledgment. I would argue that all forms of knowledge deserve not only recognition but also protection from "abusive" IPR regimes. This protection should come in the form of inclusion in the patent system as prior art with the legitimate ability to destroy patent novelty and inventive step claims.

6.2.2 *Traditional knowledge as prior art and the debate over geographical limitations*

In light of the above definition of prior art, it follows that much of what has been treated as TK also falls within the ambit of what is considered prior art in the EU – and should be considered prior art everywhere. This should hold true whether or not such TK has been documented. Meanwhile, TK holders are increasingly wary of the misappropriation of their knowledge and its patenting in countries that do not recognize undocumented TK held abroad as prior art.[192]

[191] *Traditional Knowledge as Prior Art and the Use of the Patent System as a Defensive Measure Against Misappropriation*, www.southcentre.org/publications/occassional/paper09/paper9-04.htm, 4.

[192] G. Dutfield, *Developing and Implementing National Systems for Protecting Traditional Knowledge: Experiences in Selected Developing Countries* (New York, United Nations, 2004). Until recently, Japan, like the US, did not recognize foreign undocumented TK as prior art.

There have been instances of patents being issued on "inventions" that were long known to traditional peoples due to the geographical limitations on prior art within the US patent system.[193] The US patent code (specifically paragraph 102 (a)) distinguishes between prior (unwritten) knowledge and use in foreign countries and prior knowledge and use in the US. US patent law does not recognize as prior art foreign knowledge and use of a practice that is not contained in a printed publication. It is not true, as some may believe, that the US patent system only accepts instances of novelty-destroying prior art generated within the US borders. While it may be more common for patents to be rejected based on US prior art, there are many cases of rejection of patent applications based on prior art originating outside the US as well. The problem arises with unpublished knowledge and use.[194] The implication of this geographical limitation for TK holders outside of the US is significant, for this limitation allows for the patenting by essentially any applicant of numerous TK-based innovations that have been passed down orally in the traditional community. The limitation denies local communities the opportunity to bring their unwritten knowledge, practices, and innovations that demonstrate obviousness or lack of novelty to the attention of US patent examiners. Generally, these communities are unaware of the appropriation until after the patent has been granted.

The case of the neem tree patent provides examples of both TK as prior art and the geographical limitation controversy. The neem tree has been long revered in India for its properties as a cure for many ailments. As a result of its importance to Indian communities, it has acquired a momentous cultural significance. Indeed, some Indian communities worship the neem tree as a god. In addition to its pharmaceutical properties, the neem tree has also been used as a natural pesticide by soaking its crushed seeds in water or alcohol and applying the resultant blend to crops. This TK was limited, however, in that the blend did not retain its insect-repelling

193 The most well known cases of alleged "biopiracy" are the neem tree, the enola bean, and the hoodia cactus. With regard to the neem patent controversy, E. Marden, "The Neem Tree Patent: International Conflict over the Commodification of Life", (1999) 22 *Boston College International & Comparative Law Review* 279. An ongoing encyclopaedic project on the internet is currently being undertaken, www.en.wikipedia.org/wiki/Talk: Biopiracy. This website presents ongoing updated information on the topic, and anybody can give his/her contribution.

194 *Tri-Collar, Inc. v. Reamco, Inc., A Division of Sun Oil Co.* 538 F. Supp. 669, 686 (D.C.La., 1982) (where a US patent application was denied on account of domestic prior art). For an extensive list of examples of novelty and nonobviousness destroying prior art in the US; J. Rydstromnot, "Comment Note – Application and Effect of 35 US CODE paragraph 103, Requiring Nonobvious Subject-matter in Determining Validity of Patents", (2005) 23 A. L. R. Fed. 326.

properties for long. In the early 1990s, the international company Grace developed a method for stabilizing the neem seed mixture, thus making it marketable as a packaged commodity. Grace subsequently obtained US and European patents on a pesticide based on the neem seed formula. The patents were contested in both the US and Europe as being obvious, therefore failing the novelty requirement of a patent. One of the patents granted in Europe was consequently invalidated on those grounds. In the European view, the TK related to the use of the neem tree seed was considered novelty-destroying prior art. The patents granted in the US, however, were unscathed by the attacks. These attacks were due to the geographical limitation of US patent law as contained in the aforementioned section 35 of US Code paragraph 102(a) which provides that foreign TK can only defeat a US patent's novelty claim if that foreign TK appeared in a printed publication before the invention or patent application by the US applicant. In the case of the neem tree, there was no readily available publication.[195] There are cases, however, where prior art from outside the US related to TK has been able to bar a patent application in the US. In the turmeric case,[196] the GR was not just pre-existing in its wild state but also quite extensively known both in its structure and defining characters and in its actual and potential uses. Turmeric is a ubiquitous Indian plant used for centuries to heal wounds and rashes. Two expatriate Indian scientists in Mississippi were granted a patent over turmeric as a healing agent for wounds. The Indian Council for Scientific and Industrial Research consequently instituted re-examination proceedings seeking to cancel the patent. They argued that the healing properties of turmeric had been known and used for centuries in India. The USPTO subsequently revoked all six patents that had been granted, finding that the turmeric-based healing agent lacked novelty.[197] The distinguishing factor between these two cases is that whereas in the neem case the opponents were not able to provide documentation of prior art, in the turmeric case the Indian Council for Scientific and Industrial Research was able to produce the necessary printed publications.

It is increasingly common for DC interest groups to assert the TK of DCs as prior art against the patents or patent applications of foreign corporations that have taken allegedly undue advantage of them. My main focus in this section will be to discuss cases in which companies

[195] Bagley, "Patently Unconstitutional".

[196] P. Cullet, C. Germann, A. Nascimiento and G. Pasadilla, "Intellectual Property Rights, Plant Genetic Resources and Traditional Knowledge", in T. Cottier and S. Biber-Klemm (eds.), *Rights to Plant Genetic Resources and Traditional Knowledge: Basic Issues and Perspectives* (CABI, Geneva, 2006) 135.

[197] Bagley, "Patently Unconstitutional", footnote 23.

have acquired patents for inventions that draw on TK related to PGRs. These cases include the patenting of quinine as a cure for malaria taken from Peruvian TK of the properties of cinchona tree bark[198] and a Japanese corporation's patenting of sweetening proteins derived from *katempfe* and serendipity berry, two plants that have long been used in African communities as a natural sweetener.[199]

Another example of a patent based on TK occurred in 1990 when an American scientist obtained a patent for colored cotton.[200] Latin American indigenous groups, who spent centuries breeding and cultivating the cotton, received no recognition or recompense. These cases demonstrate that patent examiners were not apprised of, did not seek, or where not permitted to consider the available prior art contained in the TK from the knowledge systems of indigenous and local communities.[201]

It has been argued that this geographical limitation was originally sought in the US to facilitate patenting of technology that had been copied from England.[202] By the same token, many authors consider that today's US patent law is a mechanism for legitimizing acts of biopiracy through misappropriation of prior art within the public domain.[203] Supporters of the current geographical limitation, on the other hand, argue that the law maximizes economic benefits arising from TK because it allows TK to be transformed and used in an industrial innovative process much more rapidly than if the patent examiners had to trace very hard to access information in order to assess whether the innovation is non-obvious and novel in comparison with its prior art. The dispute over the geographical limitation of US patent law has thus polarized into two conflicting views: the unconstitutionality of patenting prior art argued by Bagley,[204] and the utilitarian ideal, as argued by Nard.[205]

198 United States Patent 6,844,356 (January 18, 2005), www.pharmcast.com/Patents100/Yr2005/Jan2005/011805/6844356_Hemorrhoid011805.htm.

199 Roht-Arriaza, "Of Seeds and Shamans", 923. 200 *Ibid.*, 924.

201 Bagley, "Patently Unconstitutional", 682.

202 Dutfield, *Developing and Implementing National Systems for Protecting Traditional Knowledge.*

203 A. Gupta, *Intellectual Property, Traditional Knowledge and Genetic Resources Conserving Biodiversity and Rewarding Associated Knowledge and Innovation Systems: Honey Bee Perspective* WIPO/ECTK/SOF/01/3.8,16,(May 29–31,2001), www.iimahd.ernet.in/-anilg/ ("There is a tremendous amount of knowledge, which is available only in oral form and has not yet been documented. There have been cases when such knowledge communicated in good faith by local people has been used without acknowledgement or reciprocity to claim intellectual property on the same.")

204 Bagley, "Patently Unconstitutional", 697.

205 C. Nard, "In the Defense of Geographic Disparity" (2003) 88 *Minnesota Law Review* 221.

6.2.2.1 Arguments against geographic limitation

Bagley and other authors[206] consider that US geographical patent law subverts the IP clause of the Constitution[207] because it rewards patent holders for use of knowledge that is already in the public domain. This argument follows from a belief that a central principle behind the IP clause was the avoidance of abusive monopolies as experienced in England, and that the principle should continue today. For example, going back to the neem tree case, a US patent on a neem-derived pesticide would, in effect, introduce a beneficial product to the US market. However, because the patent was revoked in Europe, that same product would be introduced to the European market at a competitive price, whereas the consumers in the US would be at the mercy of the patent holder's monopolistic pricing.[208] Bagley argues that while Section 35 of US Code paragraph 102 may have been constitutionally sound at its inception – when the technological limitations on travel and information access seemed to justify the exclusion of unpublished foreign prior art – the same cannot hold true today.[209] This assertion relies on an evolutive interpretation of the IP clause of the US Constitution. Due to the technological advancements in today's world and the ease of sharing information over long distances, what is public domain in one area of the world must be considered public domain in another.

[206] J. Golden, "Biotechnology, Technology Policy, and Patentability: Natural Products and Invention in the American System" (2001) 50 *Emory Law Journal* 101, 104–5 ("[B]ecause patents provide this spur to progress through a monopoly grant, there is an ever-present concern that they will overreach – granting property rights beyond what inventors legally deserve, or (of more fundamental concern) beyond what best promotes the development and dissemination of technological products."); R. Merges and G. Reynolds, "The Proper Scope of the Copyright and Patent Power", (2000) 37 *Harvard Journal on Legislation* 45, 46 ("[T]here are limits on Congress's power to create and extend intellectual property interests. Such limits are 'internal' in the sense that they are the result of the very same constitutional provision giving rise to Congress's power in the first place, the [Intellectual Property] Clause of the Constitution"); D. Chisum, "Foreign Activity: Its Effect on Patentability Under United States Law", (1980) 11 *International Review on Industrial Property and Copyright* 26, 36.

[207] US Constitution, Article I, paragraph 8, cl. 8 declares that Congress has power "[t]o promote the Progress of Science and useful Arts, by securing for limited Times to Authors and Inventors the exclusive Right to their respective Writings and Discoveries."

[208] The enola bean case is a perfect example of the effects of the patent monopoly. In 1996 an American filed for a patent on a bean taken from Mexico after modifying the bean's color to a particular yellow hue. The patent holder subsequently sued another company importing the beans within the patented color range. The result was a huge blow to Mexican farmers and exporters and to American consumers who were no longer able to obtain the beans at a competitive price. *Mexican Bean BioPiracy*, RAFI Geno-types (January 17, 2000), www.etcgroup.org/en/materials/publications.html?pub_id=339.

[209] Bagley considers that the Supreme Court's holding in *Gayler v. Wilder*, 51 US (10 How.) 477, 497 (1850) is no longer relevant: Bagley, "Patently Unconstitutional", 698, 733.

Furthermore, the legislative history of 35 US Code paragraph 102 reveals little justification behind the law besides "perceived evidentiary difficulties" in proving foreign, undocumented prior art.[210] Therefore, in Bagley's view, geographical limitation on prior art is no longer justified even if it does expedite the introduction of beneficial advancements within the US borders.[211] Besides being unconstitutional, Bagley also argues that paragraph 102 is bad policy because indigenous TK holders are rarely rewarded for sharing their knowledge with individuals that subsequently become patent holders. The result is the creation of a trade barrier where DCs are precluded from marketing their product to the US, because the US company markets a product that contains the patented TK, that would have otherwise served the indigenous community to develop its own independent product. Thus, according to Bagley, not only are the constitutional constraints of the IP clause violated, the resultant hampering of foreign TK holders' rights is clearly unjust. In support of her proposal that paragraph 102 is obsolete, Bagley points out that Japan, the US's long-time sole supporter of paragraph 102(b) modified its law to follow the European model of inclusion of unpublished knowledge as prior art.[212] Her argument now appears vindicated because a bill to amend Section 35 was introduced in the House of Representatives in April 2005. The amended paragraph 102(a)(1)(B) reads in relevant part:

> A patent for a claimed invention may not be obtained if the claimed invention was patented, described in a printed publication, *or otherwise known* before the effective filing date of the claimed invention [...]. (italics added)[213]

Apparently, the proposed new language would include unpublished TK as prior art.[214] It remains to be seen whether this bill will ever become law, but the bill represents a step towards recognition that the world is growing smaller in the field of IP and that substantive harmonization is necessary in order to address the issue of unpublished knowledge as prior art.

6.2.2.2 Arguments in favor of geographic limitation

In contrast to Bagley's view, Nard's utilitarian view of patent law allows for the patentability of undocumented prior art to reward investments made to commercialize useful knowledge that would otherwise remain inaccessible. If patents were denied on the basis of unwritten foreign prior

210 *Ibid.*, 699. 211 *Ibid.*, 684. 212 *Ibid.*, 732–33.

213 S. 3818 to amend Title 35, US Code, to provide for patent reform, in the Senate of the US (August 3, 2006), www.thomas.loc.gov/cgi-bin/query/z?c109:s3818.

214 The Committee Print is www.promotetheprogress.com/ptpfiles/patentreform/houseoversight/committeeprint.pdf.

art, there would be no incentive for investment-backed researchers to make possible improvements on TK. Under this view, the driving force behind the IP clause of the US Constitution is the enhancement of public welfare, and not an overriding concern for abusive monopolistic practices.[215]

In the utilitarian view, Nard reasons that undocumented foreign knowledge is generally inaccessible to the public, as there are no means of direct retrieval of such knowledge. Because it is inaccessible, such undocumented knowledge remains outside the public domain. It follows that public welfare within the US and abroad can only be advanced by providing patents on TK. Nard proposes that the development and exploitation of undocumented TK can serve to enrich both the patent owners as well as the TK holders through bilateral benefit-sharing agreements that would otherwise not exist, leaving both the traditional community and the US patentee without the resulting revenue.[216]

6.2.2.3 How to solve the dilemma

The debate between Nard and Bagley[217] provides a good look at the opposing scholarly views on geographical limitation in US patent law. Similar discussions can be undertaken in other systems as well. However, a discussion that limits itself to domestic law and does not consider applicable principles of international law does not seem to provide an effective solution to the controversial relationship between patents based on foreign GR, TK, and prior art. These issues are going to be addressed separately in order to find a legal basis in international law against geographic limitation.

Nard's arguments are mainly economic and certainly affect the policy approach of common law. In this sense his utilitarian approach is too willing to sacrifice the inherent rights of TK holders to achieve the higher objective of making profit, even if for the whole international business society. Section 5.2.1 above discussed the major difficulties related to benefit-sharing contracts (such as those Nard proposes) due to unequal negotiating positions between indigenous and industrial parties. So the benefits arising from the commercial exploitation of GRs and related TK enrich the patent holder rather than the provider State or the indigenous/local community. Nard seems to assume that such benefit-sharing systems will arise spontaneously. The utilitarian view neglects some underlying questions of law by pointing to attractive benefit-sharing possibilities arising out of the patenting of TK.

[215] Nard, "In the Defense of Geographic Disparity", 223. [216] *Ibid.*, 224.

[217] Bagley, "Patently Unconstitutional"; Nard, "In the Defense of Geographic Disparity"; Bagley, "Still Patently Unconstitutional: A Reply to Professor Nard", (2003) 88 *Minnesota Law Review* 238.

The first question of law is whether a patent application based on GRs and TK actually meets the basic legal standards for patentability. The question of whether to include TK as prior art should be examined. This question can only be fully addressed when all relevant knowledge is available to the examiners as prior art. A patent right should only be granted by a patent examination system that is attentive to the abuses of overly extensive and illegitimate monopoly rights. Nard's utilitarian approach eschews these legal issues. In my view, the legal justification of geographical limitation should be based on the uncertain legal status of undocumented non-industrial TK as prior art, rather than on bald economic considerations, including non-obviousness and novelty.

On the other hand, whereas Bagley is no doubt correct in asserting that the US geographical limitation of paragraph 102 poses a problem to indigenous communities, criticizing it solely on the grounds of unconstitutionality may be too reductive. Indeed, one can argue that paragraph 102 is not unconstitutional because the Constitution expresses a very general principle that the patent rights of the inventors are for "promot [ing] the progress of ... useful arts."[218] The IP clause of the US Constitution does not define how the principle of territoriality interacts with prior art and other patent law principles. In other words, Bagley may be exaggerating the conflict between the Constitution and current patent law. She fails to demonstrate how a general principle (*lex generalis*) of the Constitution conflicts with paragraph 102 that, in turn, sets forth the *lex specialis*. By their very nature, *lex specialis* and *generalis* are complementary rather than conflicting. It is worth noting that neither international patent law nor the US Constitution defines prior art, nor do they address the admissibility of geographical limitations.

6.2.2.4 An international law approach resolving geographic limitations

A truly international prior art search has the potential to include TK as prior art. However, the question arises of how international law can influence US patent law and other patent systems in this direction. International law may offer an additional legal basis against geographic limitations and may provide a more specific analysis of TK as a possible source of prior art that could destroy novelty or non-obviousness. International law is rarely used as justification for modifying US law, though sometimes harmonization of international patent law is persuasive in the US so as to justify a modification of practice – as is currently occurring in the push to change US patent law from "first to invent" to

[218] US Constitution, Article I, paragraph 8, cl. 8.

"true inventor." One may doubt whether international treaty and customary law will generate an immediate feeling of compliance among US and other domestic legislatures. Given the lack of an inter-state dispute settlement mechanism,[219] the current task of international legal doctrine and of international organizations will be one of reminding the States to comply with their international law obligations. The CBD has therefore instituted committees to assure that States comply with the obligation to cooperate, ensuring that the IPRs "are supportive of and do not run counter to [the CBD's] objectives."[220] However, the US is not bound by the CBD, except for certain possible customary principles that, even in case of recognition as such on the part of the US administration, are unlikely to influence patent law directly.

The analysis of the relevant international patent treaty law and customary norms stemming from the CBD principles leads to a conclusion that favors the institution of international prior art search in domestic patent examination.

6.2.2.4.1 International treaty law

Applicable international treaty law obligations offer a more solid legal ground on which to base the necessity of amending the geographical limitation in paragraph 102 of US law. As observed in section 6.2.1.1 above, Article 102(a) provides that knowledge or use abroad is not able to destroy novelty.

The geographical limitation can be challenged on two main grounds:

The first is based upon Article 4 of TRIPS Agreement that states that "with regard to the protection of intellectual property, any advantage, favor, privilege or immunity granted by a Member to the nationals of any other country shall be accorded immediately and unconditionally to the nationals of all other Members."[221] It is evident that paragraph 102(a) of the US Code is in violation of the national treatment obligation in the Paris Convention and the most favored nation treatment principle enshrined in Article 4 because it discriminates against foreign knowledge; hence it discriminates against foreign residents of WTO Members.

The second ground is based upon Article 15 of the PCT mandating that (i) each international application shall be the subject of international search, and that (ii) the objective of the international search is to discover relevant prior art.

[219] The CBD provides for arbitration and conciliation in Part 1 of Annex II and judicial dispute settlement through the ICJ in Article 27.

[220] Article 16.5 of the CBD.

[221] M. Ricolfi, "Patent Harmonization: First to File v. First to Invent", in D. S. Chisum, C. Nard, H. F. Schwartz, P. Newman and F. Scott Kieff, *Principles of Patent Law* (3rd edn., Foundation Press, New York, 2004) 516–21.

An "international-type" search should be undertaken as the term is defined under the PCT[222] for all patent applications. Indeed, this international search is required in USPTO's current written policy, although it reportedly is not being carried out.[223] According to the aforementioned Article 15.2 of the PCT, the objective of an international search is to discover relevant prior art from around the world.

Currently USPTO examiners perform the "international-type" searches only for applications that enter the national stage from international applications. However, a look at 37 Code of Federal Regulations (CFR) paragraph 1.9 indicates that "national applications," for which examiners are required to perform international-type searches under 37 CFR paragraph 1.104(a)(3),[224] includes any US application,[225] and not only applications entering the national stage from international applications. Consequently, the international-type search must be performed on all US patents filed on and after June 1, 1978.[226]

6.2.2.4.2 International customary law

I refer to the analysis on the international customary normative value of certain principles of the CBD outlined in section 6.1.1.2 above. CBD substantive norms enshrined in Articles 15 and 8(j), read in conjunction with CBD Article 16(5), have crystallized into the basic international customary norm that GRs and TK may be accessed and commercially exploited only *after* PIC and benefit sharing. These are universally accepted principles that the US, as a non-party to the convention, supports (sections 6.1.1.2 ff. above). In case the theory of persistent objector is valid under international law, the US may qualify as a persistent objector to the eventual emerging customary norm mandating the modification of patent law in this sense (see section 6.1.1.2.5 above).

As already observed, the relevant customary norm of protection of TK in the IP system is not specific. It has a higher degree of vagueness than the introduction of the certificate of origin which, as opposed to the specific

222 PCT.

223 Center for International Environmental Law, *Recommendations on Traditional Knowledge Relating to Biological Diversity Submitted to the United States Patent and Trademark Office* (August 2, 1999) 11–2, www.ciel.org/Publications/IdentificationofPriorArt.pdf.

224 37 CFR paragraph 1.104(a)(3) reads: "An international-type search will be made in all national applications filed on and after June 1, 1978," www.uspto.gov/web/offices/pac/mpep/documents/appxr_1_104.htm.

225 Any US application for a patent filed under 35 US Code paragraph 111.

226 37 CFR paragraph 1.9(a)(1) reads: "A national application as used in this chapter means a US application for patent which was either filed in the Office under 35 US Code 111, or which entered the national stage from an international application after compliance with 35 US Code 371."

inclusion of TK as prior art, is expressly mentioned in the Bonn Guidelines. This lack of specificity in the norm may face the obstacle that courts pose for granting a binding effect in a particular case. In our case the customary norm is not yet able to produce binding effects in the field of IP law.[227] In sum, there is no customary norm as yet that obliges States to include the disclosure requirement in the patent system.

There is another way in which such customary principles of PIC and benefit sharing are relevant to the present discussion of prior art. States are obligated to exercise due diligence in preventing GRs and TK from being misappropriated from provider countries. Compliance with this international customary norm will be achieved more easily if States implement policies and measures controlling allegedly misappropriated GRs and TK submitted for patenting. Short of implementing such measures, along with other measures outside patent law, States may be found liable for a breach of customary international law through failing to exercise due diligence.

Patenting misappropriated TK and GRs may become a *contributory factor* in the establishment of the wrongful act. For instance, given the US unwillingness to carefully apply the applicable PCT and PLT guidelines and regulations on international prior art searches in patenting of allegedly misappropriated GRs and TK, the US can easily be found responsible for breaching a customary norm of international law.

The fact that US legislation does not allow the use of foreign unpublished knowledge as prior art does not disengage the US from State responsibility. On the contrary, the USPTO, by rejecting orally transmitted but unpublished TK, is essentially weakening its ability to rebut any presumption of non-compliance with the due diligence obligation with regard to the customary PIC and ABS norms. In sum, these CBD-crystallized customary norms oblige the entire international community to employ its best efforts against misappropriation of GRs and related TK for the purpose of commercial exploitation (including patenting the same; see CBD Article 16.5 and its derivative law). Although this customary norm does not have the effect of mandating modification of the patent

[227] On the requirement of specificity of the customary norm, for instance the US Supreme Court states that the customary norm "expresses an aspiration that exceeds any binding customary rule having the specificity I require"; *Sosa v. Alvarez-Machan*, 6. 542 US (2004) 49. One of the judges, Scalia, in the same decision, issued an opinion that supports part of the US doctrine and reveals certain tendencies of the US Supreme Court: "the fact that a rule has been recognized as [customary international law], by itself, is not an adequate basis for viewing that rule as part of federal common law." D. J. Meltzer, "Customary International Law, Foreign Affairs, and Federal Common Law", (2002) 42 *Virginia Journal of International Law* 513, 519; *Sosa v. Alvarez-Machan* 542 US 692 (2004).

system, it is precisely by modifying it to comply with the precise norms of the CBD that a State can prevent any allegation of responsibility for due diligence vis-à-vis the CBD obligations.

For all these reasons as well as the reasons invoked by Bagley, it would be advisable for the US to modify its legislation to allow foreign TK to be duly searched as prior art during the examination of biotech inventions based on GR and related TK.

In an objectivist approach to international law it is inherent to the international community to provide legitimate patents that avoid international conflicts. It is vital for a new globalized patent system to be based on common rules that bind the international community as a whole and that are necessary for the proper functioning of the larger international law system. By excluding TK as prior art, the US is granting bad patents on inventions that take unfair advantage of prior art derived from TK in the use of PGRs.

The utilitarian view of the patent system should not be ignored since it arguably provides a mechanism for maximizing the value of TK, provided that an adequate system of benefit sharing is implemented. Putting products on the market may be a noble goal, but it can hardly stand as a legal justification for the appropriation of undocumented knowledge.[228] While not disregarding Nard's utilitarian approach, economic benefits should not displace the fundamental principles of patentability. Patents should only be granted for inventions that represent real innovation, and determining whether they do requires the presence of all relevant knowledge, including TK, as prior art. Consequently, striking a balance between economic interests and the rights of traditional communities is in order. By advocating such a balance, one can maximize the contributions that patent applicants can make while considering the interests of TK holders. The additional provisions with regard to TK included in the PCT are to be welcomed. If properly implemented, an effective system of TK database filing may ensure that novelty-destroying prior art is globally available.

6.2.3 International prior art searches and the role of traditional knowledge databases

The role of prior art searches will be facilitated once TK databases are in place and function in an efficient way. A patent office like the USPTO may consider TK as prior art only if it is documented in a written form that is

[228] Nard's initial response to the monopolistic nature of the patent system is "at least there is a product on the market": Nard, "In the Defense of Geographic Disparity", 224.

accessible to the public. As observed above, the international, regional, and national patent systems do possess the legislative tools to proactively ensure the defensive protection of TK through an international prior art search of undocumented TK. But it must be noted that the US did not implement clear laws mandating this type of search; on the contrary, its laws do not allow materials found in such a search to be used as novelty-destroying prior art.

On the one hand, industrialized (usually recipient) countries are reluctant to introduce anti-biopiracy mechanisms (such as prior art searches) that would oblige patent applicants to provide relevant information pertaining to the use and sources of TK or genetic material for their inventions. On the other hand, provider (usually developing) countries argue that patent offices in industrialized recipient countries should, indeed, undertake such international searches more seriously because they will decrease the number of patents based on an ambiguous appropriation of TK. If the search uncovers an invention that takes unfair advantage from its use of TK prior art, such patent applications could be rejected rather than requiring indigenous communities to challenge the patent after it is granted. The problem is that, in order to qualify as novelty- or inventive-step-destroying prior art, the description of TK in the databases must meet certain requirements. For instance, the allocation or the entitlement of prior art must be defined with a certain level of clarity and disclosure. The objective of this section is to discuss how the description of TK can qualify for novelty-destroying prior art.

6.2.3.1 How traditional knowledge should be documented to become novelty- or inventive-step-destroying prior art[229]

This matter can be approached by first raising a very basic question: if TK information concerning hoodia[230] and turmeric[231] were published in a database, would it suffice to constitute novelty-destroying prior art? The answer depends on the law and practices of the domestic patent office at stake. But, in general, the answer to this question would depend on how the TK was described. In some countries, even if published, TK could not challenge

[229] S. Biber-Klemm, "Documentation and Registration", in Cottier and Biber-Klemm (eds.), *Rights to Plant Genetic Resources and Traditinal Knowledge* 254–64.

[230] For more information about this patent case www.en.wikipedia.org/wiki/Biopiracy and www.williams.edu/go/native/hoodia.htm. Wikipedia is the popular community-driven online encyclopedia, but it is doomed to "amateurism" and should be complemented by more serious research efforts. For a more accurate legal analysis of the case study, Cullet *et al.*, "Intellectual Property Rights, Plant Genetic Resources and Traditional Knowledge", 135.

[231] *Ibid.*

some disputed patents because it was not disclosed in a way that would teach someone to come up with an invention similar to or exactly as described in the specification of the actual patent.[232]

Without delving into the complex problem of the description of relevant TK in the database,[233] it is important to note that national laws vary with respect to how information or material in the public domain should be presented or described in order to constitute novelty-defeating prior art.[234] On this point, Lord Hoffmann stated in a decision of the EPO Technical Board of Appeal: "the concept of novelty must not be given such a narrow interpretation that only what has already been described in the same terms is prejudicial to it [...]. *There are many ways of describing a substance*"[235] (italics added). To describe and make TK easily accessible would encourage the patent applicant to be very cautious about using TK and about finding the most appropriate legal relationship with the TK holders proving the underlying TK; such descriptions would economically empower the TK holders. Such a database would also ensure that the accessibility standards are met for prior art of the various national systems. As Bently and Sherman indicate, prior art in the UK can be novelty- or inventive-step-destroying only when it is obtained by a person skilled in the art without "undue burden or without the need to exercise any additional inventive effort."[236]

Although written disclosure is a *sine qua non* condition for information to become relevant prior art for the purposes of an international search,[237] the PCT allows that "[t]he date on which the written disclosure was made available to the public may have been after the filing date of the international application."[238] This proviso means that while orally transmitted TK would not qualify as prior art, it would qualify once it was collected in a database, even if these data were not registered in that form until after

[232] *Merrell Dow v. HN Norton*, (1996) 33 *Intellectual Property Law* 11.

[233] Biber-Klemm, "Documentation and Registration" in Cottier and Biber-Klemm (eds.), 254–55.

[234] Different answers have been provided by the governments and regional patent offices to the questionnaire carried out by WIPO's SCP (2001), "Information provided by members of the Standing Committee on the Law of Patents concerning the definition of prior art. Brief summary. Prepared by the International Bureau," SCP/6/INF/2. For instance, in Japan, "novelty-defeating disclosure [...] has to be enabling, i.e. it teaches those skilled in the art how to make and use the claimed invention. If novelty-defeating disclosure fails to provide such information, the disclosure will not be a novelty-defeating bar"; M. Morneault and B. F. Rademaker, "A Maze of Laws and Exceptions: Examples of Novelty Around the World", (2001) 4 *Journal of World Intellectual Property* 1, 28.

[235] Bently and Sherman, *Intellectual Property Law*. [236] *Ibid.*, 420.

[237] Rule. 31.1(b); *PCT International Search Guidelines*, PCT Gazette, ch. VI paragraph 1.2 (Special Issue No. 06/1998, Oct. 8, 1998), www.wipo.org/eng/main.htm.

[238] *Ibid.*, ch. VI paragraph 1.2.

the patent application had been filed.[239] This highlights the importance of database collection or printed publication of TK to be undertaken at the national level in DCs. There are still unresolved questions about the creation of TK databases: in the field of plant variety patents, it is hard to see how a database can destroy the patent claim if the information in the database does not describe all existing landraces. There is also the question of whether the databases should be private or public, with the risk that in the latter case a public database might even be counter-productive since it could also provide opportunities for further biopiracy cases.[240] For example, while the documentation of TK in databases may protect TK that is already known, it may be harmful for TK that is not known outside its geographical region of origin. In cases where TK is protected by a trade secret, the aforementioned documentation may destroy the confidentiality of the information by granting access to information "harvesters." A solution can be found by making accessible only the well-known TK, while restricting the access to as yet unknown TK. If the knowledge is well known, then there are probably already published documents describing it, and so it is already available as prior art. In my view, a database would only function to collect that information in central locations; it would not, however, function to protect unwritten TK.

Such hybrid open and closed access databases have already been implemented by the Tulalip Tribes in the US. The Tulalip Tribes are a federally recognized Native American tribe located in the ecologically rich Puget Sound area of Washington State. While compiling a database of their TK, they distinguished between "Type A knowledge," reserved exclusively for members of the tribe, and "Type B knowledge" available to the public at large. A patent examiner will be given just enough information concerning "Type A knowledge" to make a prior art search. Should bioprospectors need more information than they can get from the Type B database, they can get more information only with the consent of the tribe. Furthermore, when users log on to the database, their activities within the system are monitored to ensure against "harvesting" activities. This multi-layered database system safeguards the sovereign rights of the tribe with respect to the divulgence of their knowledge and secrets.[241] Prior art, by definition,

[239] M. Leistner, "Analysis of Different Areas of Indigenous Resources" in S. Von Lewinski (ed.), *Indigenous Heritage and Intellectual Property* (Kluwer, The Hague, 2004) 60.

[240] WIPO, *Intergovernmental Committee on Intellectual Property and Genetic Resources, Traditional Knowledge and Folklore, Third Session*, Draft report prepared by the Secretariat, WIPO/GRTKF/IC/3/17 (June 13–21, 2002)

[241] *Intergovernmental Committee on Intellectual Property and Genetic Resources, Traditional Knowledge and Folklore* WIPO/GRTKF/IC/8/7, 3 (June 6–10, 2005), www.wipo.int/edocs/mdocs/tk/en/wipo_grtkf_ic_8/wipo_grtkf_ic_8_7.doc.

must be accessible to the public. The EPO considers that in order to be public the information must be available and comprehensible to any single member of the public with no obligation to maintain secrecy,[242] and the USPTO has a similar standard.[243] For the purposes of an international search, the Tulalip Type A database contains information that is, in effect, not available to the public and thus not available as prior art.

The question concerning the access to this documented knowledge remains troubling. On the one hand, databases need to be accessible to examiners so that they do not grant patents in error; on the other hand, their access should be restricted to prevent further abuses. Part of the rationale for giving printed publications a privileged status is that they serve as a reminder of their registration function, but that is only true if they are publicly accessible. The question here arises: if these publications are accessible to both the bioprospector and the patent examiner, will the examiner not be able to reject attempted misappropriations made possible by the accessibility?

On the other hand, industry can see these TK databases as valuable sources of knowledge that can bypass painstaking and time-consuming research. Dutfield cautions against only creating TK databases, affirming "that without other reforms to the patent system databases would be useful only for the most egregious cases of TK misappropriation, and not even all of these."[244]

6.2.3.2 The Patent Cooperation Treaty perspective on traditional knowledge databases

Since there is still no universal international definition of novelty- and inventive-step-destroying prior art, it is now appropriate to look at what has been done at the international level to provide guidance on the matter. The PCT has procedural rules that are relevant to the status of TK as novelty-destroying prior art. Article 27(5) of the PCT expressly states

> Nothing in this Treaty and the Regulations is intended to be construed as prescribing anything that would limit the freedom of each Contracting State to prescribe substantive conditions of patentability as it desires. In particular, any provision in this Treaty and the Regulations concerning the definition of prior art is exclusively for the purposes of international procedure and, consequently, any

[242] *Ibid.*

[243] USPTO Manual of Patent Examination Procedure paragraph 2128.01(III) (Internal Documents Intended to be Confidential are not Printed Publications), www.tess2.uspto.gov/tmdb/tmep.

[244] G. Dutfield, *Protecting Traditional Knowledge and Folklore, A Review of Progress in Diplomacy and Sustainable Development* (UNCTAD, ICTSD, Geneva, 2003) 37.

Contracting State is free to apply, when determining the patentability of an invention [...] the criteria of its national law in respect of prior art.

Thus, while the PCT provides procedural guidelines for member States, it leaves the substantive conditions for patentability to national law. Despite the mere procedural role of the PCT, it does provide significant direction on international examination and its role in the treatment of TK as prior art.

Article 15 of the PCT stipulates that every patent application shall be subject to an international prior art search. Prior art for the purposes of Article 15.2 is defined in Rule 33(1)(a) of the Regulations under the PCT as consisting of:

everything which has been made available to the public anywhere in the world by means of written disclosure (including drawings and other illustrations) and which is capable of being of assistance in determining that the claimed invention is or is not new and that it does or does not involve an inventive step [i.e., that it is or is not obvious], provided that the making available to the public occurred prior to the international filing date. Unfortunately, the application of this provision is not always the same in all national patent systems. A question that needs to be explored (and that falls outside the scope of the present study) is how much attention and effort is made by USPTO (and EPO) examiners to find relevant TK domestic and foreign technical literature. Indeed, at the patent application stage, most patent offices may not survey foreign literature (and much less give attention to orally transmitted TK) in which descriptions of TK appear if an invention is already part of the prior art and therefore not novel.[245] Given the problem of insufficient prior art searches of foreign literature, it comes as no surprise that prior art searches of orally transmitted TK are equally if not more neglected even though the PCT has provisions on how to deal with orally transmitted TK.

PCT Rule 33(1)(b) considers oral TK in the following way:

when any written disclosure, refers to an oral disclosure, use, exhibition or other means whereby the contents of the written disclosure were made available to the public, and such making available to the public occurred on a date prior to the international filing date, the international search report shall separately mention that fact and the date on which it occurred if the making available to the public of the written disclosure occurred on a date which is the same as, or later than, the international filing date [...] [and] the date on which the written disclosure was made available to the public may have been after the filing date of the international application.[246]

[245] *Intergovernmental Committee on Intellectual Property and Genetic Resources, Traditional Knowledge and Folklore, Second Session,* Progress Report on the Status of Traditional Knowledge as Prior Art WIPO/GRTKF/IC/2/6 (December 10–14, 2001).

[246] *PCT International Search Guidelines* chapter VI, paragraph 1.2.

This provision refers to public presentations etc. that were written down, not to oral communications within a closed group. Oral disclosure is thus not sufficient for prior art to destroy novelty according to the PCT, nor can it establish prior art in international searches. However, where a written disclosure attests to an oral disclosure preceding the filing date of a challenged patent, the written disclosure may serve to corroborate the prior art in a manner sufficient to destroy the novelty claim.[247]

As earlier observed, judicial and practical problems surface where public domain knowledge is not easily accessible through the normal search procedures that patent offices use. The judicial problem consists of deciding how novelty and prior art must be adapted to accommodate TK holders. The practical problem consists of creating an efficient means of identifying pre-existing knowledge given that vast amounts of TK are undocumented and given that existing databases are neither widely available nor expansive enough in their cataloguing of TK. Additionally, database searches are not sufficiently integrated into the procedures of patent offices.[248]

Part of the practical problem is that existing database systems are not suitable for efficient prior art searches. This is because the liaison between IP offices and TK documentation initiatives has not as yet become functional, either because IP offices do not know about the databases, or because the databases are not very easy to be accessed or be searched.[249]

WIPO has implemented measures improving the defensive protection of TK. Proposals have been made to the WIPO IGC on IPGRTKF that leave the following options open:

(1) The Committee could compile an inventory of existing periodicals, databases and other information resources which document disclosed GRs, with a view to discussing a possible recommendation that certain periodicals, databases and information resources may be considered by International Search Authorities for integration into the minimum documentation list under the PCT;[250]
(2) The Online Portal of Registries and Databases which was established by the Committee at its third session, could be extended to include

[247] Leistner, "Analysis of Different Areas of Indigenous Resources", 69, 70.
[248] *Ibid.*, 60. *Survey on Existing Forms of Intellectual Property Protection for Traditional Knowledge – Preliminary Analysis and Conclusions*, WIPO/GRTKF/IC/2/9 (December 3, 2001).
[249] *Ibid.*
[250] This has already been successfully accomplished for periodicals concerning disclosed TK, as provided in *Inventory of Existing Online Databases Containing Traditional Knowledge Documentation Data* WIPO/GRTKF/IC/3/6 paragraphs 41–45 (June 13–21, 2002) WIPO/GRTKF/IC/2/6.

existing databases and information systems for access to information on disclosed GRs (additional financial resources would be required to implement this option);[251]

(3) The Committee could discuss a possible development of recommendations or guidelines that existing search and examination procedures for patent applications take into account disclosed GRs as well as a recommendation that patent granting authorities also make national applications which involve GRs subject to "international-type" searches as described in the PCT Rules.[252]

6.2.4 *Conclusion*

This section has explored the ways in which TK can be defended through establishing it as prior art. Opposing interests (nourished by cultural biases in the patent system) are at stake in this discussion: on the one side, most TK stakeholders do not want their TK to be patented; on the other side, patentees want inventions based on GR and TK to be readily patentable. These two groups are often in different countries, so the tension can be defused only at the international level, within WIPO or WTO.

Patent offices of recipient countries such as the USPTO are still trying to implement existing international prior art searches.[253] The path to full consideration of TK as prior art presents obstacles regarding the way the TK is described and how easily it can be accessed. Without efficient database systems, patent examiners cannot be aware of all the TK that could possibly be at the heart of an invention. The turmeric and neem tree cases[254] provide striking examples of the problems likely to be perpetuated unless these database systems are streamlined and constantly improved upon. As the definition of TK continues to be clarified, these databases must also develop as well. Prior art is a constantly evolving legal concept that will increasingly serve to check novelty claims as more and more information moves into the public domain.

My legal analysis lends credence to the findings of Graham who states that under the current legal situation in the US and EU, TK databases only prevent the granting of patents that blatantly rely upon the prior art described in the TK database and, thus, constitute clear cases of patent

[251] *Ibid.*, paragraph 15. [252] *Ibid*, paragraph 52.

[253] Center for International Environmental Law, *Recommendations on Traditional Knowledge Relating to Biological Diversity.*

[254] Cullet *et al.*, "Intellectual Property Rights, Plant Genetic Resources and Traditional Knowledge", 135.

applications lacking novelty or non-obviousness. There are many doubtful cases in which the connection to TK as prior art is less clear even when the TK is included in databases; thus, in such instances, the TK is not likely to be considered prior art and it will not affect the patent application.[255]

In conclusion, there is the need to elevate the bar of the requirement of novelty to absolute novelty. At the same time, the crucial element is the establishing of international coordination for the ongoing creation of databases cataloguing TK. Transparency and public access are key policy issues that must be resolved in order for the whole patent system to be functional. Unrestricted access can create abuses of misappropriation that are difficult to correct. At the same time, the utilitarian approach of Nard should facilitate contacts between the biotech industries and the TK holders' communities to bring about further progress and benefit sharing of profits arising from commercializing such knowledge. The creation of the clearing house mechanism (CHM) for the administration of the IP issues related to GRs and TK can be an extremely useful tool to facilitate innovation and benefit sharing based on TK when (i) TK becomes legally available as prior art to be taken into account by patent offices, and (ii) guidelines are followed on CHMs' management and their accessibility by patent offices for the search of prior art to make TK effectively accessible to patent examiners.

6.3 *Ordre public* and morality as exception to patentability

This section examines the structures in Europe and the United States for addressing fundamental patent policy questions, particularly (as an exemplary issue) since they relate to inventions and their consequences for the protection of biodiversity and the environment in general. The relevant legal doctrines have been weakened to the point of providing little room for policy analysis. Next we will examine the administrative and political vehicles for examining patent policy questions. There is more potential here for serious examination of patent policy, but for a variety of reasons, such examination often does not take place. In many countries the judicial branch looks to the political branch, the political to the judicial, and in the end, both sit with their arms crossed.

The concept of *ordre public* has already been analyzed in order to justify the compatibility of the certificate of origin with the TRIPS Agreement as a possible implementation of the CBD concept of PIC in IP law. In

[255] Dutfield, *Protecting Traditional Knowledge*, 37.

addition to the disclosure of origin/source (section 6.1 above) and the TK as novelty-destroying prior art (section 6.2), the CBD concepts become relevant to the patentability of biotechnology as mandated by Article 27.1 of TRIPS when a biotech-patent can be challenged by reference to the general concepts of *ordre public* and morality. So my analysis here enlarges its scope to consider how patent law can include environmental concerns and also TK within these concepts.

Through the adoption of Article 27.2 of TRIPS, these concepts have come to encompass the protection of "human, animal or plant life or health" and the prevention of "serious prejudice to the environment." The interpretation of these concepts offers a potential reconciliation between TRIPS and the CBD. But here the "classical" and the "radical" approaches to biotech-patentability must be confronted.[256] The classical approach views the patent system as autonomous and neutral on issues such as ethics and the environment which are not related to traditional patentability requirements. The radical approach argues that national patent systems must carefully use the *ordre public* and morality exceptions to patentability in order to take into account the numerous and complex ecological and ethical implications of granting these patents.[257] Under the radical view, patent law should not be autonomous, but should be integrated and reconciled with international environmental law and with other legal and ethical considerations.

For Europe, this will involve the patentability exceptions of *ordre public* and morality of EPC Article 53(a) (largely inspired by Article 27.2 of TRIPS). The EPO case law currently interprets narrowly concepts of "*ordre public* and morality" of Article 53(a) of the EPC. This narrowing has largely been the triumph of what I have called the "classical" over the "radical" approach to patentability.

The analysis of the US legal system is more prospective than that of the European system, as there are currently few ways to challenge the patentability of biotechnology. However, as patent law becomes increasingly global, the US government and patent applicants will increasingly need to deal with *ordre public* and morality common in other parts of the world. Harmonizing national patent law systems will likely be a major issue on

[256] The labels of "classical" and "radical" are inspired by what Ricolfi respectively calls "mainstream or standard legal literature" and a "more radical approach": Ricolfi, "Biotechnology, Patents and Epistemic Approaches", 77.

[257] R. Pavoni, "Brevettabilità genetica e protezione della biodiversità: la giurisprudenza dell'Ufficio europeo dei brevetti" (2000) 83 *Rivista di diretto internazionale* 429–80. His study relies on the opinions of authors such as P. Drahos, R. Steinbrecher, D. Beyleveld and R. Brownsword, D. Wirth, V. Walker, B. Bergmans, G. Winter, A. Wells.

the international IP stage in the coming decades. I then move beyond the US and EU to analyze how Article 27.2 can be used by a WTO Member State to make exceptions to patentability for inventions that it finds contrary to the principles of the CBD, through a "mutually supportive" interpretation of the TRIPS Agreement and the CBD and other MEAs. I will show how the EPO interpretation of EPC Article 53(a) offers various policy options regarding the implementation of the similar provision of TRIPS Article 27.2.

In any discussion of the patentability of biotechnology, one must acknowledge that patent law cannot control whether an invention is practiced or commercialized, but it does have a significant effect on the economics of the field because patents undeniably encourage the flow of money for R&D.

In the interest of such harmonization, the classical and radical approaches are discussed, after which a possible reconciliation of the two approaches is presented: a board within national patent offices that addresses the concerns of the radical approach while preserving the efficiencies of the classical approach.

This analysis concludes with the proposal to create a mechanism with the competence and agility to address questions of fundamental patent policy as they arise: boards responsible for public policy operating within the national and international patent offices. These boards would not create law or decide cases. Rather, they would bridge the policy gap left open in the current system, serving as a resource and a starting point for collecting and creating the information needed for patent policy questions to be addressed. The aim would be a patent system that encourages innovation in a way that best serves the public interest, in the broadest sense.

6.3.1 *Ordre public* and European intellectual property legal order

This section examines the role that public policy concerns can take in the patent system and the evolution of the interpretation of EPO case law to the current narrow interpretation of the concepts of "*ordre public* and morality" in Article 53(a) of the EPC. This narrowing has largely been the triumph of what may be called the "classical" over the "radical" approach to patentability.

Ordre public and morality have, though to different degrees, a limited role in the exercise of IPRs. Vivant stated that "*l'ordre public se mêle manifestement à la propriété intellectuelle d'une manière non univoque*" and concludes that its role is "*bien ambigu*" when it constitutes a patentability exception, to the point that the *"recours à la notion ne paraît plus, tout simplement, avoir d'utilité."*[258]

These doctrines are of general applicability, and even in very morality-free copyright law, *ordre public* and morality have traditionally played a certain role. Any original creation is copyrighted, regardless of its compliance with standards of ethics or morals. However, the extent of the exercise of this right has a limit. For instance, pornographic works are protected and, in case of infringement, reparation of damages can be granted. However, when the pornographic work involves children, children sex-slavery, pedophiliac acts etc., the work is copyrighted but cannot claim reparation in damages.[259]

In distinction to copyright law, where rights begin at the moment of creation, patents (and trademarks) involve government examination, and thus *ordre public* and morality can provide an earlier opportunity to exercise scrutiny, and there are specific provisions allowing governments to do so.[260]

The concept of *ordre public* can be associated with the "general interest" or "public policy." European case law outside the field of IP has found that the scope of *ordre public* includes the fundamental values of the international community.[261] A number of questions can arise or can be raised when this concept is associated with patent law: should the EPO integrate these values in the concept of *ordre public* of the EPC? Should it take into account the evolving interpretation of *ordre public* in accordance with general European law? What interest is pre-eminent in the current EPO interpretation of Article 53(a)? These questions shall be discussed in the following sections.

[258] M. Vivant, 'Propriété intellectuelle et ordre public', *Jean Foyer – auteur et législateur* Leges tulit, jura docuit – *écrits en hommage à Jean Foyer* (Presses Universitaires de France, 1997) 310.

[259] *Ibid.* [260] *Ibid.*

[261] 61977J0030 Judgment of the ECJ in *Régina v. Pierre Bouchereau* (October 27, 1977) commented by M. Doppelhammer, (1999) 1 *Revue du marché unique européen* 238. For an author in favor of this view, P. Benvenuti, *Comunità statale comunità internazionale e ordine pubblico internazionale* (Giuffrè, Milano, 1977); G. Barile, "Ordine pubblico" (dir. int. privato) in *Enciclopedia del diritto, Vol. XXX* (Milano, Giuffrè, 1980) 1106, 1110–114. The classical doctrine of *ordre public* is anchored to its national notion: Morelli G., *Elementi di diritto internazionale privato italiano* (Napoli, 1986) 83–6. Opposing this vision is J.-C. Galloux, "Ethique et brevet ou le syndrome bioéthique" (1993) *Recueil Dalloz Sirey* 83 and 88–9 who states: "il n'existe pas, au sens technique d'ordre public commun à tous les pays signataires de la CBE auquel un examinateur de l'OEB puisse se référer afin de rendre effectives les prescriptions de l'art. 53 a. Cette disposition demeure ainsi largement inapplicable."

6.3.1.1 The classical approach of the European Patent Convention

Article 53(a) of the EPC prevents the patenting of inventions "contrary to *ordre public* or morality." This Article has historically been applied, using our distinction, in a "classical" rather than a "radical" way. In the classical approach, patent examination focuses on the type of monopolistic right to exclude others (*ius excludendi*). The patent has nothing directly to say about the permissibility of the research that lead to the patented invention or the commercialization of the invention afterwards.[262] The permissibility of commercial exploitation of an invention is left to other parts of the government and other bodies of law with appropriate authority and adequate instruments to assess the risk, in this case, to harm the environment.[263] This procedure is completely separated from the patent examination.

The Guidelines of the EPO indicate that inventions can be excluded by invoking Article 53(a) only in "rare and extreme cases" that, broadly speaking, boil down to two situations: either (i) the invention has no foreseeable lawful use whatsoever and is likely to induce riot or public disorder (e.g. a letter bomb); or (ii) there has been a previous law or decision of a competent organ that completely bans the exploitation of the invention.[264] It would seem that under such a classical interpretation these two applications of *ordre public* and morality offer very little protection against patents that may violate morals or harm the environment.

6.3.1.2 The Biotechnology Directive and radical interpretations of ordre public *and morality*

Opposed to this classical interpretation stands the more radical one that seeks to integrate into the patent system an analysis of the potential health and environmental risks of inventions, e.g. the growth of plant species (superweeds) resistant to human control or the existence of plant or animal recombinant micro-organisms that can cause adverse effects on human health or biodiversity. These preoccupations are most forcefully expressed by a strident minority of patent scholars, and are slowly becoming more common among consumers, international NGOs, and local farmers.

[262] U. Schatz, "Patentability of Genetic Engineering Inventions" (1999) 29 *International Review of Industrial Property Copyright Law* 2–6 (1999) and the *Enlarged Board of Appeal "Plant Genetic Systems"* 545 OJ EPO 1995 (February 21, 1995).

[263] C. Noiville, *Ressources génétiques et droit. Essai sur les régimes juridiques des ressources génétiques marines* (Pédone, Paris, 1997) 437; M. Ricolfi, "La brevettazione delle invenzioni" *Rivista di diritto industriale* (Giuffre Editore, Milan, 2003). C. M. Mazzoni (ed.), *Bioethics Markets and Morals: The Case of Biotechnological Patents, A Legal Framework for Bioethics* (Kluwer, Dordrecht, 1998) 131.

[264] EPO Guidelines C-IV, 3.1.

The Biotechnology Directive generally supports the traditional limited application of *ordre public*. Article 6 of the Directive provides that the exploitation of an invention shall not be deemed contrary to *ordre public* and morality "merely because it is prohibited by law or regulation in some or all of the Contracting States."[265] However, the Directive has also been applied in a more radical way. Advocate General Jacobs, in *Kingdom of the Netherlands v. European Parliament and Council of the EU*, moved toward the radical approach when he affirmed that the granting of a patent for an invention which constitutes "a sufficiently serious threat to the environment" would be in violation of the concept of *ordre public* in Article 6.1 of the European Directive on Biotechnological Inventions.[266] There is a significant and unanswered question as to what constitutes a "sufficiently serious threat." Advocate General Jacobs also stated that moral safeguards within the framework of the patent system go beyond the mere application of traditional criteria of patentability. He reasoned that the system embodies certain generally unquestioned social and moral values, thereby acting as a "social and moral filter," allowing certain things into mainstream commercial life while blocking others.[267] These statements are hard to reconcile with the classical approach.

In light of various existing MEAs, Article 53(a) of the EPC inspired Article 27.2 of the TRIPS Agreement which states: "Members *may* exclude from patentability inventions, the prevention within their territory of the commercial exploitation of which is necessary to protect *ordre public* or morality, including to protect human, animal or plant life or health or to avoid *serious prejudice to the environment*, provided that such exclusion is not made merely because the exploitation is prohibited by their law" (italics added).[268]

Thus, there is a legal principle, taking form in Europe and available to other WTO countries, which may provide a basis for using *ordre public* and morality to bring environmental concerns into the patent examination process.

[265] The provision as it was amended in 2000 to be in compliance with Article 6 of the Biotechnology Directive.

[266] *Opinion of Advocate General Jacobs*, paragraph 109. S. J. R. Bostyn, "One Patent a Day Keeps the Doctor Away? Patenting Human Genetic Information and Health Care" (2000) 7 *European Journal of Health Law* 242; Ricolfi, "Biotechnology, Patents and Epistemic Approaches"; D. Beyleveld and R. Brownsword, "Is Patent Law Part of the EC Legal Order? A Critical Commentary on the Interpretation of Art. 6(1) of Directive 98/44/EC in Case C-377/98" (2002) *Intellectual Property Quarterly* 97–98.

[267] *Opinion of Advocate General Jacobs*, paragraph 227.

[268] *Resource Book on TRIPS and Development*, 375–83.

6.3.1.3 European Patent Office decisions: moving away from a purely classical view

The interplay between biotechnological patents and environmental law has prompted the EPO to weigh in on the subject of morality and *ordre public*. I will take into account the most important opposition proceedings, such as *Onco-Mouse* and *Plant Genetic Systems*, where the EPO has given *ordre public* and morality a progressively narrower place in patent law.

6.3.1.3.1 Cost-benefit analysis in the *Onco-Mouse* case

In 1985, Harvard University applied to the EPO for a patent on a type of mouse, dubbed the "Onco-Mouse", which was genetically engineered to have enhanced susceptibility to cancer, for use in cancer research. The application was eventually allowed.[269] An opposition proceeding was then started by certain environmental and other interest groups. The principal question was "whether the potential benefit to cancer research justifies the use of genetically engineered animals with an increased sensitivity to carcinogens,"[270] based on Article 6.2(d) of the Biotech-Directive that denies patents for "processes for modifying the genetic identity of animals which are likely to cause them suffering without any substantial medical benefit to man or animal, and also animals resulting from such processes." Article 53(a), the *ordre public* and morality exception of the EPC, was also invoked on the grounds that the release of genetically manipulated animals into the environment might entail irreversible adverse effects through the spread of the gene giving increased cancer susceptibility and thus cause more animal suffering.[271]

The Board of Appeal gave guidelines for weighing these moral considerations when it remitted the case. The Board of Appeal made clear that genetically engineered life forms are not a "*domaine réservé*" (a domain completely reserved from patenting).[272] In the Board's view, patentability depends on the careful "weighing" of animal suffering and possible risks to the environment on the one hand and the usefulness of the invention on the other.[273] After carrying out the balancing exercise (a cost-benefit analysis), the Board concluded that the risk of animal suffering and oncogenes escaping into wild organisms compared with the reduction in animal testing that Onco-Mouse provided[274] did not justify an exception to patentability.

[269] *Onco-Mouse/Harvard II*, T 19/90. M. Bagley, "Patent First, Ask Questions Later: Morality and Biotechnology in Patent Law" (2003) 45 *William and Mary Law Review* 469, 517–30.

[270] *Onco-Mouse/Harvard II*, T 19/90, 476, paragraph 5. [271] *Ibid.*

[272] Noiville, *Ressources génétiques*, 403.

[273] *Onco-Mouse/Harvard III* (OJ EPO 1992, para 4(iv)).

[274] Due to its increased susceptibility to cancer, fewer Oncomice than ordinary mice need to be used in a given cancer experiment.

The cost-benefit analysis introduced in the *Onco-Mouse* case has many potential applications in international environmental law.[275] However, commentators are split as to the desirability of the test. The classical approach has criticized the integration of moral considerations into the patent system on two grounds: (i) because of the technical character of patent law, patent examiners are not competent to examine environmental and ethical concerns;[276] and (ii) because the concepts of *ordre public* and morality are vague and variable, and thus hard to apply with predictability. A more moderate classical view represented by Noiville welcomes the Board of Appeal's decision because it introduces a reasonable consideration of environmental and moral dimensions of patents on biological subject-matter.[277] The moderate radical view represented by Beyveld and Brownsword sees the *Onco-Mouse* case as evidence that the EPO has neglected its role as a "social and moral filter" responsible to keep inventions against public policy from gaining patent protection.[278]

6.3.1.3.2 The *Plant Genetic System* case and the problem of scientific uncertainty

In the *Onco-Mouse* proceedings, ethical and environmental issues – as a part of the morality and *ordre public* exception – were not analyzed very extensively. The following year, in the *Relaxin* case, the Opposition Division retreated somewhat to the classical position when it observed that "the EPO is not the right institution to decide on fundamental ethical questions."[279] This classical approach to Article 53(a) was reaffirmed by the Board of Appeal when it stated that it "repeatedly found that such exceptions are to be narrowly construed."[280]

[275] P. A. Nollkaemper, "What You Risk Reveals What You Value and Other Dilemmas Encountered in the Legal Assaults on Risks, The Precautionary Principle and International Law" in D. Freestone and E. Hey (eds.), *The Precautionary Principle and International Law: The Challenge of Implementation* (Kluwer, The Hague, 1996) 73–94.

[276] A. Gallochat, "Le brevet et l'éthique ou le mélange des genres" (Dossiers brevets, 1993) 7; S. Mayer and D. Alexander, "Mice, Morals and the Environment" (1992) 514(II) *Propriété Industrielle – Bulletin Documentaire* 10; P. Brandon and J. Dunnet, "E.P.O. Case Law; Ignore It at Your Peril! A Review of the Recent Board of Appeal Decisions" 52 (1993) *Patent World* 39.

[277] Noiville, *Ressources génétiques*, 405.

[278] D. Beyleveld and R. Brownsword, *Mice, Morality and Patents* (Common Law Institute of Intellectual Property, London, 1993) 33–46; P. Drahos, "Biotechnology Patents, Markets and Morality" (1999) 21(9) *European Intellectual Property Review* 441–9.

[279] *Hormone Relaxin*, 0J EPO 6/1995, 403.

[280] *Hormone Relaxin*, 0J EPO 6/1995, 398. The patentability of medical treatments has also led courts into discussions of patents and ethics: Eli Lilly 7 Company's Application (UK, Patents Appeal Tribunal, 1975, RPC 438); *Wellcome Foundation v. Plantex Ltd and*

However, the concepts of *ordre public* and morality have continued to play a role in judicial analysis of patentability. The relationship between the concept of morality and *ordre public* was analyzed in two opinions by the Board of Appeal on patent applications from Plant Genetic Systems (PGS). PGS and Biogen in 1990 held a patent for an herbicide-resistant plant. Greenpeace challenged the patent before the EPO on ethical and practical grounds, warning about the negative impact on biodiversity protection that can reveal itself at different levels, including transfer of the engineered genes into wild plants and adverse effects on agriculture.[281] Greenpeace hoped the court would revoke the patent given the potential ecological risks.

The first objection raised by Greenpeace – that granting a patent for higher life forms was intrinsically immoral – was immediately rejected by the Opposition Division on the ground that this argument was merely "philosophical" in nature and lacking any objective criteria to evaluate it. It also stated that:

> The development of [genetic engineering] technology allows a better understanding and control of the natural phenomenon linked to plants. However, in the Board`s view, this does not render activities in this technical field intrinsically wrong. Indeed, in the Board's judgment, plant biotechnology per se cannot be regarded as being more contrary to morality than selective breeding because both traditional breeders and molecular biologists are guided by the same motivation.[282]

Although the Opposition Division seemed willing to make an "objective attempt" to assess the potential risks of this invention,[283] it did not see any possibility of evaluating the "invention on the basis of what amounts to a risk/benefits assessment"[284] due to the lack of scientific certainty concerning the risk of ecological harm due to transgenic plants.[285] As the court wrote:

> there is still no agreement concerning the extent of these risks and the Opponent has indeed conceded that the risks are impossible to determine with certainty. Scientific expertise thus does not provide a sufficient basis to conclude that the risks [...] preclude any application of this technology.[286]

Pharmap Lantex Ltd (Israel, Supreme Court, 1974, R P C 514); *Commissioner of Patents v. Wellcome Foundation* (New Zealand, Supreme Court, 1979, 2 NZLR); *Anaesthetic Supplies v. Rescare* (1994, 28 IPR 383 Australia, Federal Court).

281 R. Steinbrecher, 273. Decision T356/93. *Plant cells / Plant Genetic Systems* OJ EPO 1995 545.

282 *Ibid.*, paragraphs 3.4 and 3.10. 283 *Ibid.*, paragraph 3.12.

284 *Ibid.*, paragraph 3.16. 285 *Ibid.*, paragraph 3.12 and 3.13.

286 *Ibid.*, paragraph 3.13.

The court, thus, did not even deem it necessary to undertake a cost-benefit analysis as a balancing exercise where the alleged costs or harms had not been demonstrated. Where scientific data are controversial or uncertain, the EPO default position is that the application is not unpatentable:

> The Board observes that such a "balancing exercise" is not the only way of assessing patentability with regard to Article 53(a) EPC but just one possible way, perhaps useful in situations in which an actual danger and/or a disadvantage (e.g. suffering of animals in the case of [the Onco-Mouse]) exists.[287]

From this quote, one can infer that the Opposition Division intends to perform the cost-benefit analysis only when dealing with a risk that is known to exist, such as animal suffering. Perhaps the test will only be applied when there is a risk of animal suffering; the drafts of the EU Biotechnology Directive mention balancing interests in Article 6.2 only in relation to the suffering or physical handicaps caused to animals.[288]

Although the Board of Appeal did not seriously analyze the environmental impact of the PGS invention, important *obiter dicta* analyzed the scope of *ordre public* and morality as stipulated in Article 53(a) of the EPC:

> It is generally accepted that the concept of *ordre public* covers the protection of public security and the physical integrity of individuals as part of society. This concept encompasses also the protection of the environment [...].[289]

The concept of morality is related to the belief that some behavior is right and acceptable whereas other behavior is wrong, this belief being founded on the totality of the accepted norms which are deeply rooted in a particular culture. For the purposes of the EPC, the culture in question is the culture inherent in European society and civilization. Accordingly under Article 53(a) EPC inventions the exploitation of which is not in conformity with conventionally accepted standards of ethics pertaining to this culture are to be excluded from patentability as being contrary to morality.[290]

It is clear that protection of the environment can constitute an exception to patentability.[291] But the EPO deals with environmental matters in a very ambiguous way: on the one hand the EPO states that it is not the right place to fully consider the environmental impact of inventions. On the other hand, it sets out the level of evidence of an environmental threat that it requires in order to revoke a patent:

287 *Plant Genetic Systems*, paragraph 18.8.
288 R. Pavoni, "Brevettabilità Genetica", 451.
289 *Plant Genetic Systems II*, paragraph 5.
290 *Ibid.*, paragraph 6.
291 Article 27.2 of the TRIPS; Article 6.2 of the Biotech-Directive; Case C-377/98, *Kingdom of the Netherlands v. European Parliament*, paragraphs 35–39, 61–62, 76.

In the Board's judgment, the revocation of a European Patent in Article 53(a) on the grounds that the exploitation of the invention for which the patent has been granted would seriously prejudice the environment presupposes that the *threat to the environment be sufficiently substantiated* at the time the decision to revoke the patent is taken by the EPO [...].[292] (italics added)

This quote suggests that the EPO analyzes the merits of environmental issues only when the risk to the environment is sufficiently certain. The "mere possibility" of environmental damage represents too low a degree of risk to warrant such an endeavor. In two ways, the EPO showed that the exception would be applied in a very narrow manner: (i) it required Greenpeace to present scientific evidence regardless of the inherent uncertainty of this field of research; and (ii) it sets a high standard for the level of proof necessary to reject a patented invention that prejudices the environment.

I suggest a moderate classical approach that follows Llewelin who proposed that the EPO, when confronted with these matters, should have established beforehand the degree of certainty of the scientific evidences expected by Greenpeace in accordance with the specific paradigms of the field in which the negative impact of genetically modified organisms (GMOs) on biodiversity can be ascertained.[293]

6.3.1.3.3 Concealed cost-benefit analysis

Pavoni asserts that the Board of Appeal in *PGS II* performed a "concealed balancing exercise."[294] The Enlarged Board of Appeal referred to "possible risks," "negative potential effects," and "actual disadvantages" for biodiversity and for the ecosystem. Although the Opponent submitted a series of documents containing an analysis of the possible risks, the Opposition Division stated that "he has not been able to prove, or at least to render highly plausible, his allegations;"[295] he has not been able "to prove the *extent of the risks*"[296] (italics added). Despite these phrases that clearly admit the existence of risks, the Opposition concluded that Greenpeace "has not been able to prove the *existence* of risks and indeed the true *extent of the risks* is impossible to determine with any degree of accuracy."[297] Although admitting that harmful effects "*may occur to some extent*," the documentary evidence "is not sufficient to substantiate the existence of a threat to the environment such as to represent a bar to patentability under Article 53(a) EPC."[298]

292 *Plant Genetic Systems II*, paragraph 18.5.
293 M. Llewelyn, "Article 53 Revisited", (1995) 17 (10) *European Intellectual Property Review* 506, 511.
294 Pavoni, "Brevettabilità Genetica". 295 *Plant Genetic Systems I*, paragraph 3.12.
296 *Ibid.*, paragraph 3.13. 297 *Ibid.*, paragraph 3.16. (italics added)
298 *Ibid.*, paragraph 18.6.

Therefore, the EPO seems to be weighing the sufficiency of the evidence on the one hand. On the other hand, the EPO seems to weigh the legitimacy of the technology. The Opposition Division held that any objection to the patentability of a biotechnological invention must also be considered "*in the light* of the fact that several member States of the EPC subsidize plant molecular biology research."[299] Hence, the EPO could not "reasonably declare immoral the results of research financially supported by some of its member States."[300] This financial support gave legitimacy to this type of patentable technology and weighed strongly against the uncertain evidence of the environmental risk of GMOs. The official recognition of GMO research in EU member States, despite its political rather than legal nature, seems to have led to the grant of the patent.

The approach and methodology of the EPO decisions remain anchored in the classical approach to patent law. However, the EPO seems to have moved towards considering environmental protection as a legitimate application of *ordre public.*

6.3.1.4 International instruments, subsequent practice, and their effect on ordre public *and morality*

The EPC exceptions of *ordre public* and morality should be viewed in light of subsequently adopted MEAs that also relate to the exercise of IPRs. The EPC, as an international treaty, must be interpreted in accordance with the rules of interpretation of Article 31.3 of the VCLT,[301] i.e. in light of subsequent practice, especially when analyzing the evolving meaning of its terms. As the ICJ stated: "Interpretation cannot remain unaffected by the subsequent development of law [...]. Moreover, an international instrument has to be interpreted and applied within the framework of the entire legal system prevailing at the time of the interpretation."[302]

[299] *Plant Genetic Systems I*, paragraph 3.8. [300] *Ibid.*

[301] Article 31.3 of the VCLT, provides that in interpreting a treaty, there shall be taken into account, together with the context: "(a) any subsequent agreement between the parties regarding the interpretation of the treaty or the application of its provisions; (b) any subsequent practice in the application of the treaty which establishes the agreement of the parties regarding its interpretation; (c) any relevant rules of international law applicable in the relations between the parties."

[302] The Appellate Body of the WTO (cited in *US–Import Prohibition of Certain Shrimp and Shrimp Products*, WT/DS58/AB/R, paragraph 143 (November 6, 1998); *Aegean Sea Continental Shelf Case* (Greece v. Turkey) December 19, 1978, *International Court of Justice Reports* 3; and E. Jimenez de Arechaga, "International Law in the Past Third of a Century" (1978) 159 *Recueil des cours de l'académie de droit international*, 1, 49.

It should be noted that this interpretative practice has already been implemented in *Novartis*,[303] where the EPO took into account the relevant provisions of international IP conventions ratified by all EPC Contracting States such as the TRIPS Agreement, EU Directives, the Agreement on Community Patent, and the UPOV Convention.[304] The EPO has taken subsequent practice into account in interpreting Article 53(a) of the EPC.[305] When a conflict of norms arises, the EPO has treated the commercial or environmental character of each instrument or norm as irrelevant. This treatment implies that all the applicable rules are applied, and the instruments are on the same level (see section 1.2.3 above).

International judicial organs adopt this approach more and more frequently in order to avoid the creation of self-contained legal systems totally independent from general norms and from each other. In this sense, the European patent system can be compared to the WTO legal system, another highly technical body of law often found in isolation from the rest of international law. However, the WTO Appellate Body has increasingly interpreted WTO treaties in light of the subsequent legal practice of its Member States.[306]

The concept of mutual supportiveness is at stake in this discussion. The judicial organs of the EPO should interpret Article 53(a) of the EPC in light of Articles 16.5, and 22 of the CBD, since all EPC Member States are also parties to the CBD. Article 16.5 of the CBD provides that "patents and other IPRs may have an influence on the implementation" of the provisions of that convention; therefore, Member States are under an obligation of cooperating to ensure that IPRs are "supportive of and do not run counter to its [the CBD's] objectives" (Article 16.5 of the CBD). *Ordre public* and morality provide

303 The patent application refers to the well-known genetic technology that empowers a plant with resistance characteristics to pathogenic agents. All the claims including plant varieties have been rejected on the same grounds of Article 53(b) as in the *Plant Genetic Systems* paragraph 19. An appeal to the Enlarged Board of Appeal of the EPO has been dismissed; R. Nott, "The *Novartis* Case in the EPO" (1999) *European Intellectual Property Review* 33.

304 It seems self-evident for the EPO to decide that the EPO decided that the principle of interpretation of treaties was applicable to the EPC, *In re Eisai Co. Ltd*, Enlarged Board of Appeal, case no. Gr 05/83, P.J.E.P.O. 1985, 64 paragraphs 1–6 (December 5, 1984).

305 The most common finding has been that while the TRIPS Agreement and the UPOV Convention for the Protection of New Varieties of Plants were fully compatible, there was a legal conflict between a radical interpretation of Article 53(b) of the EPC and Article 4.2 of the EU Biotechnology Directive. Article 4.2 seemed to support the "more than a single variety approach" in that "inventions which concern plants or animals shall be patentable if the technical feasibility of the invention is not confined to a particular plant or animal variety" (*Novartis* Case paragraph 2.1 and in particular paragraph 78).

306 M. Lennard, "Navigating by the Stars: Interpreting the WTO Agreements" (2002) *Journal of International Economic Law* 17–89.

a natural place to implement the subsequent legal practice of Article 22 of the CBD, which indicates that the CBD's provisions will not affect the rights and obligations of countries to other "existing international agreements, except where the exercise of those rights and obligations would cause a serious damage or threat to biological diversity." In other words, the EPO judicial bodies should also indirectly apply the relevant CBD provisions.

The next section uses the concepts of subsequent practice and mutual supportiveness to develop an interpretation of Article 53(a) that best fulfills the objectives of international instruments relating to IP and to the environment.

6.3.1.5 *The interpretation of the concept of* ordre public *and morality in the relationship between Article 53(a) of the European Patent Convention, TRIPS, and the EC Biotechnology Directive*

The Biotechnology Directive 98/44[307] declares in recitals 55–56[308] that it is compatible with the Convention on Biodiversity[309] and the TRIPS Agreement. These recitals remind States that in applying the Directive, they should respect certain norms of the CBD.

The concept of *ordre public* is addressed in Article 6 of the EU Directive, and its meaning is provided by Recital 36 which reaffirms that concepts of *ordre public* and morality include preserving the environment by quoting verbatim Article 27.2 of TRIPS:

> Whereas the TRIPS Agreement provides for the possibility that members of the WTO may exclude from patentability inventions, the prevention within their territory of the commercial exploitation of which is necessary to protect *ordre public* or morality, including to protect human, animal or plant life or health or to avoid serious prejudice to the environment, provided that such exclusion is not made merely because the exploitation is prohibited by their law.

However, the protection of biodiversity or the environment in general is notably absent from the list of patentability exceptions in Article 6.2 of the Directive. Some authors consider this list as non-exhaustive, and criticize this omission as incompatible with the facultative exception of Article 27.2 of TRIPS that all EU countries have ratified – which includes protecting

[307] Article 1.2 of the *Directive No. 98/44.* expressly provides that "this Directive shall be without prejudice to the obligations of the Member States pursuant to international agreements, and in particular the TRIPS Agreement and the Convention on Biological Diversity."

[308] Recitals 55 and 56 of the Biotechnology Directive 98/44.

[309] Decision of the Council 93/626/EEC (October 25, 1993) with respect to the Conclusion of the Convention on Biodiversity, GUCE December 13, 1993, L 309, 1.

the environment within the concepts of *ordre public* and morality.[310] The Directive does include "serious prejudice to the environment" as within the scope of *ordre public* in Recital 36.[311] Thus, I conclude that subsequent legal practice approves of the use of *ordre public* and morality to protect the environment.

It can be also argued that Member States that are party to the CBD, and thus bound by its provisions, are responsible to flesh out the Directive with more specific rules of implementation, as Recital 55 of the Directive acknowledges. This responsibility is in line with one of the fundamental principles of the TRIPS Agreement set forth in Article 8: "Members may, in formulating or amending their national laws and regulations, adopt measures necessary to protect public health and nutrition [...] provided that such measures are consistent with the provision of this Agreement."[312]

The historical interpretation of Article 27.2 of TRIPS at the international level shows that the scope of the concepts of *ordre public* and morality is broad and encompasses the exclusion of inventions contrary to public order, law, generally accepted standards of morality, public health, or the basic principles of human dignity or human values.[313]

6.3.1.6 The standard of preponderance of the evidence

In order to revoke a patent that allegedly has a "*serious* prejudice to the environment," the EPO Board linked the concept of Article 27.2 of TRIPS with the standard of evidence of "sufficient substantiation."

310 Pavoni, "Brevettabilità Genetica", 468. The same criticism has been addressed to the lack of more ample environment protection provisions or exceptions in the GATT; F. Francioni (ed.), *Environment, Human Rights and International Trade* (Hart Publishing, Oxford, 2001) 177. The Netherlands challenged Directive 98/44/EC before the ECJ on the grounds *inter alia* of incompatibility with international obligations, including the CBD.

311 *Plant Genetic Systems II*, paragraph 18.5.

312 Whether this environmental exception, unique in the GATT/WTO system, serves to effectively protect the environment or is merely formalistic because it is to be interpreted in a very restrictive way, will depend on the WTO Dispute Settlement Body. Indeed the facultative nature of this provision imposes only a character of "necessity" similar to the one established in Article XX.b of *GATT* which is interpreted restrictively. In other words, this kind of phrasing encourages judges to exclude certain biotechnological subject-matter from patentability only if it consists of an indispensable measure, namely when the harm to the environment caused by such an invention is far greater than the commercial benefits thereof. In my view, the EPO will substantially follow the reasoning construed by the GATT/WTO system. Francioni, *Environment*, 161–2.

313 Draft of Article 27.2 of *TRIPS*, reads as follows: "4. The following [shall] [may] be excluded from patentability: 4.1. Inventions, [the publication or use of which would be], contrary to public order, [law,] [generally accepted standards of] morality, [public health,] [or the basic principle of human dignity] [or human values]. 4.2. Scientific theories, mathematical methods, discoveries and materials or substances [already existing] [in the same form found] in nature."

Pavoni – a supporter of the radical approach to patent examination – maintains that this decision casts more doubts than clarity on this matter.[314] He wonders about the level of evidence of environmental risk that is sufficient to reject a patent or whether the rejection should only occur in extremely rare cases. "Sufficient substantiation" may require that a scientific authority has already ascertained the environmental danger that the patentable invention would pose once it is commercialized because, as the Board held in *PGS*, it is not sufficient to present some scientific articles denouncing the environmental risks related to the invention.[315] This rejection of the patent would require either that the product be so similar in its details to an existing product or the invention itself already so commercialized that a scientific organization could evaluate it. This is not likely to be a frequent occurrence.

Noiville justifies the high standard of evidence to reject a patent on three grounds: (i) the lack of previous widespread use of the invention to test its real risk for the environment; (ii) the lack of specialization of the EPO in environmental risk assessment; and (iii) the EPO's reluctance to invade the territory of other authorities specialized in this type of evaluation.[316] In the same line of thought, Ricolfi acknowledges that patent examiners – technical officials – have little if any background to perform the necessary review of the concepts of *ordre public* and morality in the realm of environmental risk, as the radical approach would have them do.[317]

Ricolfi puts forth a major argument against the entire range of radical approaches. He states that *a priori* arguments against the patentability of biotechnology often fail to consider the negative effect of *exclusion* of this science from patent protection. It must be clear that, if biotech inventions are excluded from patentability, this does not mean exclusion from legal protection altogether. One of the major drawbacks to denying patent protection is the risk of secrecy under trade secret law.

A ban on patenting may therefore fail to discourage innovative activity in the relevant field while inducing innovators to seek the alternative protection provided by trade secret law, which is hardly a desirable outcome to the extent that society wants to increase rather than decrease public scrutiny of novel technologies, and especially those which concern

[314] Pavoni, "Brevettabilità Genetica", 455–60. [315] *Ibid.*

[316] Noiville, *Ressources génétiques*, 405 and 417–20. F. Kernaleguen, "Les principes fondamentaux des lois 'bioétique'", in B. Le Mintier (ed.), *Les lois "bioétique" à l'épreuve des faits* (Presses universitaires de France, Paris, 1998) 35.

[317] Ricolfi, "Biotechnology, Patents and Epistemic Approaches", 78.

the building blocks of life and which may seriously prejudice the environment.[318]

Thus, Ricolfi favors patenting as a means of maintaining the benefits of disclosure and public scrutiny, with environmental concerns still having a role in patent law.[319]

In evaluating the narrow approach taken by the Board of Appeal and endorsed by Noiville and Ricolfi, I find the reasoning of the Board not entirely cogent. The *ratio decidendi* of this case indicates that, where the invention has not been presented to the EPO in its final form as a commercialized product – and therefore no scientific authority has likely evaluated it beforehand – its potential harm cannot be deemed sufficiently proven even if some scientific evaluations determine that the invention could seriously prejudice the environment. Thus, although the EPO standard provides the possibility of denying patent protection in order to protect the environment, there are few if any realistic circumstances when the prohibition would be applied.

In my view, similar cases should be decided on the basis of "preponderance of the evidence" which is closer to the cost-benefit analysis already applied by the EPO. The concept of sufficient substantiation of scientific certainty is too vague and can be interpreted in too restrictive a sense so that it falls short of the concept of "absolute certainty." In order to be intellectually appealing, this preponderance of the evidence must be construed as a coherent chain of pieces of evidence.

This standard calls for an ultimate decision that is based on weighing the amount and quality of submitted circumstantial evidence in favor or against granting a patent.[320] It would need to be demonstrated that the evidentiary weight in favor is greater than the weight opposing the claim.[321]

6.3.1.7 The relevance of a full cost-benefit analysis

There has been criticism of the non-transparent or opaque balancing exercise in the *PGS case* and the recommendation that a real cost-benefit

318 *Ibid.*, 80. 319 *Ibid.*

320 The modern evidence rules in court trial do not draw a distinction between circumstantial and direct evidence. This division can be applied only in some cases. Direct evidence is easier to admit, provided it is relevant, because there is no inference that needs to be drawn as with circumstantial evidence.

321 T. J. Martens, "The Standard of Proof for Preliminary Questions of Fact under the Fourth and Fifth Amendments" (1988) 30 *Arizona Law Review*, 119–33. For these reasons, I urge the EPO to define in clearer terms the level of certainty necessary to deny a patent allegedly harming the environment.

analysis becomes a normative standard "occurring on a case-by-case basis at the court level."[322]

However, even where a full cost-benefit analysis is performed, the analysis would be somewhat arbitrary. The benefits as well as the risks of GMOs and other biotechnology are uncertain. Some have argued that the benefits arising from these inventions are rather exclusively in favor of the biotech-industries, whereas the "costs," i.e. environmental harms, are borne by society as a whole.[323]

In the cases described above, it is clear that the real impact of biotechnological inventions on the environment, on public safety, and on public health is generally better characterized by *ignorance* rather than *uncertainty*. The epistemological relationship between ignorance and uncertainty in cost-benefit analyses has been elegantly explained by Ricolfi who states that our society arrives, from time to time, at the extreme frontiers of our past experience:

> I certainly have good reasons to challenge prevailing, entrenched attitudes in evaluating the implications of the unprecedented steps we are taking. In dealing with complex, non linear systems, it may well be that the final outcome of decisions taken is rather likely to be explained by catastrophe models as illustrated by the corresponding, specialized branch of mathematics than by the more familiar costs-benefits analysis – and even their more modern game-theoretical variants – which I employed in the past. However, it seems to me that I should not too easily discard the conventional tools with which I have been working in the past, to the extent they may still assist us in clarifying issues and sorting them out analytically.[324]

When science is uncertain about possible harm to the environment, two elements can help assess more fully the potential risks that the innovation may provoke in the environment. The first, the allocation of the burden of proof, is an important procedural element, and it may depend on the substantive precautionary principle.

6.3.1.8 Allocation of the burden of proof upon the defendant

Crucial in proceedings to determine possible harm to the environment are the allocation of the burden of proof and its degree of certainty. That the burden

[322] J. P. Trachtman, "Trade and Problems, Cost-Benefit Analysis and Subsidiarity" (1998) 9(1) *European Journal of International Law* 45. It is also probable the challenges on Article 53(a) may have somewhat and paradoxically contributed to turn the scale in favour of the rejection "more than a single variety approach" presented by the patent applicant. In other words, the EPO decided that a product which embraces within its subject-matter "plant varieties" is not patentable," *Plant Genetic Systems II*, paragraph 24.

[323] Pavoni, "Brevettabilità Genetica", 96.

[324] Ricolfi, *Biotechnology, Patents and Epistemic Approaches*, 81.

of proof lies with the claimant is a well-established principle of international law. Moreover, it is the claimant who appeals an exception to a legal right (to a patent), further placing the burden on the claimant's shoulders. Clearly, in a case like *PGS*, in which the environmental risk was characterized by a high level of scientific uncertainty, allocation of the burden of proof can determine the outcome of the Board of Appeal's decision.[325]

After Greenpeace produced a series of documents to demonstrate the irreversible damage that the patented invention could potentially cause, the Opposition Division noted that "the Proprietor [...] does not dispute that such risks may exist."[326] The Board of Appeal thus marked its distance from the radical approach when it granted the patent notwithstanding Greenpeace's undisputed assertion. Pavoni maintains that the Board of Appeal merely paid lip service to the procedural rules governing the burden of proof; he believes the rules, if correctly applied, would have led the Opposition Division to require a demonstration of measures to guarantee the safety of the invention. Instead, the Opposition Division considered the defendant's acknowledgment of the risks to be proof according to the principle of non-contestation.[327] I must disagree with Pavoni: this case involves an exception to an established right;[328] therefore, the burden of proof lies squarely with the claimant and not the defendant. Only application of the precautionary principle would have shifted the burden of proof.

6.3.1.9 The application of the precautionary principle

Authors who favor the radical approach to patent examination often advocate introducing the precautionary principle[329] into patent law: patents for inventions that present serious and not well-understood risks to health, the

[325] *Plant Genetic Systems I*, paragraphs 3.12, 3.13, 3.16 for clear indications of how a failure in the burden of proof results in rejection of an objection based on Article 53(a) of the EPC. D. Wirth, "The Role of Science in the Uruguay Round and NAFTA Trade Disciplines" (1994) 27 *Cornell International Law Journal* 817; V. Walker, "Keeping the WTO from Becoming the 'World Trans-science Organization': Scientific Uncertainty, Science Policy, and Factfinding in the Growth Hormones Dispute" (1998) 31 *Cornell International Law Journal* 258.

[326] *Plant Genetic Systems I*, paragraph 3.11. *Plant Genetic Resources II*, paragraph 18.6.

[327] Pavoni, "Brevettabilità Genetica", 457–58.

[328] *Ordre public* and morality are exceptions to the right to patent protection for an invention, rather than being requirements for patentability themselves.

[329] Article 15 of the *Rio Declaration*, elaborates the precautionary principle: Where certainty shall not be used as a reason for postponing cost-effective measures to prevent environmental degradation. For a general understanding of the application of this concept in international environmental law, D. Hunter, J. Salzman and Z. Durwood, *International Environmental Law and Policy* (University Casebook Series, New York Foundation Press, 1988) 360–61.

environment, etc., should be rejected until their safety is proven. In other words, the burden of proof for safety is placed on the applicant. Pavoni argues that the precautionary principle is enshrined in the Biosafety Protocol to the CBD.[330] This principle states that scientific uncertainty should not be an excuse for not undertaking measures apt to prevent or minimize serious and irreversible environmental harms. It is very probable that the precautionary principle may be increasingly invoked in European patent oppositions.

The EPO does not apply the precautionary principle, distinguishing itself from other judicial fora that have dealt with environmentally controversial biotechnology. For example, the WTO Appellate Body had to adjudicate a case regarding GMOs admitting that "a risk assessment has not come to a monolithic conclusion" on possible environmental threats or threats to human health. It noted that "the very existence of divergent views presented by qualified scientists who have investigated the particular issue at hand may indicate a state of scientific uncertainty."[331] Despite these findings, the Body applied the precautionary principle.

The value of the precautionary principle is that it allows regulatory bodies to address potential dangers in highly uncertain technological

[330] The *Carthagena Biosafety Protocol to the CBD* (January 29, 2000) provides for the precautionary principle (in Articles 10.6 and 11.8) that allows for restrictions on introducing GMOs into the environment despite lack of scientific certainty as to the hazard; S. D. Murphy, "Biotechnology and International Law" (2001) 41(1) *Harvard International Law Journal* 47. Article 2.2 of the *Agreement of Sanitary and Phytosanitary Measures*, Annex 1A to the WTO Agreement allows for import-restricting measures intended to protect human, animal and plant life, and health only if they are "based on scientific principles … and not maintained without sufficient scientific evidence." The decisions of the WTO panels that are based on the Agreement on the Sanitary and Phytosanitary Measures presuppose that any national measure able to block imports of environmentally dangerous products must be based on scientific principles and not applied without sufficient scientific evidence. Article 3.3, indeed, allows States to adopt a superior level of protection than the one established by international standards only if there is a "scientific justification." Article 5.2 states that the risk on which the decision/measure is made has to be based on available scientific evidence. The precautionary principle was raised in the *Hormones* case. The Appellate Body considered that the EU applied precautionary measures to protect human health that had too severe an impact on international commerce. Since the end of the 1980s the precautionary principle has been used and been developed in national and international environmental law. While accepting scientific uncertainty and the mere possibility of damage, a precautionary measure requires a serious threat to the environment. The legal status of the precautionary principle in international law is difficult to determine because its definition lacks precision in international treaties and jurisprudence. However, it is certain that the precautionary principle is on the way to becoming a general principle of international law: see *Reports of the Panel*, WT/DS26/R/USA (August 18, 1997) and *EC Measures Concerning Meat and Meat Products* (Hormones case) WT/DS48/R/CAN (January 16, 1998); D. Wirth, "International Decisions. European Communities – Measures Concerning Meat and Meat Products" (1998) *American Journal of International Law* 755–59.

[331] *EC Measures Concerning Meat*, paragraph 194.

fields. Various authors who have commented on WTO decisions concerning GMOs in international trade have clearly stated that "scientific uncertainty pervades the empirical sciences and virtually all current estimates of risk"[332] because "uncertainty and lack of data [...] characterize much of the scientific basis for regulation."[333] Indeed, disagreement among scientists is inherent in many fields.[334] It is commonly believed that adverse consequences to the environment caused by GMOs can be detected only many years after their release.[335]

In the eyes of those who advocate the radical approach, it is absurd to refuse to apply patentability exceptions to protect the environment solely because scientific opinion is not unanimous on the environmental risk. Pavoni, for example, would be eager to integrate the most sophisticated scientific assessment methodologies into biotech patent examination, when as yet these methods are not even integrated into the case law on international environmental risk.

When the precautionary principle is applied, the standard of evidence necessary to revoke the patent, as Wirth explains, should be no higher than a "minimum level of scientific rationality"[336] or a scientific plausibility instead of a certainty.[337] The Appellate Body of the WTO in the *Hormones* case applied a comparable standard when it affirmed that "the results of the risk assessment must sufficiently warrant – that is to say reasonably support – the [health] measure at stake."[338] The precautionary principle thus recognizes that scientific uncertainty does not mean absence of risk:[339] "the continuum between a merely speculative risk and a conclusively demonstrated one lies on a vast stretch of undemonstrated, unquantified, but scientifically plausible risks. Within that zone, the risk of harm is real so long as safety is unproven."[340]

A fortiori, when the precautionary principle is applied to controversial biotechnology (such as GMOs), the lack of scientific evidence on potential adverse effects on the environment is a perfectly good reason for imposing importation and other restrictions. Had the EPO Board of Appeal applied the precautionary principle in *PGS* or *Onco-Mouse*, would it have come to a different conclusion? The answer is a clear yes.

332 Walker, "Keeping the WTO from Becoming the 'World Trans-science Organization'", 258.
333 Wirth, "Role of Science", 837.
334 *Ibid.*; Walker; Pavoni, "Brevettabilità Genetica", 459.
335 Friends of the Earth, www.foe.co.uk/resource/evidence/environmental_audit_committee.html.
336 Wirth, "Role of Science", 855–6; *EC Measures Concerning Meat*, paragraph 193.
337 Walker, 262–63 and 179–85.
338 *EC Measures Concerning Meat*, paragraph 193.
339 Pavoni, "Brevettabilità Genetica", 459.
340 Walker, 304–05.

The precautionary principle implies a shift of the traditional burden of proof allocation from the claimant to the defendant, and there is no evidence in the current case that the defendant could have met such a burden. This inability of the defendant to prove safety, would certainly be true of most biotech patents with potentially harmful environmental or health effects in the US, the EU, and in other patent systems. The very uncertainty that characterizes such risks makes it difficult to prove that the invention will be safe.

6.3.1.10 The difficulty of application of the precautionary principle

However, there are two main reasons why it would be inappropriate to apply the precautionary principle during patent examination.

First, there is a notable difference between interpreting a provision in the light of subsequent legal practice and the direct application of provisions contained in other than EPC treaties. The EPO applies its constitutive treaties and not other treaties. The rules of interpretation of Article 31 of the VCLT do not require that the EPO *directly* applies binding international environmental norms when granting a patent, since it is not part of its constitutive instrument. Hence, the concepts of *ordre public* and morality should be interpreted in light of the rest of conventional and customary norms resulting from the evolution of international environmental law that refer to those concepts. The interpretative effort of expanding the concept of *ordre public* has been made regarding the ban on patentability of inventions that seriously prejudice the environment. I maintain that the interpretation of the concept of *ordre public* as including the environment correctly follows the rules of interpretation of treaties. However, I do not think that this interpretation should include the application of the precautionary principle. The concept of *ordre public* in the EPC is construed as an exception to patentability that must be interpreted in a restrictive manner.

The second main reason that the precautionary principle should not be applied during patent examination is that the EPO judicial organs cannot directly apply the norms contained in the CBD and in its Biosafety Protocol (e.g. the precautionary principle).[341] The provisions of this treaty are destined to be implemented in the domestic legal systems of

[341] It must be noted that the precautionary principle is not a customary principle as yet: "The status of the precautionary principle in international law continues to be the subject of debate among academics, law practitioners, regulators and judges [...]. [I]t is unnecessary, and probably imprudent for the Appellate Body in this appeal to take a position on this important, but abstract question" (*EC Measures Concerning Meat*).

EPC members and are to be applied when the invention is exploited rather than when the patent is examined.[342]

The fact that the EPO should not apply the precautionary principle, does not prevent it from adopting and developing a real cost-benefit analysis as a fundamental method of decision when confronted with allegations of environmental harm possibly caused by certain biotechnological inventions. But, in this case, the claimant would still carry the burden of proof.

Thus, it is very doubtful that there is a legal basis in the EPC for applying the precautionary principle in patent law. The EPO has no competence to undertake a risk assessment or to apply the precautionary principle.

6.3.1.11 The European Court of Human Rights

An innovative proposal to deal with the concepts of *ordre public* and morality has linked Article 53(b) of the EPC to the European Convention of Human Rights (ECHR). This relationship has been analysed at various levels.[343] Beyleveld and Brownsword suggest that no decision of the EPO organs should conflict with any right enshrined in the ECHR.[344] They propose that the EPO form a special plural membership Ethics Board to hear matters of *ordre public* and morality when an appeal is made to oppose the grant of a patent. Ethics committees of diverse membership are nowadays commonly formed for decision-making in the field of, for instance, medical experimentation.[345] Under this proposal, the Ethics Board's decisions would be appealable to the ECHR.[346]

Pavoni goes so far as to propose a judicial review of the ECHR over the States' implementation of a decision of the EPO. Although the European Court of Human Rights is not competent *ratione personae* to examine proceedings before the decisional bodies of the EPO and since the latter

[342] *Cartagena Protocol on Biosafety to the Convention on Biological Diversity* (January 29, 2000) 39 *ILM* 1027 has the objective "to contribute to ensuring an adequate level of protection in the field of the safe transfer, handling and use of living modified organisms resulting from modern biotechnology that may have adverse effects on the conservation and sustainable use of biological diversity, taking also into account risks to human health, and specifically focusing on transboundary movements" (Article 1). Report Cartagena Protocol on Biosafety: *A Report on the Investigation into the Scientific Issues, Protocol Mechanisms and Proposals* (August 20, 2004), www.columbia.edu/cu/mpaenvironment/pages/projects/f2004/CARTEGENA_FINAL_REPORT.pdf.

[343] J. Frowein, "The European Convention on Human Rights as the Public Order of Europe" (1990) *Collected Courses of the Academy of European Law*, I-2, 267; F. Sudre, 'Existe-t-il un ordre public européen?', in P. Tavernier (ed.), *Quelle Europe pour les droits de l'homme?* (Bruylant, Bruxelles, 1996) 39.

[344] Beyleveld and Brownsword, *Mice, Morality*, 68–70 and 89–90. [345] *Ibid.*, 90.

[346] Pavoni, "Brevettabilità Genetica", 473–4.

is not party to the ECHR,[347] it does have jurisdiction over individual States that implement the decisions of the EPO. States are still responsible to fulfill the law, even when they transfer authority to do so in part to an international organ such as the EPO.[348] Under this view, the ECHR would be competent to reject a patent from a national register.[349] Although the ECHR has even held that it is a legal "instrument of European public order (*ordre public*) for the protection of individual human beings,"[350] there are three major impediments to the ECHR's involvement with the exception of *ordre public* in patent law.

First, Article 53(a) of the EPC has not yet generated enough cases of EPO oppositions and appeals to make it useful for the ECHR to get involved (although the involvement of the ECHR might stimulate additional litigation).

Second, the ECHR does not contain any provision that expressly guarantees the right to a clean environment. This right is only indirectly dealt with through the protection of privacy and family life in Article 8 or through the right not to be submitted to a degrading treatment in Article 3.[351]

Third, it seems inefficient to require judicial review of an exclusive right to a biotechnological invention that allegedly infringes the right to a clean environment. Instead, the ECHR can be used to start an action directly against the State that has authorized exploitation of the biotechnological invention.

It is hard to support this radical approach as an efficient and effective solution.

6.3.1.12 The defensive protection of traditional knowledge through the concepts of ordre public *and morality*

Because of the reach and scope of the ethical and environmental considerations of patent law, the concepts of *ordre public* and morality may serve as an additional legal basis for formulating arguments against blatant cases of biopiracy: if a patent application does not comply with the CBD concepts of PIC and benefit sharing, can it be banned because it violates the EPC concepts of *ordre public* and morality or the US concept of "moral utility" (see section 6.3.2 below)? In this section, I consider whether TK can be protected by refusing to grant patents for inventions based on a GR that has

[347] ECHR Case 21090/92 *Heinz* v. *Contracting States also Parties to the Europeant Patent Convention* (1994) 76A DR 125.

[348] Pavoni, "Brevettabilità Genetica", 474. [349] Drahos, "Biotechnology Patents", 449.

[350] ECHR Case 15318/89 *Loizidou v. Turkey*, Preliminary Objections 10 (March 23, 1995).

[351] Article 3 of the *European Convention of Human Rights* (November 4, 1950), www.echr.coe.int/ECHR/.

been "misappropriated" from its provider country, which is prohibited by Article 15 of the CBD. Experts disagree on what misappropriation means, so that what exactly is meant by misappropriation is somewhat ethically or morally controversial. I will now define the scope of the legal basis on which a patent can be rejected when GRs or TK have been misappropriated. Opponents of protection for TK (who take the classic approach) note that most "misappropriated" patents claim only improved versions of TK. While I have noted instances where this is not true,[352] I agree that most of these patents would not prevent indigenous peoples from continuing their traditional practices, even if the patent protection extended to their local jurisdiction.[353]

However, such patents can prevent further innovation by the indigenous society because such continued innovation would infringe on the broad scope of the patent protection obtained by industrialized countries. Once patent protection is obtained, the "raw materials" will very probably be transformed and commercialized into final products at unaffordable prices. It strikes many commentators as blatantly inequitable that societies that have bred plants and animals over centuries and have kept the knowledge do not share the benefit of that resource and may even have to pay a premium to multinational corporations for its use. However, most commentators believe that *ordre public* and morality cannot be extended to the class of patents based on GRs and related TK.[354]

The international subsequent practice may contribute to this classic interpretation. Ricolfi seeks to push the boundaries of the interpretation of *ordre public* and morality by viewing it in light of Article 15 of the CBD. He submits that a patent based on a biological material that, in turn, has been accessed and appropriated without the CBD-required PIC of the provider State is "against the internationally mandated standard of fairness and, therefore, invalid as against morals or *ordre public*.[355]

Moreover, a WTO Member may refuse to grant a patent under *ordre public* and morality to preserve natural resources against misappropriation if the patent fails to share benefits with the provider State.[356] Ricolfi notes,

[352] Cullet *et al.*, "Intellectual Property Rights, Plant Genetic Resources and Traditional Knowledge", 135–37.

[353] Dutfield, *Intellectual Property Rights, Trade and Biodiversity*, 61; Leistner, "Analysis of Different Areas of Indigenous Resources", 67.

[354] R. Moufang, "The Concept of *Ordre Public* and Morality in Patent Law", in G. Overvalle (ed.), *Patents, Ethics and Biotechnology* (Katholieke Universiteit Brussel, Bruxelles, 1998) 65–77; T. Roberts, "Patenting Plants around the World" (1996) 18 *European Intellectual Property Review* 531–36.

[355] Ricolfi, "Biotechnology, Patents and Epistemic Approaches", paragraph 4.6; Spada "Liceità dell' invenzione", 5, 18.

[356] Moufang, "The Concept of *Ordre Public*", 69.

however, that this interpretation is complicated by the last part of Recital 27 of the Biotech-Directive, which provides that applicant's failure to comply with disclosure obligations "is without prejudice to the processing of patent applications or the validity of rights arising from granted patents."[357]

Thus, there are indications in the CBD and Bonn Guidelines (subsequent international law) that *ordre public* and morality could be used to deny patent protection for biodiversity-based inventions lacking the necessary PIC. However, the Directive makes it difficult to do so in the EU. The evolution of soft law and customary law[358] on the protection of TK against misappropriation calls upon all patent offices and legal systems to reconsider the morality of granting a patent on GRs without the proper PIC of the host community. However, given the constraints on the concept in current EU law, a different approach may be more useful in protecting TK. In light of these considerations, strengthening prior art search practices and developing a Board of public policy to scrutinize patent applications will help achieve the objective of granting patents that do not breach internationally recognized standards of protection of TK and GRs.

6.3.2 *The concept of moral utility in the US patent system*[359]

The US has a doctrine that bears some similarity to *ordre public* and morality: moral utility. The utility requirement (of which moral utility is a part) stems from the statutory imperative that an invention be "useful",[360] and corresponds to the industrial application requirement in the EU and other countries.

357 Ricolfi, "Biotechnology, Patents and Epistemic Approaches".

358 Bonn Guidelines; *Overview of the Activities and Outcomes of the Intergovernmental Committee*, WIPO/GRTKF/IC/9, www.wipo.int/edocs/mdocs/tk/en/wipo_grtkf_ic_11/wipo_grtkf_ic_11_9.doc; *Genetic Resources and Patent Disclosure Requirements: Transmission of Technical Study to the Convention on Biological Diversity* WIPO/GRTKF/IC/6/9 (December 12, 2003), www.wipo.int/documents/en/meetings/2004/igc/doc/grtkf_ic_6_9.doc. *Fourth Session of the Working Group on Reform of the Patent Co-operation Treaty* PCT/R/WG/4/14 (19–23 May, 2003), www.wipo. int/pct/en/ffieetings/reform_wg/pdf/pc~_r_wg_4_14.pdf; *Fifth Session of the Working Group on Reforming of the Patent Co-operation Treaty* PCT/R/WG/5/13 (November 17–21, 2003), www.wipo.int/edocs/mdocs/pct/fr/pct_r_wg_5/pct_r_wg_5_13.doc; *Draft Report: Ninth Session of the SCP* SCP/9/8 (May 12–16, 2003) www.wipo.int/edocs/mdocs/scp/fr/scp_9/scp_9_8.doc; *Doha Ministerial Declaration.*

359 This section including the one on the Board of Public Policy proposal has been developed and drafted with Peter Bradford, currently associate at the intellectual property law firm of Sughrue Mion, PLLC, in Washington, DC, and graduate student in International Science and Technology Policy at The George Washington University.

360 35 U.S.C. 101.

The moral utility doctrine dates back to *Lowell v. Lewis* in 1817 during the formative period of US patent law. Justice Story defined the utility requirement as encompassing requirements of morality and policy:

> All that the law requires is that the invention should not be frivolous or injurious to the well-being, good policy, or sound morals of society. The word "useful," therefore, is incorporated into the [patent] act in contradistinction to mischievous or immoral. For instance, a new invention to poison people, or to promote debauchery, or to facilitate private assassination, is not a patentable invention.[361]

This US concept of moral utility as a doctrine to protect "the well-being, good policy, or sound morals of society" sounds much like the European concept of *ordre public* and morality. Moral utility was used to find various inventions unpatentable between *Lowell* and the mid-twentieth century, most often gambling devices[362] and inventions to facilitate deception.[363] The case law during this era seemed to balance the invention's benefit with its harm, similar to the EPO analysis in *Onco-Mouse*.

However, with time, the analysis shifted to evaluating whether such inventions had any use that was not injurious; even if inventions were primarily useful for immoral purposes, they would be patentable if they did have some utility that was not immoral.[364] (This shift is similar to the EPO's shift to a largely "hands off" attitude to morality and social policy in the *PGS* and *Relaxin* cases.) In the latter half of the twentieth century, moral utility retreated from prohibiting gambling[365] and deception.[366] In *Juicy Whip*, overruling the application of moral utility to an invention that a lower court had found unpatentably deceptive, the Federal Circuit[367] put the ball squarely in the court of the legislative branch:

[361] 15 F. Cas. 1018, 1019 (US Court of Appeals 1817).
[362] *National Automatic Device Co. v. Lloyd*, 40 F. 89 (N.D. Ill. 1889); *Reliance Novelty v. Dworzek*, 80 F. 902 (N.D. Cal. 1897); *Schultze v. Holtz*, 82 F. 448 (N.D. Cal. 1897); *Brewer v. Lichtenstein*, 278 F. 512 (7th Cir. 1922).
[363] *Rickard v. Du Bon*, 103 F. 868 (2d Cir. 1900); *Hall v. Duart Sales Co.*, 28 F. Supp. 838, 42 USPQ 354 (N.D. Ill. 1939).
[364] *Fuller v. Berger*, 120 F. 274 (7th Cir. 1903); *Koppe v. Burnstingle*, 29 F.2d 923 (D.R.I. 1929); *Chicago Patent Corp. v. Genco*, 124 F.2d 725, 728, 52 USPQ 3 (7th Cir. 1941) (gambling); *In re Corbin*, 6 F. Cas. 538 (No. 3224) (C.C.D.C. 1857); *Klein v. Russel*, 86 US (19 Wall.) 433 (1873); *Naylor v. Alsop Process*, 168 F. 911 (8th Cir. 1909).
[365] *In re Murphy*, 200 U.S.P.Q. 801 (PTO Bd. App. 1977).
[366] *Juicy Whip, Inc. v. Orange Bang*, Inc., 185 F.3d 1364, 1366–1368 (Fed. Cir. 1999).
[367] The Court of Appeals for the Federal Circuit is the US appellate court that reviews all patent cases.

the requirement of 'utility' in patent law is not a directive to the PTO or the courts to serve as arbiters of deceptive trade practices. Other agencies, such as the Federal Trade Commission (FTC) and the Food and Drug Administration, are assigned the task of protecting consumers from fraud and deception in the sale of food products [...] As the Supreme Court put the point more generally, "Congress never intended that the patent laws should displace the police powers of the States, meaning by that term *those powers by which the health, good order, peace and general welfare of the community are promoted*" [...] Of course, Congress is free to declare particular types of inventions unpatentable for a variety of reasons, including deceptiveness [...] Until such time as Congress does so, however, we find no basis in section 101 to hold that inventions can be ruled unpatentable for lack of utility simply because they have the capacity to fool some members of the public.[368] (emphasis added)

While the broad pronouncement here was not necessary to resolve this case, in light of the lack of an application of the moral utility doctrine in decades, this case seemed to close the door on moral utility. However, it has not disappeared completely. A later Federal Circuit decision defined utility using the above quote from Lowell, including the reference to the "well-being, good policy, or sound morals of society." This quotation was an *obiter dictum*, as the Court did not have occasion in that case to apply the "moral" part of the utility requirement. Also, a US patent office commentary[369] (although predating *Juicy Whip*) and at least one patent examiner's action[370] have indicated that a patent application that would encompass a human being would be rejected under the moral utility doctrine. It is unclear if there are any other inventions to which moral utility could be applied.[371]

The sidelining of moral utility has been a necessary consequence of removing moral or policy assessments from patent adjudication. This process began in the US Supreme Court case of *Diamond v. Chakrabarty*, in which the Court stated: "It is argued that this Court should weigh these potential hazards [to health, the environment and respect for life] in considering whether respondent's invention is patentable subject-matter

[368] *Juicy Whip*, 1368, quoting *Webber v. Virginia*, 103 US 344, 347–48, 26 L. Ed. 565 (1880).

[369] USPTO media advisory, *Facts on Patenting Life Forms having a Relationship to Humans* (April 1, 1998), www.uspto.gov/web/offices/com/speeches/98–06.htm ("It is the position of the PTO that inventions directed to human/non-human chimera could, under certain circumstances, not be patentable because, among other things, they would fail to meet the public policy and morality aspects of the utility requirement.").

[370] U.S.S.N. 08/993,564, the Chimera application of Dr. Newman, Office Action of 29, 2003.

[371] The extent of what "human being" refers to here is not even clear, as the USPTO said this would refer to some human-animal chimeras, and which chimeras would be covered is only vaguely addressed in the USPTO's actions in the case of Dr. Newman.

under 35 US Code section 101. I disagree."[372] It was for Congress to act if morality was to enter the patent law.[373]

A few commentators have suggested reviving the moral utility doctrine to deal with contemporary ethical and environmental concerns, especially with biotechnology patents. Smith has argued that the most consistent application of moral utility has been to inventions that would solely or predominantly encourage illegal behavior;[374] that is to say, the law gives content to the morality of moral utility. This position is congruent with the prohibition of patents on humans: human cloning is an activity that is illegal under executive order[375] and in the laws of many States, and several bills have been introduced to prohibit cloning nationwide (although disagreements on scope have thus far prevented passage).[376] Activities which are essentially never permitted by law are a reasonable indicator of what our society rejects as immoral or against public policy, and, as such norms change, the law changes as well.

The moral utility doctrine has significant pros and cons as a tool to solve such problems. As a general doctrine, it allows patent examiners and courts to address issues that arise on which there is no clear guidance. It already exists (though significantly narrowed) in US law. Because moral utility is similar to *ordre public* and morality, this would allow the US to give meaning to and take meaning from *ordre public* and morality provisions enacted in the patents laws of the EU, Japan, and other countries with developed patent systems.[377] Using the moral utility doctrine might allow for a fuller and more nimble evolution of the doctrine that would respond to changing needs. The executive branch (the patent office), the courts, and Congress could each contribute as the need arose.

However, it must be noted that there is no express statutory requirement concerning morality in US Patent law. There seem to be no ethical safeguards, the important ethical values being left unaddressed by US Patent law in this field. Indeed, the 1998 USPTO guidelines on the

[372] *Diamond v. Chakrabarty*, 447 United States 303, 309 (1980), reported also in F. Abbott, T. Cottier and F. Gurry (eds.), *The Intellectual Property System; Commentary and Materials* (Kluwer, The Hague, 1999) 25.

[373] *Ibid.*

[374] A. R. Smith, "Monsters at the Patent Office: The Inconsistent Conclusions of Moral Utility and the Controversy over Human Cloning" (2003) 53 *DePaul Law Review* 159.

[375] US Presidential Executive Order, *Prohibition on Federal Funding for Cloning of Human Beings* (March 14, 1997), http://grants.nih.gov/grants/policy/cloning_directive.htm.

[376] E.g. Human Cloning Ban and Stem Cell Research Protection Act of 2005, S.876.

[377] E.g. Japan (patent law section 32, www.wipo.int/clea/docs_new/en/jp/jp006en.html); Peru (Ley de Propiedad Industrial Decreto Legislativo N° 823, Article 102, www.indecopi.gob.pe/ArchivosPortal/estatico/legislacion/osd/DECRETOLEGISLATIVO823.pdf).

patentability of biotechnological inventions makes no mention of moral, social, or public policy-related issues, except to state that they are not a basis for rejecting biotechnology patents.[378]

The US Patent Office rejected the Chimera patent application for failing to recite statutory subject-matter, including enablement and best mode of disclosure requirements and for anticipation but, perhaps surprisingly, not for lack of moral utility. Thus only general and specific utility must be satisfied for such an invention to be patentable subject-matter. The Chimera applicants could easily have demonstrated that they would produce transplantable organs to satisfy the utility requirement. The application was instead rejected because it might encompass a human being.[379] The somewhat ad hoc prohibition against patenting a human being would seem to fit naturally into a broader ethical framework such as moral utility might provide. Lacking such a framework, ethical issues (and similarly economic and social considerations) related to patents are addressed only in rare cases that grab attention, such as the potential patenting of a human being.

Given the current approach of US law, it seems unlikely that general applications of moral utility, even limited to clearly illegal behavior, would find a place in the US legal system. Indeed, illegality cannot be used to reject a patent either under current US law, EU law or TRIPS.[380] As the former content of moral utility has been largely removed, the use of moral utility as a framework for evaluation of patents would require a major reworking of US law or of a judicial application of the doctrine within current jurisprudence, which does not presently appear likely.

While the applicability of moral utility in modern patent law is questionable, inventions are required to have both general and specific (or practical) utility. General utility refers to the type of usefulness of a class of inventions; specific utility requires a showing that the specific invention claimed has that usefulness.[381]

6.3.3 *Reconciliation of the classical and radical approaches*

The classical approach has carried the day, for the most part, in both EU and US law (as well as the other major patent systems). This is due in part to the power of the classical approach. It is highly efficient, because it limits patent examination to an analysis of the technology and the classical

[378] USC S.218(a) 1982.
[379] US patent application number 10/308135, http://portal.uspto.gov/external/portal/pair.
[380] TRIPS Art. 27.2; EPC Art. 53(a).
[381] Manual of Patent Examination Procedure paragraph 2107.01.

requirements for patentability, fairly narrow issues in which patent examiners can be rigorously trained. However, the radical approach is powerful as well and is having increasing influence on the patent systems of the developing world. The radical approach recognizes the power of the patent in shaping crucial economic sectors, such as determining the availability of pharmaceuticals and agricultural products. From this point of view, the patent system is not inherently just or good – it is a powerful system that needs to be actively adjusted to promote the public good.

Can the values of the classical and radical approaches be reconciled? At first glance, the answer would seem to be no: one approach seeks to include public order and morality in patent examination; the other seeks to exclude it. However, as described below, we believe there is a possibility of reconciliation between the "radical" and the "classical" approach.

6.3.3.1 The technical character of patent law

A common objection to including assessments of morality or public policy in patent examination is that these questions fall outside the competence of patent examiners. Like sports or banking law, patent law is considered highly technical because it contains numerous peculiarities that must be mastered to practice it. To a certain degree, any branch of law has its own technicalities. However, efficiency is undoubtedly important in the patent examination process. This is the power of the classical approach. Examination is limited to the technology at hand and the classical elements of patentability, in which examiners can be rigorously trained. With the issues thus limited, patent examination can proceed much more quickly than it could if the process were burdened with a multitude of policy and morality issues. In an era of ever-increasing numbers of patent applications and wait times, such efficiency is valuable.

6.3.3.2 The inherent vagueness of ordre public *and morality*

Furthermore, when one analyzes policy issues in patent examinations, one inevitably confronts ambiguity; there are few applications of the *ordre public* and morality exceptions that would not be highly controversial. Genetically Modified Organisms (GMOs) provide a good example. To take the European example, laws on the experimentation, control, and diffusion of GMOs in European countries do not forbid the development of genetically modified animals; on the contrary, research is encouraged, especially in oncology. However, EU public opinion is fairly negative toward GMOs. Can the result of such state-supported research be contrary to European values? For the vast majority of issues, determining what is immoral or against public policy will depend on the individual making the determination.

Although Article 53 of the EPC does not contain any reference to the morality of scientific research, from the radical point of view, *ordre public* and morality may not be solely confined to very harmful applications. In national systems, for instance, the judge has to consider the dominant opinion at a given moment and foresee the effect that a certain legal decision will have on a society. However, these variable concepts cover subjective values not unanimously shared; yet it is because of their imprecise nature that they need to be assessed by a judge.[382] When such concepts are instead applied in the first instance by patent examiners whose expertise is confined to one technical area of the law the variability and uncertainty of outcomes increase.

The EPO has only applied *ordre public* with legal instruments too limited to enable a serious analysis of these concepts. Some authors say that it is because a national judge can more easily define the concepts of national morality and *ordre public*; however, there is no European benchmark for *ordre public* and no homogeneous understanding of the concept of morality in biotechnology.[383] Thus, one can see the reasoning behind the classical rejection of *ordre public* and morality: both lack of expertise by the patent examiner and lack of uniformity of the concept within Europe make the concept hard to apply. Widespread application of public policy exceptions to patentability could lead to widespread uncertainty in the patent process.

6.3.3.3 The concerns of the radical approach

While recognizing the legitimacy of the classical critique, its approach does not put an end to the discussion in a satisfactory way. A patent is an individual government monopoly granted with the public good in mind. The thought that patents should not be granted for inventions goes back at least to *Lowell v. Lewis*.[384] Concerns that certain patents or classes of patents are against the public interest should be addressed squarely rather than brushed off.

Regardless of the ambiguity of *ordre public* and morality, there is a clear legal basis for their recognition in the EU and some other countries. The EU Treaty defines *ordre public* as those values that might justify certain limitations on economic liberties while the ECHR refers to *ordre public* and

[382] Noiville, *Ressources génétiques*, 424.

[383] J. Straus, "Ethische, rechtliche und wirtschaftlische Probleme des Patent und Sortenschutzes für die biotechnologische Tierschützung und Tierproduktion" Geschäftsstelle der Deutschen Vereinigung für gewerblichen Rechtsschutz und Urheberrecht (December 12, 1990) 913; Galloux, "Ethique et brevet", 87–88.

[384] 15 F. Cas. 1018, 1019 (US Court of Appeals 1817).

morality as safeguards of the fundamental values of democracy. Article 17 of the UPOV Convention expressly provides for a limitation in the exercise of rights for "reasons [...] of public interest." Rather than elimination, *ordre public* and morality need the harmonization of national laws to provide a clear European standard.[385] While the US does not have a similar doctrine, the radical approach can still be integrated into US law.

6.3.4 *A Board of Public Policy at the Patent Office for the reconciliation between the classical and the radical approaches*

A key to the reconciliation between the classical and the radical approaches to the environmental and ethical concerns in the patent system could be solved by the formation of a Board of Public Policy within national (e.g. USPTO) or regional (e.g. EPO) patent offices. The Board of Public Policy would review patent applications referred to it as implicating public policy concerns. Such referrals could be made by anyone with knowledge of the application number or title, including the inventor or assignee (however unlikely that might be), the patent examiner, and third parties.[386] The application would be referred in parallel rather than taken out of technical examination, and thus the patent examination would not be slowed. The Board of Public Policy would not have the power to reject patent applications, so the applicant would not have additional uncertainty as to the patentability of the invention. Review would be discretionary; that is, if the Board of Public Policy did not think that a given application implicated sufficiently significant public policy concerns, it could set it aside without further action. The Board of Public Policy would have access to all the resources of the national patent office, and some additional resources that are discussed below.

6.3.4.1 The composition of the Board

The membership of the Board should be as broad and representative as possible so that various viewpoints could be integrated: a member (or members) with long experience at the national patent office, a bioethicist, an expert on environmental law and policy, perhaps a theologian knowledgeable in the various religious streams representing the population of the country in question, and perhaps a representative from industries that rely heavily on patent protection. Public policy includes economics. The

385 R. Schulze, "Le droit privé commun européen" (1995) 1 *Revue Internationale de Droit Comparé* 7.

386 Allowing referrals from anyone would be optimal. However, if the volume of referrals became unmanageable, restrictions could be imposed on permitted parties.

economic analysis of biotech-patents is crucial to achieve the main objectives of patent law which is to foster innovation in society and thus we recommend the membership of an economist.[387] Finally, we suggest the membership of an expert in bioinformatics.[388]

6.3.4.2 The powers of the Board in the EU

The charge of the Board of Public Policy would be to make recommendations to ensure that the patent system is serving the public interest. I suggest three main activities of the Board: (i) it would recommend legislation; (ii) it would comment on how current judicial doctrines are applied (and perhaps suggest new doctrine); and (iii) it would make recommendations to the national patent offices as to rule and policy making, internal procedures, and examiner training. In essence, the Board of Public Policy would serve as an ombudsman for the public interest, broadly defined, in the patent system. The Board of Public Policy would add several elements to the current commentary and recommendations on the patent system made by academics, NGOs, etc. It would add focus, speed, competency, and prestige. The Board of Public Policy would deal exclusively with the patent system and how it interacts with public policy; this focus would allow the Board of Public Policy to understand the breadth and depth of the issues. This focus would allow the Board of Public Policy to respond quickly to ethical and policy concerns as they arise. This Board located in the patent office could inform the patent courts, such as the Federal Circuit in the US, and the legislature. Indeed, the Board of Public Policy would have access to information and people that would allow it to understand the breadth of the considerations behind current and

[387] The US FTC, the regulatory body charged with preventing illegal monopolies, made a number of recommendations to the USPTO and the Federal Circuit (*The Proper Balance of Competition and Patent Law and Policy: A Report by the Federal Trade Commission*, October 2003, www.ftc.gov/os/2003/10/innovationrpt.pdf). Among these was a recommendation to "expand consideration of economic learning and competition policy concerns in patent law decision making," *ibid.*, 17 (Recommendation No. 10). This recommendation was opposed by the USPTO and the American Intellectual Property Law Association (AIPLA). AIPLA wrote that "injecting economic theory into the interpretation and application of clearly defined statutory criteria will simply result in greater uncertainty. AIPLA believes that Congress, and not the USPTO or the courts, is the proper authority to consider economic theory and competition policy-oriented principles." AIPLA Response to the October 2003 FTC Report: "To Promote Innovation: The Proper Balance of Competition and Patent Law and Policy," www.aipla.org/Content/ContentGroups/Issues_and_Advocacy/Comments2/Patent_and_Trademark_Office/2004/ResponseToFTC.pdf. Clearly, sophisticated economic analysis is beyond the competence of the patent examiner and administrative judge.

[388] Chin has used bioinformatics to argue that patents on DNA impede the discovery of new genetic material, A. Chin, "Research in the Shadow of DNA Patents" (2005) 87 *Journal of Patent & Trademark Office Society* 846.

proposed law and policy. With a legislative mandate to perform its duties, the Board of Public Policy would have the power to command real consideration of its suggestions.

Such a Board could function in all the major patent offices of the world. In the EU, Japan, and other countries that have *ordre public* and morality exceptions to patentability, the Board of Public Policy could serve an additional function: it could act as a resource for examiners faced with questions on how *ordre public* applies to a specific patent application. This function is essentially what Beyleveld and Brownsword have proposed.[389] This role would be more limited than the policy recommending role of the Board: here, rather than speculating on what the content of the law should be, the Board would confine itself to applying the law as it *is*. In assessing what the law is, the Board would provide much expertise on ethics and policy that the examiner simply does not have. The recommendations of the Board might be obligatory or hortatory. In addition, the Board might alternatively adjudicate intermediate appeals on *ordre public*. The recommendations or decisions of the Board should be appealable, as the question of *ordre public* is ultimately a legal question, and the Board's main focus would be policy.

6.3.4.3 The Board as a reconciliation of the classical and radical approaches

In keeping with the classical approach, the creation of the Board of Public Policy would not necessitate any new requirements to be introduced into the patent law. The efficiencies and competencies that the classical approach seeks to protect would remain. However, the frequently voiced ethical, environmental, economic, and other concerns with the patent system would be considered and then crafted into legislative or departmental recommendations if appropriate, or would be discarded with adequate reasons when such a course would be appropriate. Such a system would allow the public to play a role in the patent system that it is not able to do now. Given time to do its job, the Board would contribute to the formation of a patent system better tuned to the public interest and the wider world of concerns that patent law affects.

The Board would also coordinate the roles of the various players in the formation of patent law. The courts and the USPTO usually defer to Congress on policy issues. Congress, however, often defers to the expertise of the USPTO and the Federal Circuit. Advocacy as to the proper distribution of ideas is divided.[390] With its full view of the relevant policy

[389] Beyleveld and Brownsword, *Mice, Morality*, 68–70 and 89–90.

[390] 130 Cong. Rec. 28,071, reprinted in 1984 U.S.C.C.A.N., 57–58 (statement of Professor Herbert F. Schwarz, advocating that Congress defer to the Federal Circuit to create appropriate precedent on joint inventorship).

issues, the Board could recommend not just what improvements to the law are appropriate, but also which entity should take the relevant steps. The work that the Board could produce includes (i) legislation to address a policy issue; (ii) specific rules and practices that the patent office could use to implement the law; and (iii) an analysis of issues that could arise in adjudication. While some of these recommendations might not be adopted, such a comprehensive approach would give law- and policy-makers a better idea of the consequences of potential legal and policy solutions.

Such a Board would eliminate the main objections of the classical approach regarding application of morality, *ordre public*, moral utility, or similar doctrines at the patent offices. The examiners would not make decisions based on ambiguous policy doctrines; they would focus on their core competencies of technology and legal standards of patentability. Furthermore, we envision such a board facilitating the formation of specific rules to address specific problems, rather than broad doctrines like *ordre public* and morality.

In this approach, the Board would go beyond the function of reconciliation between the classical and radical approaches to patent law. It would focus not only on concerns about health, the environment, and ethics, but also on how to make patents more friendly to business and economic opportunities.

6.3.5 *Concluding observations on the methods of assessing* ordre public *and morality*

In the analysis of the role of *ordre public* and morality and moral utility in EU and US law, there have been a number of unsuccessful attempts to apply the *ordre public* and morality exceptions to prevent the patenting of biotechnology inventions. The EPO has interpreted the concepts of *ordre public* and morality beyond the narrowest classical approach. In *PGS* and the *Onco-Mouse* cases, the Opposition Division has brought post- and pre-IP issues of environmental protection and animal suffering inside the generic umbrella of *ordre public* and morality. However, I have suggested that entire categories of inventions cannot be *a priori* excluded from patentability under the concepts of *ordre public* and morality. Each application must be considered individually. It is evident that patent law has been applied in a generally classical manner, although several legal statements, *obiter dicta*, and academic analyses demonstrate that there is a place in *ordre public* and morality to address inventions that are harmful to the environment, human health, etc. In the US, there is currently little room for ethical and policy concerns to be addressed, but I have examined several ways in which such concerns could be integrated. In both the US and European systems, the classical approach tends to currently dominate.

6.3.5.1 The necessary reconciliation between radical and classical approaches

Both the classical and radical positions are fundamentally legitimate; however, they are for the most part addressing separate concerns and speaking past one another. The radical position is right in saying that patents should be in the public interest, and the public interest is not only represented by the patent applicant. Numerous scholars, activists, and others have pointed to various ethical, environmental, and other public policy concerns that are not taken into account in determinations of novelty, inventive step, utility, and adequate disclosure.

However, advocates of the classical position correctly respond that patent examiners are not equipped to address these public policy concerns. Patent offices around the world struggle to keep their offices stocked with technically proficient examiners. The flow of patent applications is on the rise, and delay times are increasing. Complaints are common about the quality of examination on the basis of patentability standards (whether the best prior art is found, whether the examiners appreciate all the relevant legal issues) and costs of examination. In short, patent offices struggle to do their "classical" work proficiently, and adding other tasks to the examination process would lead to higher costs, longer delays, and distractions from these core competencies.

Furthermore, most of these ethical and policy concerns are vague, and many are controversial. Giving patent examiners the responsibility of determining whether a given patent is in the public interest would add significant uncertainty to the process of obtaining a patent. These problems, together with the other concerns, would make companies less inclined to patent in general, not only for inventions that are likely detrimental to the public good. Thus, the patent examiner is not the right individual to perform the task of determining ethical and policy questions. The classical approach points to national legislation as the solution to any perceived public policy concerns.

Unfortunately, the legislative solution has significant drawbacks as well. It is slow, especially on issues that are not in the public eye, which is usually the case with patent law. For example, the EU Biotechnology Directive was ten years in the making. Numerous parties have spoken of the need to adjust various elements of the traditional US patent system for some time now, but it was not until 2005 that a bill was introduced into the House of Representatives. Patent reform acts were introduced in 2006 and 2007, but there is no sense of urgency to pass such a bill, and no immediate prospects for such passage. Political negotiations and the numerous interests represented usually ensure that on such issues national legislatures act on time scales of several years or even decades. Furthermore, the patent applicants typically are strongly represented in such fora, as they are before patent offices; their interests are represented, but to what degree are other concerns? Also, although patent legislation is vital, it is perhaps currently ill-suited to deal with many of the public policy concerns implicated by the patent system. In a sense, the national legislature paints with too broad a brush; the patent examiner with a brush that is too technically narrow. To analogize to the tale of Goldilocks and the Three Bears, the legislature is too big; the patent examiner is too small. We propose the creation of a Board of Public Policy that can propose solutions that reconcile the two extremes, to be "just right".

6.3.5.2 The cost-benefit analysis and the preponderance of evidence

I encourage the adoption of the cost-benefit analysis as it may also be used to question certain fundamentals of patent law: since a patent is an exceptional reward for the creative efforts of the inventor through a temporary monopoly right as opposed to the normal market competitive use of the knowledge, the inventor cannot abuse its exceptional monopoly by creating prejudice to a legally protected public interest, i.e., the preservation of the environment. As a consequence, such an analysis cannot be simply an absolutistic or formalistic analysis of the costs and benefits. This balancing exercise must not evaluate the nominal benefits as opposed to the costs. The concept of public order requires a full-fledged analysis of who benefits from the exercise of the patent right and who bears the costs. If the patent holder or the manufacturer of a patented invention – which represent a limited category of society – benefits most prominently at the expense of one of the most fundamental legally protected public interests such as the environment – which is also a right *erga omnes* at the national and international level – then this patent application should be deemed abusive and contrary to the *ordre public*.

The cost-benefit analysis is strictly connected with the concept of preponderance of evidence as it functions as a method of performance of the cost-benefit analysis. A few general considerations on the methods of application of the preponderance of evidence are in place. The preponderance is realized when "contemporary, primary evidence of a number of related matters all point in the same direction, and the evidence so accumulated leaves no doubt in the reader's mind that only one reasonable conclusion can be drawn from it." Then it is appropriate to say that piece or element of evidence is established by the preponderance of the evidence (which is the standard for a defendant in court).

"However, if there is found a piece of evidence that points in a different direction and if it is not possible to show clearly that this piece of evidence is wrong,"[391] then the standard calls for an ultimate decision that is based on weighing the amount and quality of submitted evidence in favor or against. The evidence should be admitted after careful examination of the compliance of each piece of evidence with the present guidelines. It must be demonstrated that the evidentiary weight in favor is greater than the one opposing these claims.

Two qualitative elements should be taken into account in this weighing exercise:

(i) *Force and exclusivity*. The superior evidentiary weight is determined by the convincing force rather than by the number of pieces of evidence. This convincing force must be "sufficient to free the mind wholly from all reasonable doubt" and "sufficient to incline a fair and impartial mind to one side of the issues rather than the other."[392]

(ii) *Quality and quantity*. The preponderance of evidences is established if the circumstantial and strong pieces of evidences are more significant in quantity and quality than the contrary pieces of evidences. I maintain that in this particular context, the quality should supersede the quantity. The methods of appreciation of the quality and quantity vary from court to court.

The banning of the patent will vary upon the importance that a certain legal system will attach to the preservation of biological resources vis-à-vis the importance that it attributes to human technological inventions and their commercialization. A society that attributes an absolute priority to the preservation of the environment may ban a patent as an incentive to a certain technique even if only one claim brings prejudice to the environment.

6.3.5.3 Insights on the performance of the cost-benefit analysis

In conducting the so-called "balancing exercise/cost-benefit analysis," courts and patent examiners are being called on to pronounce judgments on issues of great social and environmental concern. International law related to patent and environmental law (i.e. the TRIPS Agreement, EU Biotech-Directive, and the subsequent international practice of the CBD) has led us to conclude that:

[391] Walter Lee Sheppard, Jr., quoted in S. Guinn, *Preponderance of Evidence*, www.grapevine.net/~swguinn/evidence.html.

[392] B. A. Garner (ed.), *Black's Law Dictionary* (8 edition, Thomson West, 2004). US Federal Rules of Evidence 104(a), www.law.cornell.edu/rules/fre/rules.htm#Rule104 for the general admissibility of evidence in court proceedings.

(i) The protection of the environment is not a distinct exception but is part of a more general concept of *ordre public* and morality.
(ii) European patent law accepts that an invention that seriously harms the environment is not patentable.[393]

How to determine when an invention is contrary to *ordre public* and morality remains to be seen. The classical approach views the examination efforts performed by the Board of Appeal in *Onco-Mouse* and *PGS* as largely sufficient, whereas a more radical approach wishes to apply the most updated methodologies used by environmental international bodies on risk assessment, including the precautionary principle.

This study has demonstrated that it can be appropriate to ban patenting on grounds of protection of the environment under the concepts of morals and especially *ordre public* as long as the patent examiners bear in mind that these concepts logically belong to a "level of scrutiny which pertains to exclusivity over ideas rather than permissibility of commercial exploitation."[394] Thus, the methods of assessment of a patent cannot be the same as those of specific competent authorities for commercialization of the final product.

A reasonable policy suggestion would consist in fully applying the concepts of *ordre public* and precautionary principles at the moment of the delivery on the market instead of at the patent application stage where the form and the composition of final product cannot be easily foreseen. A final biotech product is generally made of several patents. For instance, the current form of *Goldenrice* includes up to 44 patents which, although related to the same product, have different functions. While not altogether impossible, it can be very difficult for examiners to assess the ecological risk of each inspected patent. Such a risk will be much better assessed once the final product is prepared, before delivery on the market, and by a competent authority other than the EPO. This is a "post-IP" matter that should not be examined at the patent stage.

There are two main possibilities for dealing with a patent that may pose problems potentially falling within the scope of *ordre public* and morality. The first is when the patent's commercialized product impact is unforeseeable, which happens in most of the cases. The cost-benefit analysis should then be the usual method of decision about the patentability of an application. The balancing test will usually consist of granting the patent

[393] Thanks to this clause, the EPO does not need to overelaborate new concepts such as the one of "imperative exigencies" infamously used by the ECJ in order to recognize (the legitimacy of the limits to the fundamental liberties of the Treaty) when a general interest (for instance, environment protection) has to be pursued.

[394] Ricolfi, "Biotechnology, Patents and Epistemic Approaches", 78.

because of the beneficial effect of disclosing an innovative process for society as a whole. The costs will be inherently not provable because the possible damage is unforeseeable.

In certain cases, the patentable invention will clearly result in a product whose manufacture is likely to prejudice the environment. In such cases, the concepts of *ordre public* and morality assume paramount importance, and two particular measures should be adopted:

(i) The EPO should require a shift of allocation of the burden of proof from the *opponent* to the *claimant* in order to demonstrate that the invention is environmentally friendly;

(ii) The EPO should adopt what Ricolfi calls a "pre-emptive action"[395] instead of the "precautionary principle." This Ricolfian concept of pre-emptive action can be used *mutatis mutandis* to ban the patentability only when adverse effects on the environment are clearly assessable and the patentable invention consists of a marketable product. In such a case, the ban on exploitation may be precisely "offered by the device of coupling the prohibition on the manufacture, sale and use of the threatening device with a parallel exclusion from patentability."[396]

6.3.5.4 Economic policy

Economic policy plays a role in the cost-benefit analysis. An economic consideration presents sound advantages flowing from a more careful analysis of the concepts of *ordre public* and morality. If patent protection were available for environmentally unfriendly inventions, by allowing such protection, society would be encouraging capital flow into R&D of "undesirable" technical solutions. In other words, if the ban is just the

[395] Ricolfi uses the term in a different context, and I expand its scope: "society at large may have strong convictions on the undesirability of resorting to a particular sort of technical solution and fear that the corresponding ban might gradually be bent, maybe first at the margins, as the time goes by. A good precaution against this kind of risk of erosion of the ban on exploitation may precisely be offered by the device of coupling the prohibition on the manufacture, sale and use of the threatening device with a parallel exclusion from patentability. The latter measure may well reinforce and stabilize the former; and can be seen as a form of legitimate 'pre-emptive action' against future changes in the rules": Ricolfi, "Biotechnology, Patents and Epistemic Approaches", 79.

[396] *Ibid.*, 80. It has also been proposed that the issues of compatibility with *ordre public* and morality should be left to national judges, the EPO granting the patent for a test-period of ten years; then national judges would examine the patent after its exploitation has occurred. Also, in the course of ten years, the concepts of *ordre public* and morality will have evolved in a permissive way; therefore, what was not permitted today will be very probably permitted in ten years. However, I maintain that this option should be discarded on the simple ground that, as Gallochat puts it, the law should be applied to solve problems when they arise instead of deferring to a period when the law will change for multiple reasons; Gallochat, "Le brevet et l'éthique", 8.

manufacture of unfriendly products, then inventors are going to keep trying to obtain patents on them. The innovator may hold an exclusive right on know-how that he can transfer and license, nourishing a hypothetical hope that the ban might eventually be relaxed or lifted. Obtaining the monopoly right of a patent gives investors confidence in their ability to bring about a change in the rules concerning commercial exploitation which may eventually have the effect of rewarding their outlay.

It is finally submitted that where the patentable invention is clearly determined by the examiners to be environmentally unfriendly, it will be especially important for the inventor to cease further investment of financial resources in the development of a product that might prove unfit to be marketed. If the inventor knew of possible environmental risks at an early stage of product development, the inventor might direct his/her research efforts in another, more environmentally friendly direction.

7 Positive protection of traditional knowledge

7.1 Positive protection of plant genetic resources and related traditional knowledge in provider countries

While defensive protection seeks to prevent the granting of patents based on misappropriation of TK, positive protection seeks to grant exclusive rights on TK.[1]

There are various options to protect TK subject-matters through IPRs existing in TRIPS or newly crafted *sui generis* IPRs. This action calls for a certain urgency at the domestic level since international negotiations in the relevant fora (in order to recognize new types of IPRs and to clarfy the concepts thereof) progress at a slow pace. Certain countries have been considering particular legislative measures to protect TK with national IP legislation.[2] Nevertheless, treaties harmonizing IP concepts on this matter can grant protection beyond the boundaries of a certain State because of the principle of territoriality of IPRs (i.e. their enforcement stops at national boundaries).

There are two major types of protection of TK. The first is the holistic that intends to create one right to protect all the different aspects of a TK subject-matter. This protection seems to satisfy the needs and expectations of most of TK holders (section 4.1 above). The second type of protection divides the different aspects of a TK subject-matter according to the types of suitable IPRs. This section shall focus only on the latter as holistic protection requires too deep an anthropological study of TK and it is not immediately clear how this can relate to the existing IP system for

[1] WIPO IGC on IPGRTKF: *Matters Concerning Intellectual Property and Genetic Resources, Traditional Knowledge and Folklore* (April 30–May 3, 2001) WIPO/GRTKF/IC/1/3, www.wipo.int/edocs/mdocs/tk/en/wipo_grtkf_ic_1/wipo_grtkf_ic_1_3.doc

[2] OAU, the Andean Community, and India as explained in T. Cottier *et al.*, "The Current Law of Plant Genetic Resources and Traditional Knowledge", in T. Cottier and S. Biber-Klemm (eds.), *Rights to Plant Genetic Resources and Traditional Knowledge* (CABI, 2006) 83–92.

an effective form of protection in the market economy and globalized world of intellectual property.

These attempts to protect TK by IP and the required infrastructure will be taken up for legal analysis. In this section, proper distinctions among IPRs apt to protect various TK subject-matters (traditional agriculture, TMK, informal breeding, etc.) will be considered. Protection can be attained through existing IPRs such as low-cost patents, trademarks, trade-secrets and GIs, or through new IPRs or quasi-IPRs, such as traditional IPRs and liability regimes.

Furthermore, indigenous communities might find it useful to use several different forms of IP protection in an overlapping way to ensure that various elements are protected, i.e. the use of GIs does not exclude the simultaneous protection of trademarks and vice versa in the same way in which some jurisdictions grant a double protection through a patent and a PBR that largely overlap their scopes of protection. In other words, just as software designers use patent and copyright protection to cover different technical aspects of their product, trademark and perhaps trade-dress and trade-secret protection as well, so too might a practitioner of traditional medicine rely on overlapping forms of protection to protect his plant variety through a *sui generis* right, or medicinal formula with a trade-secret, or designs and ritual chants with copyright.[3]

There is a further very important advantage in fragmenting TK into various IPRs: to protect the integrity of a certain TK subject-matter, its holder would simply need to prove the violation of one of these IPRs that cover a single element of the whole TK subject-matter. In contrast, had its holistic nature been adopted within an IP protection framework, its holder must prove the total infringement of all parts of the TK element. The following brief example illustrates this advantage. Imagine a shaman who holds the knowledge of a traditional healing method. He has protected various elements of this TMK by various IPRs: the traditional medicinal formula is protected by a petty patent; the traditional chant is protected by a copyright,[4] etc. If someone were to reproduce the traditional medicine in a chemical process without reproducing the traditional chant, the shaman (or the TK holder), in order to protect the whole traditional healing method, would only need to prove infringement of the chemical process protected by the petty patent.

If biodiversity provider countries adopted a *sui generis* IP regime for TK it would require the country to set up appropriate rules for access

[3] US intervention at WIPO/GRTKF/IC/3.

[4] *Survey on Existing Forms of Intellectual Property Protection for Traditional Knowledge – Preliminary Analysis and Conclusions*, WIPO/GRTKF/IC/2/9 (December 3, 2001).

to protection, scope of protection and ownership. The danger in choosing *sui generis* protection for TK and PGRs instead of existing IP regimes is that recipient countries would not have the corresponding legislation to acknowledge these rights and their enforcement in another country, and they would remain void through the principle of territoriality of IPRs.

7.2 Protecting traditional knowledge through the implementation of Article 27.3(b) of TRIPS[5]

This section discusses whether the international legal instruments of the CBD and ITPGRFA make it possible for the concerned Member State to shape legislation implementing Article 27.3(b) of the TRIPS Agreement in such a way as to give protection not only to what has been termed as technological innovation but also and simultaneously to this kind of other-than-technological and non-laboratory knowledge, i.e. TK. I will adopt a two-pronged approach.

This discussion will focus on the fundamental concepts sketched in sections 4.2.2 and 4.2.4 above, namely (i) the plant GR itself, (ii) the TK referring to it, and (iii) the technology applied to it. Although I will try to differentiate between them, they actually lie along a continuum in relationship to the various possibilities of implementation of national or regional legislation and their compliance with international legal instruments.

Section 3.3.2 above examined the principal content of farmers' rights according to the FAO treaty; this section examines in more detail the impact that the UPOV Convention could have on DCs if they had to adopt UPOV as the *sui generis* system under 27.3(b) of TRIPS. These following subsections consider the different opinions of both the positive and negative aspects of this regime for protecting new varieties of plants. The main focuses are breeders' rights and the effects of UPOV on traditional grassroots innovators in developing and LDCs. In light of the definition of farmers' rights in FAO treaty, I will finally elaborate some proposals to be incorporated in domestic or regional laws aimed at creating "effective *sui generis*" protection for plant varieties in compliance with Article 27.3(b) of TRIPS.

[5] The ideas contained in this chapter were developed in collaboration with Professor Marco Ricolfi, University of Turin, in the Brasil–Italy project sponsored by Istituto Agronomico per l'Oltremare.

7.2.1 *The two-pronged approach: protecting traditional knowledge by implementing Article 27.3(b) in provider countries*

Article 27.3(b) TRIPS provides that

> 3. Members may also exclude from patentability: [...] (b) plants and animals other than micro-organisms, and essentially biological processes for the production of plants or animals other than non-biological and microbiological processes. However, Members shall provide for the protection of plant varieties either by patents or by an effective *sui generis* system or by any combination thereof.

The following considerations need to be taken into account in order to determine how provider countries, mainly developing ones, may use this Article to kill two birds with one stone, i.e. appropriately protect TK by implementing Article 27.3(b):

(i) Article 27.3(b) of TRIPS primarily deals with technology and technological innovation applied to plant (and animal) GRs. The mandated protection clearly applies regardless of the place of origin of the resource itself.

(ii) Through Article 27.3(b), industrialized countries have successfully extended, as far as possible, most of their types of protection to DCs in order to solve the problem of territoriality of IPRs. Article 27.3 (b) has bridged the gap of this legal framework in the field of biotechnological innovation. Here the question arises whether WTO members, including DCs, which are the biodiversity providing countries, should interpret this provision in line with the whole spirit of the TRIPS Agreement, i.e. the function of the entire Agreement is to create a legal framework whereby industrialized countries can exercise and enforce their IPRs in DCs and vice versa (especially in the future). In light of this teleological interpretation, I will explore the question whether the expression "effective protection" in this Article means that the type of protection should correspond to more or less patent or plant variety (in existence in the UPOV area).

(iii) The mechanisms for protecting technological innovation concerning "plant varieties" are patent laws. Moreover, PVP, along the lines of the 1991 UPOV, is also compliant. And this latter is the option that most of DCs are choosing either because it is a system already in place or because the DCs have entered into bilateral or regional investment treaties with (usually) industrialized countries that mandate UPOV standards of PVP.

(iv) Other TRIPS-compliant options can be also envisaged. First, utility models and petty patents may supply variations on the basic patent paradigm.[6] In addition, the idea of fashioning some kind of *sui generis* protection has been explored by various authors.[7]

(v) I note that, although the recognition of farmers' rights is still vague, it can be combined to the entitlement of the grant of a PBR. These paradigms are understandably proposed to provide protection not for formal technological innovation but rather for TK. In connection with these alternatives, some hard questions arise: can the "effective protection" of Article 27.3(b) be translated into an exclusive right that may provide appropriate TK protection and at the same time comply with the TRIPS requirement to also protect formal technological innovation (thus being TRIPS-compliant)?

(vi) Another consequent question arises: what are the pros and cons and costs and benefits of implementing Article 27.3(b) of TRIPS in such a way that it covers technological innovation and TK at the same time?

(vii) The "two-birds-with-one-stone" approach undoubtedly offers a very positive outcome for TK-providing states since TK-based intellectual contributions would benefit from the same kind of protection that is granted to foreign technological innovation. Certainly, this approach will inevitably entail costs. The conclusion that the features mandated by Article 27.3(b) are appropriate to protect both technological innovation and TK depends on the degree of modification of the regime applicable whenever the subject-matter is TK rather than technological innovation. These modifications consist, for instance, in the scope of "fair use" which is different. As far as TK is involved, it consists in a "fair use" as against the protection of the main regime for technological innovation. Before presenting the major issues of the "two-birds-with-one-stone" approach, I will identify the conditions under which this legislative approach can function.

7.2.2 *The criteria for implementing the two-pronged approach*

A country may be at the same time a biodiversity provider and the recipient of the GR since it has the capacity to generate a substantial amount of intellectual added value pertaining to living matter in the form of TK that is worth protecting by IPRs. To comply with TRIPS,

[6] A. K. Gupta, "Conserving Biodiversity and Rewarding Associated Knowledge and Innovation Systems: Honey Bee Perspective", in T. Cottier and P. Mavroidis (eds.), *Intellectual Property, Competition and Sustainable Development* (University of Michigan Press, 2003) 148.

[7] M. Ricolfi, "The Interface between Intellectual Property and International Trade: The TRIPS Agreement" (2002) 1(1) *Italian Intellectual Property* 42.

Ricolfi has identified some criteria in line with the western IP approach that views TRIPS as a harmonizing tool of IPRs among States that will avoid the territoriality problem. In this regard, certain requirements that any implementation of protection of Article 27.3(b) must meet can be spelled out, paraphrasing Ricolfi, as follows:

(i) the access requirements must be similar – and not higher – than those provided for under current patent or plant variety legislation;
(ii) the scope of protection must be similar – and not lower – than the one provided for under current patent or plant variety legislation;
(iii) the exceptions and limitations must be similar – and not broader – than the ones provided for under current patent or plant variety legislation;
(iv) the term of protection must be similar – and not lower than – the one provided for under current patent or plant variety legislation; and, finally,
(v) the protection granted in the technology-recipient country must enjoy some sort of link (similar to the Paris Convention priority)[8] to make sure that the holders of title in the technology-provider country may benefit from extending the protection from the country of origin to the State of the grant.[9]

Within these criteria, I will describe the boundaries of effective PVP that takes into account the considerations proposed by the legal doctrine already sketched in section 3.3.2 above.

The following sub-sections discuss the suitability of a regime of protection of plant varieties. I will sketch the evolution of the UPOV system and the potential negative impact that it can have on farmers' rights in DCs (section 3.3.1 above). However, a *sui generis* protection system for plant varieties is important for the innovation in this field. Its effectiveness resides in the way the PVP is implemented and in the exceptions that are granted so that various types of farmers can be motivated to innovate in this field.

7.2.2.1 The evolution of the UPOV Convention for developing countries

The first effective protection is UPOV because this is a system that is already in place at the international level. The International Union for the Protection of New Varieties of Plants, UPOV, which was signed in Paris in 1961 and entered into force in 1968, is being promoted by some industrialized countries as the benchmark of the "effective *sui generis* system" for

[8] www.en.wikipedia.org/wiki/Priority_right.

[9] The expression "current patent or plant variety legislation" refers to the major conventions on these IPRs.

PVP required by Article 27.3(b) of TRIPS. UPOV was established by the International Convention to protect new varieties of plants. The purpose of the UPOV Convention is to ensure that the Member States of the Union acknowledge the achievements of breeders of new plant varieties by making available to them an exclusive property right, the PBRs, on the basis of a set of uniform and clearly defined principles. These principles include the stability, uniformity, and distinctiveness of the new plant variety giving rise to the right.

The Convention was revised in Geneva in 1972, 1978, and 1991. During these revisions, genetic engineering has contributed to bring substantive manipulations of plant genetic codes; consequently, plant breeders have progressively sought to soften the distinction between UPOV and patent regimes.

Until recently most countries allowed farmers and other traditional breeders to be exempted from the provisions of such rights, as long as they did not indulge in branded commercial transactions of the varieties. Now, however, after an amendment in 1991 and the subsequent harmonization of the principles established in the Convention, UPOV itself has tightened the monopolistic nature of plant variety breeder rights by substantially removing the exemptions for farmers.

7.2.2.2 Proposals for an effective sui generis *system for plant varieties in developing countries*

During the Uruguay Round negotiation of TRIPS, DCs did not restrict their arguments to the mere fact that UPOV system was not included in Article 27.3(b) as *sui generis* protection of plant varieties. The African Group tried to insert a footnote stating that any *sui generis* law for PVP could provide for the protection of innovations of indigenous communities, the continuation of traditional farming practices, and prevent anti-competitive rights or practices which would threaten sovereignty of people in DCs. Such precise statements could effectively balance the monopoly rights granted by TRIPS; nevertheless the footnote was refused because of countering pressures by industrialized countries.

The obligation thus remains for DCs that are not members of UPOV to protect plant varieties by an "effective *sui generis* system." If, on the one hand, the inclusion of certain suggestions in a revision of UPOV is deemed highly unlikely at present, on the other hand, DCs can consider a national or regional system which I call "UPOV plus CBD plus FAO treaty." Therefore, considering the UPOV model, these countries can create an acceptable "effective *sui generis* protection" of plant varieties by amending its current provisions to keep pace with the interests of farmers and the environment. The goal of the following suggestions is to grant rights and privileges to farmers so that they can produce, save, and exchange seeds freely. In a wider sense, farmers' rights may be extended to farming communities.

In my view, the "two-birds-with-one-stone" approach (criteria (i), (ii), and (iii)) can be realized by implementing Article 27.3(b) through legislation that realizes the goal of granting rights and privileges to farmers so that they can produce, save, and exchange seeds freely. In a wider sense, farmers' rights may be extended to farming communities. In order to do so, the following legal concepts within PVP are proposed and should be developed:

(i) *Definition of "variety"*: the term "variety" defined in UPOV text should be implemented in the national legislation so that a plant variety developed by farmers shall be eligible for protection. Further, a provision should provide that a variety shall be eligible for protection only if it differs from another variety, not in one characteristic, but by a wider margin, in order to prevent a series of genetically homogenous or closely related lines in several characteristics. The protection requirements of plant varieties should be more flexible and adapted to the "country varieties."[10]

(ii) *Lowering the required level of uniformity or stability*: the requirements of uniformity and stability required by member States under the UPOV Act 1991 do not allow for heterogeneity. Acceptance of such heterogeneity would facilitate and create incentives for breeding varieties that are better adapted to the needs of indigenous and small-scale farmers; it may, therefore, be advisable to lower the required level of uniformity or stability. The background of the proposal is to switch from "distinctness," "uniformity," and "stability" to "distinctness" and "identifiability."[11] On the other hand, broadening the limits of heterogeneity within a plant variety to be protected inevitably leads to broader property claims. This has to be taken into account when defining the acts requiring the right-holders' authorization in relation to the protected variety. The scope of protection depends on the requirements for protection and the breadth of the claims.

(iii) *Extension of farmers' rights*: A core set of provisions or a single provision should protect the interests of farmers. Indeed, farmers' rights should not be confined necessarily to saving seeds for reuse on farms, they should be extended to a wider scope. These rights should be shaped as follows: (a) farmers' rights should include the right to save

[10] G. Dutfield, *Intellectual Property Rights, Trade and Biodiversity* (Earthscan, 2002) 82.

[11] D. Leskien and M. Flitner, "Intellectual Property Rights and Plant Genetic Resources: Options for a *Sui Generis* System" (IPGRI, Rome, June 1997) 6 *Issues in Plant Genetic Resources* 54–55.

and share seeds, to sell their product, as well as to have their contribution to the selection, breeding, and conservation of existing varieties duly recognized; (b) traditional farmers should be entitled not only to replant on their own farm the same seeds from one crop to the next but also to engage in what it is; (c) this right should extend to any variety "essentially derived" from a variety that was developed by traditional breeders; (d) the right should be recognized in accordance with the custom of traditional breeders; (e) the right to a plant variety is to be recognized regardless of whether or not it is registered.

This seed-sharing practice is cooperative more than profit-oriented since it is known as seed exchange "across the fence" from one neighbor to the other. It is often convincingly presented as an effective method of combining with the biodiversity conservation provisions of the CBD[12] since it contributes to genetic diversity and it helps to preserve the vitality of the cross through various generations.

In a case where the seed is protected by a patent, then the replanting of seeds and seed exchanges would constitute a patent infringement,[13] and this is so also under the UPOV 1991 regime. Moreover, even if a plant variety regime were adopted, the conclusion would not be very different. Under the latter set of rules, replanting might become legitimate under a farmers' exemption rule; nevertheless, under UPOV rules seed exchanges would constitute infringement.

The solution envisaged by Ricolfi and Cullet is that a biodiversity-rich provider country should enact legislation concerning TK that contains a sort of immunization from IP infringement action on behalf of non-commercial, non-profit, and cooperative seed exchanges.[14]

(iv) *Exclusion of certain varieties*: the requirement of extending rights to all varieties of plants should be amended. To secure public interest, certain varieties may not be registered if it appears necessary to prevent the commercial exploitation of such variety within their territory to protect *ordre public* (for instance, crops essential for the nation's food security needs such as rice, cereals, etc.) or morality, including the protection of human, animal or plant life, or health and the avoidance of serious prejudice to the environment.

(v) *Twin recognition*: twin recognition of commercial breeders' rights and farmers' rights was proposed at the international level a decade ago but how exactly to bring it about is still being discussed. Concerning farmers, the aim should be, *inter alia*, to protect

[12] Ricolfi, "Interface between Intellectual Property and International Trade", 43.
[13] P. Cullet, "Plant Variety Protection in Africa" (2001) 45(1) *Journal of African Law* 97–122.
[14] A very similar proposal is to found in Cullet, *ibid.*, 112.

farmers' current techniques or varieties and also allow them to derive benefits from any improvements they will carry out without being stopped by patents.

(vi) *Inclusion of public interest clause*: the grant of compulsory licensing should be included. The provision may require that a compulsory license to a party be granted if the reasonable requirements of the public for seeds have not been satisfied or if seed of the variety is not available to the public at reasonable prices.

(vii) *Reference to CBD*: this model should mention its compliance with the CBD's objectives and principles.

(viii) *Declaration of resource*: at the time of filing an application, the breeders must declare the name and source of all varieties used in the breeding of new varieties where a landrace or farmer variety has been used. The application for protection should specify the parental lines and their country of origin to safeguard the rights of local farmers. If one or more lines have been derived from a particular country, a royalty should be paid to farmers, communities, or research organizations who have developed them. In cases where it is not possible to find out the information required above or when the farmers do not seek such royalty, it should be credited to a fund to be established under the Convention. Additionally, the principle of "derived variety" has to be excluded in regional agreements. This is made possible by having a wider margin in the distinctiveness requirement.

(viii) *PIC*: any commercial breeder wishing to use the variety to develop other varieties must obtain the traditional breeder's PIC. In such a case, benefits sharing arising out of the use of TK and GRs should be ensured.

(ix) *Inclusion of limitation provision*: limitation provision to the breeders' right should be provided. These limitations may be, for example, acts done privately and for a non-commercial purpose, use of variety for research and experimentation not designed for commercial exploitation, and use of variety for teaching purposes.[15] Moreover, privileges should be granted to research institutions or people to use protected varieties for the development of varieties for non-profit, namely public interest uses.

(x) *Promotion of transfer of technology*: all applications for protection should be required to demonstrate the ability to promote immediate, substantial, and direct benefit to the people of the respective

[15] Ricolfi, "Interface between Intellectual Property and International Trade".

countries by the cultivator of such a new variety. The mode of developing the variety entitled to protection should be disclosed in order to facilitate the transfer of technology.

Some of these provisions can also be adapted to protecting TK in the form of farmers' rights in particular situations. All the above provisions can be interpreted in compliance with the CBD obligations.

The above general suggestions have been formulated on the basis of scholarly works[16] and personal contacts[17] on this issue, bearing in mind that, as Cullet puts it, "relatively little conceptual work" has been done on defining an alternative system to monopoly rights. This lack is mainly due to the pressure on DCs to join UPOV. Through these suggestions, I have shown that, without rejecting genetically engineered seeds, it is possible to conceive of an alternative system in which IPRs are allocated in compliance with other commitments under international environmental law. A PVP regime, accompanied by the aforementioned characteristics, can eventually foster positive environmental results in the interests of these countries. Finally this regime can be an inherent part of implementing the CBD in domestic or regional legislation.

7.3 The creation of new intellectual property rights for plant genetic resources related to traditional knowledge

Traditional communities are coming more frequently into contact with industrial companies that access PGRs held and bred by these communities and they are seeing how these industrial companies may potentially achieve IP protection based on their TK and PGRs. In the pages that follow, I focus on creating, *de lege ferenda*, new IPRs or alternative regimes of protection for TK related to PGRs. In doing this, I do not take into account the important distinctions between PGRs for food and agriculture and PGRs for medical purposes. I limit my efforts to sketching some

[16] Dutfield, *Intellectual Property Rights, Trade and Biodiversity*, 78–85; Leskien and Flitner, "Intellectual Property, Rights and Plant Genetic Resources". For a deeper analysis on the development of *sui generis* variety protection in Africa, Cullet, "Plant Variety Protection in Africa", 117–22. S. K. Verma, "TRIPS and Plant Variety Protection" (1995) 17 *European Intellectual Property Review* 281 and T. Roberts, "Patenting Plants around the World" (1996) 18 *European Intellectual Property Review* 531; C. M. Correa, "Biological Resources and Intellectual Property Rights" (1992) 14 (5) *European Intellectual Property Review* 154–57 (1992).

[17] These proposals are the result of joint collaboration and discussions with H. J. Arunasiri, S. A. Chowdhury and R. O. Abdel-Latif (colleagues at the LL.M in IP course organized by WIPO and the University of Turin in 2001). These suggestions are also inspired by conversations held with Professor Marco Ricolfi and with Mr. Lavignol of the UPOV Office in Geneva.

types of protection of the general concept of TK with particular emphasis on farmer's rights and their TK.

This section explores which IP regimes' concepts are aimed not only at the conservation of seeds within a community but also at the marketing of these PGRs. At the same time, I am aware that Article 9 of the ITPGRFA has a wider scope than what is outlined in the present chapter. Some of the ideas expressed in this chapter should be considered as complementary to the implementation of certain specific rights of farmers lying largely outside the scope of the present chapter. Among these rights, I include the right to define, formulate, and execute policies and programs; the right to appropriate technology and the right to participate in designing and carrying out research programs; the right to use, choose, store, and freely exchange GRs; and, finally, the right to develop models of sustainable agriculture that protect biodiversity.

In order to shed more light on the various IP-related options for implementing Article 9 of the ITPGRFA, I have identified three models. The first model, called the creative commons / open source biotechnology, will be analyzed independently since it prepares groundwork for understanding the other two models of the compensatory liability regime and the traditional IPRs.

In all the commercial relations between farmers and industrial or research entities, rights over PGRs are to be properly regulated with the aim of sharing the benefits arising from their exploitation. For this reason, other legal tools have been proposed by Reichman – whose Compulsory Liability Regime (CLR) goes beyond the simple creative commons theories and develops the compensatory liability regime, i.e. a sophisticated *domain public payant* adapted to the relations between small-scale innovators – and by Cottier, who suggests the creation of a new generation of TIPRs. This analysis will end with an overview of draft treaty provisions that have been proposed by the WIPO IGC on IPGRTKF that falls squarely in this debate of models of protection of TK. These regimes are more appropriate than patents. It is now a fact that the scope of patents and PBRs for any type of small-scale innovation is aggravating the so-called "tragedy of the anticommons."[18]

7.3.1 Creative commons for conservation communities of plant genetic resources

The strengthening of IP monopolies and its expansion to new types of inventions, including biotechnology, has provoked among legal scholars

[18] M. Heller, "The Tragedy of the Anticommons" (1998) 111 *Harvard Law Review* 621–88.

increasing skepticism that has led them to develop ways to bring conceptual unity in a re-evaluation of the public domain or intellectual commons. Starting from this critique, legal scholars have developed the theory of creative commons and its related projects on open-source biotechnology. These projects offer a different view of the incentive of creativity until now spurred on by IP protection. These projects contribute to solving the high transaction costs and the aforementioned tragedy of anticommons created by the rapid pace at which the USPTO especially has granted patents to small-scale biotechnological outcomes in various fields, but particularly in biomedical research.[19]

Developed in the field of copyrighted software, the creative commons theory has offered a possibility to share one's creativity by placing it directly in the commons of the community of users. Such a theory posits that all the individuals of the community are willing to work to produce high-quality creative works. The underlying idea is that all creative works build off other creative works. This implies that if the commons are richer and more populated, it will be easier for everyone to build newer and more interesting works. This theory has been crucial in the development of open source software.[20]

The theory offers an alternative to the trends of higher monopolistic protection of software. Software is protected by copyrights and a community of users has been able to freely use the software and access the information through interoperability, interconnectivity, and reverse engineering. While copyrighted software has rendered possible the development of creative commons in the field of software, the patentability of the software can be extremely detrimental to the open-source movement and the consequent creation of creative commons communities. Indeed, the scope of a patent has the ability to block follow-on innovations.

Creative commons models have been developing for other-than-software copyrighted works especially in the internet environment,

[19] J. H. Reichman and T. Lewis, "Using Liability Rules to Stimulate Local Innovation in Developing Countries: A Law and Economics Primer" (Paper prepared for the Conference on International Public Goods and Transfer of Technology Under a Globalized Intellectual Property Regime, Duke University Law School, Durham, North Carolina, April 2003). J. H. Reichman, "Saving the Patent System from Itself", in F. Scott Kieff (ed.), *Perspectives on Properties of the Human Genome Project* (Academic Press, 2003) 289, 291–95; P. Samuelson and S. Scothmer, "The Law and Economics of Reverse Engineering" (2002) 111 *Yale Law Journal* 1575; J. H. Reichman, "Of Green Tulips and Legal Kudzu: Repackaging Rights in Subpatentable Innovation" (2000) 53 *Vanderbilt Law Review* 1753; M. A. Lemley, "The Economics of Improvement in Intellectual Property Law" (1997) 75 *Texas Law Review* 989.

[20] D. Burk, "Open Source Genomics" (2002) 8 *Boston University Journal of Science and Technology Law* 254; E. Raymond, *The Cathedral and the Bazaar: Musings on Linux and Open Source by an Accidental Revolutionary* (O'Reilly, 2001).

whereby the creators of the works limit their exclusive right to certain usage. Generally, one of four options is chosen: (i) the author of the work has to be mentioned, (ii) the work cannot be used for profit purposes, (iii) the creation of derivate works is not allowed, or it can be allowed only under certain specific conditions.

Systems of creative commons may be useful for preserving PGRs. The value of PGRs is preserved through the planting, seed production, and continuous selection of the best-adapted farmers' varieties (landraces).[21] Such farmers generally interact among themselves exchanging seeds across the fence, thus fostering the diffusion of their varieties and further development.

Conservation communities can particularly benefit from the application of creative commons models to exchange of PGRs. Intellectual efforts have already been made by Srinivas[22] and Jane Hope[23] who have developed similar participatory models.

Srinivas considers that a plant variety should be made available to a certain community or should be in the public domain in general. Users may experiment and innovate by sharing the seed. The varieties deriving from this system will also be made available to the conservation community. An agency can coordinate the activities by creating a pool of samples of germplasm. These centers should be in contact with other public and private international collection banks like CGIAR. These foundations can buy patents to render important technologies available to farmers.

Based on similar assumptions, Jane Hope has been developing a project with more general application called Open Source Biotechnology. It extends the principles of commerce-friendly, commons-based production outlined by the open-source software movement to the development of research tools in medical and agricultural biotechnology.

Similar to the creative commons for copyrighted words on the internet and to Srinivas' proposal for plant varieties, this project is implemented through specific open-source licenses for research tools that are being actively commercialized. However, it may be that open-source licensing and commercialization strategies are not feasible or appropriate in the biotechnology context.

[21] V. Shiva *et al.*, *The Enclosure and Recovery of the Commons: Biodiversity Indigenous Knowledge & Intellectual Property Rights* (Research Foundation for Science, Technology and Ecology, Delhi, 1997).

[22] K. R. Srinivas, *The Case for Biolinuxes and Other Pro-Commons Innovations*, www.noolithic.com/IMG/pdf/09biolinux.pdf. He proposes that the General Public Licence system for open source computer software be applied in plant breeding, to keep farmers' varieties in the public domain and support further innovation.

[23] J. Hope, *Open Source Biotechnology Project*, www.rsss.anu.edu.au/~janeth/Law.html#38.

A CHM could be created to facilitate transactions pertaining to both models. If this were realized internationally, a relation between these models and the CHM of the CBD should be established. The function of a CHM is to provide information about the available TK by fostering further development of communication networks for use by traditional farmers, with an initial emphasis on information-sharing formats, protocols, and standards.

I also suggest the creation of an electronic (on-line) CHM where TK holders may display their useful innovations and potential users may easily research the appropriate GRs and examine the related know-how. Such a matchmaker tool enhances the transfer of the germplasm or of simply traditional know-how on plants. Moreover, this tool should be an institution facilitating contracts on PGR exploitation.[24]

I maintain that while creative commons models may be successful for preserving PGRs within traditional farmers' communities, this model may not be appropriate to protect all the economic and proprietary rights of traditional farmers when they have to deal with industrial parties. Application of creative commons may create a vacuum in the protection of PGR-related TK when industrial companies may freely use the PGR, manipulate its genome, and reach the stage of a patentable invention without being obliged to compensate the TK right-holders.

7.3.2 Comparative analysis of compensatory liability regime and traditional intellectual property rights for traditional knowledge on plant genetic resources

This section presents a comparative analysis of the CLR proposed by Reichman[25] and the TIPRs outlined by Cottier.[26] A thorough explanation of their models lies outside the scope of this section and the present comparative analysis relies upon a basic understanding that can be acquired only by reading their works. Article 9 of the ITPGRFA has a wider scope than that which is described in the present chapter. The spectrum of options for protecting TK is broader than Cottier's and Reichman's models, but they appear fairly complete in their legal

[24] S. Biber-Klemm and J. Curci, "Clearing House Mechanisms", in Cottier and Biber-Klemm (eds.), *Rights to Plant Genetic Resources*, 269.

[25] Reichman, "Of Green Tulips"; Reichman and Lewis, "Using Liability Rules".

[26] T. Cottier and M. Panizzon, "A New Generation of IPRs for the Protection of Traditional Knowledge in Plant Genetic Resources for Food, Agricultural and Pharmaceutical Uses", in T. Cottier and S. Biber-Klemm (eds.), *Rights to Plant Genetic Resources and Traditional Knowledge: Basic Issues and Perspectives* (CABI on behalf of Swiss Agency for Development and Cooperation and the World Trade Institute, London, 2006) 238.

construction and offer a clear choice between the two mainstream theories of IP protection (Cottier) and quasi- or non-IP protection (Reichman).

A comparative analysis allows me to outline the alternatives available and the main differences with the goal of shaping internationally agreed-upon protection that satisfies the interests of as many stakeholders as possible. It is beyond the scope of my task to analyze in depth the advantages and disadvantages that each stakeholder might derive from each model. However, this comparative analysis is necessary to lay out some principles necessary for stimulating adjustments to each system and for determining the following issues: (i) to identify which model would best suit the international protection of TK; (ii) to decide whether the two systems can be compatible; (iii) to examine which of the models is likely to create a balance of rights and obligations to all the parties involved in business transactions related to PGRs. Because CLR is already considered a sophisticated regime against unfair competition, there will be no need to expound on other regimes of unfair competition to protect TK.

7.3.2.1 General underlying issues

7.3.2.1.1 Compensatory liability regime

The economic and legal investigations on liability rules have generally focused on their application to nuisance law. However, Reichman has related the CLR to the field of innovation.[27] This should be considered as a non-IP or quasi-IP method of protection of a given subject-matter for the reasons outlined below.

Reichman first applied CLR to small-scale innovations in two articles,[28] in which he observed the routine engineers working on "common technical trajectories under the aegis of trade secret protection thus forming a *de facto* 'open source' community."[29] He criticized the response of the international community to the necessity of protecting new fields of technology through creating new forms of IPRs, such as the *sui generis* and hybrid regimes for protecting databases and the expansion of copyright beyond literary and artistic works to protect software and other applications of know-how. The CLR stands as a valid response to this

[27] I. Ayres, *Optional Law: Real Options in the Structure of Legal Entitlements* (University of Chicago Press, 2005); I. Ayres and T. Eric, "Solomonic Bargaining: Dividing a Legal Entitlement to Facilitate Cosean Trade" (1995) 104 *Yale Law Journal* 1027; P. Samuelson and S. Scotchmer, "The Law and Economics of Reverse Engineering" (2002) 111 *Yale Law Journal* 1575.

[28] J. H. Reichman, "Legal Hybrids Between the Patent and Copyright Paradigms" (1994) 94 *Columbia Law Review* 2432, 2453–2503 (1994). Reichman, "Of Green Tulips".

[29] Reichman and Lewis, "Using Liability Rules", 1057.

increasing number of commercial barriers represented by exclusive rights to the flow of the market of the economic tools. Transaction costs are, it must be recalled, the costs that society has to pay in order to benefit from the innovation.

For all these reasons, CLR seeks to achieve (i) a proper system for protecting know-how to avoid misappropriation as well as to stimulate investment in small case innovation,[30] (ii) no high social costs for market failure, (iii) enhanced follow-on applications without setting barriers that create the tragedy of anti-commons and at the same time impoverish the research commons.

Reichman has explored the applicability of CLR to TK since he considers TK as a form of know-how.[31] He argues that this model is compatible with the habits of traditional communities should they want to disclose their knowledge. He agrees with Nelson who defines TK as the oldest form of cumulative and sequential innovation known to man; thus TK fits the definition of small-scale innovation which is the subject-matter of CLR.[32]

With a registration system of the TK and a compensation mechanism for any commercial use of the registered TK, CLR can be considered as a more sophisticated version of *domaine public payant*.[33] Reichman describes accordingly the administrative and judicial aspects of all these transactions related to CLR in a similar way to the ones generally used in the system for copyrighted musical works reproduced on sound recordings.[34]

7.3.2.1.2 Traditional intellectual property rights

This regime of protection forms another plateau of protection of the same subject-matter of Article 9 of ITPGRFA, i.e. farmers' rights related to their TK on PGRs. Thinking beyond implementing traditional farmers' rights as an exception to plant-breeding rights (as described by Girsberger)[35] or as simple know-how (as envisaged by Reichman), Cottier suggests the creation of a new generation of IPRs tailor-made for the needs of TK

[30] *Ibid.*, 19.

[31] J. H. Reichman, "A Compensatory Liability Regime for Applications of Traditional Knowledge", Draft Paper Presented to the Cardozo Symposium on the Legal Protection of Traditional Knowledge, New York (February 23–24, 2000) 4, 8, on file with the author.

[32] Reichman and Lewis, "Using Liability Rules", 19; R. Nelson, "Intellectual Property Protection for Cumulative Systems" (1994) 94 *Columbia Law Review* 2678.

[33] *Protection of Traditional Knowledge, Overview of Policy Objectives and Core Principles*, WIPO/GRTKF/IC/7/5 Annex II, 32 (November 1–4, 2001).

[34] Reichman and Lewis, "Using Liability Rules", 24.

[35] M. A. Girsberger, *Biodiversity and the Concept of Farmers' Rights in International Law* (Peter Lang, 1999) 150.

stakeholders willing to participate in the IP system.[36] This is a full-fledged IP method of legal protection mainly because it comprises the concept of *ius excludendi* in its scope. Cottier considers that ITPGRFA and the CBD fail to define proprietary rights because they instead pursue the aim of free flow of PGRs based upon public funding.[37] Cottier claims that traditional farmers holding TK would obtain stronger financial incentives to conserve and use these resources if they were protected by TIPRs. At the same time, protecting TK through TIPRs would fully comply with Article 15 of the CBD that promotes the protection of assignable potential right-holders of TK in PGRFA. Unlike other public goods approaches (as described in chapter 2) and the mix of financial incentives and databases that leave private rights as contained as possible, TIPRs require enforcement of private rights over TK and related PGRs especially when defending certain rights before foreign jurisdictions.

This model offers an additional level of protection for PGRs beyond the exclusive plant variety right of UPOV and beyond the State sovereign right over PGRs contained in the ITPGRFA and the CBD. Its requirements of protection are, of course, less cumbersome than those of UPOV, mainly because the scope of the *ius excludendi* is also more limited. Cottier's proposal is instead inspired by the concepts of distributional justice and societal autonomy:

> The symbiosis of private rights law and public policies will empower farmers around the world in conserving and using TK-related PGRFA, to express esteem for their work, to validate their activities, to enhance their returns and to improve their standing when their knowledge is being used by others.[38]

[36] T. Cottier, "The Protection of Genetic Resources and Traditional Knowledge" (1998) 1(14) *Journal of International Economic Law* 559–60. Cottier refers to the major documents underpinning these conventions and in particular to the *FAO ITPGRFA*, where the public good is considered to be served best by leaving these resources in the public domain, by supporting conservation and use with public funding, such as the Food and Agriculture Organization's (FAO) Leipzig Declaration's Global Plan of Action of 1996, and its 2000–2004 Global Conservation Trust, led on behalf of the Consultative Group on International Agricultural Research by System-Wide GRs Programme (www.ipgri-pa.grinfo.net) and IPGRI/Consultative Group on International Agricultural Research's Singer-wide Information Network for Genetic Resources of 1994, the World Conservation Union's Global Biodiversity Forum of 1993, and last but not least, the WTO's Doha Development Agenda Global Trust Fund of 2002, www.wto.org/english/news_/spmm_e/spmm79_e.htm.

[37] Cottier initially pioneered the idea of TIPRs and developed this legal regime in Cottier and Panizzon, 203–38.

[38] *The Protection of Traditional Knowledge and Folklore, Summary of the Issues Raised and Points Made*, IP/C/W/370 paragraph 8 (August 8, 2002); similarly, *Trade and Development Board Commission on Trade in Goods and Services, and Commodities Expert Meeting on Systems and National Experiences for Protecting Traditional Knowledge, Innovations and Practices*, UNCTAD Document TD/B/Com.1/EM.13/2, 3–4, 9–10 (August 22, 2000).

TIPRs have the advantage of assigning TK to their right-holder with more precise rights and in certain conditions and through registration. This system is clearer than the unfair competition system. Cottier recommends addressing TIPRs based upon the conceptual work in the context of future discussions of Article 27.3(b) of the TRIPS Agreement.

TIPRs are intended to empower communities (and not merely governments) to negotiate the terms of use and exploitation of their resources with private companies in proper licensing agreements.

One of the major challenges that Cottier finds in the realization of TIPRs consists in identifying right-holders for traditional IPRs. Here lies the crucial importance of the localization of landraces.[39] There is growing evidence that such allocations and assignability to a specific territory and group is possible.[40]

7.3.2.2 The public domain

7.3.2.2.1 Liability compensatory regime

As earlier observed, Reichman maintains that classic IP regimes will not solve the problem of protecting TK.[41] Basing his conclusions on solid economic theories related to IP,[42] he argues that overprotecting IPRs in industrialized countries has already proven to hinder competition instead of fostering it. This trend should not be repeated in DCs when shaping their legal protection of grain-sized traditional innovations.

In his view, there is no qualitative difference between those who make small-scale innovations by traditional means and those who use modern technical or scientific tools "to apply know-how to industry in the form of computer programs, industrial designs, or even biologically engineered products."[43] This lack of distinction means that for the sake of legal protection by CLR, the distinction between informal/traditional and formal/industrial knowledge should not be a criterion for the type of protection.

The characteristic element of the CLR is that the knowledge is not removed from the public domain because the right-holder is only entitled to a "right to compensation" for commercial follow-on uses but not to

[39] Cottier and Biber-Klemm (eds.), *Rights to Plant Genetic Resources*, 163–64, 195–97, 185–90.

[40] M. Halewood, *et al.* "Farmers, Landraces and Property Right, Origin and Allocation of Traditional Knowledge and Landraces", in Cottier and Biber-Klemm S. (eds.), 173–202.

[41] Reichman, "Green Tulips"; Reichman, "Legal Hybrids", 2511–56.

[42] Y. Benkler, "Coase's Penguin, or, Linux and The Nature of the Firm" (2002) 112 *Yale Law Journal* 369, www.yale.edu/yalelj/112/BenklerWEB.pdf; M. Heller and R. Eisenberg, "Can Patents Deter Innovation?" (1998) 280 *Science* 698.

[43] Reichman and Lewis, "Using Liability Rules", 19.

block such follow-on uses.[44] As earlier observed, this model creates a "paying public domain."

CLR constitutes a compromise between those who want TK to remain in the public domain[45] and those willing to protect it through IPRs.[46] This model has the potential of creating a natural open-source community that generates know-how. It is based on the concept of semi-commons that extracts TK from "an inchoate public-domain status" to a legally defined temporary semi-commons.[47] These considerations form a good definitional differentiation of this model from the one of creative commons.

7.3.2.2.2 Traditional intellectual property rights

The rationale for creating a new generation of TIPRs is that leaving all TK in the public domain does not solve the imbalances and inequities of the rights of the economic actors involved in the breeding and in the biotechnological transformation of PGRs. Biotechnology has used GRs in the public domain in ways that have generated inequitable exploitation and fragmented protection mechanisms as industrial bioprospectors have used and genetically modified TK and GRs bred and preserved by rural communities. Contract negotiations also demonstrate the imbalances between the TK holders and the bioprospecting companies. Hence, TK holders should enter into contracts with industrial parties having proprietary rights over their PGR and related TK.

These inequities are caused both by powerful monopolistic rights of patent and plant variety regimes and by non-exclusive rights left in the public domain over the same subject-matter. Cottier and Panizzon are well aware of the importance of keeping the free flow of PGRFA.[48] The ITPGRFA reflects this effort at the international level, but its list of PGRs to be kept in the public domain is limited. The financing mechanisms of ITPGRFA depend on taxpayers' contributions: "since states are unwilling to part with a source of income and/or to raise taxes, in order to reimburse those who have lost revenue with the TK now in public domain, a number of important crops have not been included in the

[44] Reichman, "Of Green Tulips", 1743.

[45] C. Correa, *Traditional Knowledge and Intellectual Property – Issues and Options Surrounding the Protection of Traditional Knowledge – A Discussion Paper* (Quaker UN Office, Geneva, 2001). G. Dutfield, "The Public and Private Domains: Intellectual Property Rights in Traditional Knowledge" (2000) 21 *Science Communication* 274–95.

[46] A. K. Gupta, "Securing Traditional Knowledge and Contemporary Innovations: Can Global Trade Links Help Grassroots Innovations?", in Cottier and Mavroidis (eds.), *Intellectual Property*.

[47] Reichman and Lewis, "Using Liability Rules", 7.

[48] Cottier and Panizzon, "A New Generation of IPRs", 206–07, 218.

ITPGRFA."[49] The concept of public domain is too far from the current international market economy.

Indeed, TK may be available in the international public domain unless it is limited by international agreements such as PBRs from UPOV, patents from TRIPS, or even farmers' rights from Article 9 of ITPGRFA. The public domain can also be limited by local customary law as allowed by Article 10 of the CBD or under national sovereignty over biological resources. Because the public domain is already segmented by these treaties, TK right-holders are justified in claiming an exclusive right upon this subject-matter through TK-specific IPRs, a *sui generis* system, or a combination of both. All these types of rights imply creating a positive right of an identifiable legal person to exploit the resource and appropriate the benefits of its exploitation.

For all these reasons, Cottier and Panizzon conclude that "IPR will subject the seed under IPR protection, whereby the knowledge bearer as a right-holder will be able to appropriate the profit and the competitors will have no incentive to build upon the TK, in order to become TIPR holders themselves."[50]

7.3.2.2.3 Comparative analysis

Reichman submits that a CLR for TK would stimulate investment in commercial applications of traditional know-how without creating barriers to entry in the research commons and without otherwise impoverishing the public domain. In other words, liability rules are more apt to stimulate local innovation in DCs instead of resorting to the classic IP laws that increase the tragedy of anti-commons.[51] This positive system takes into account the ideological and practical opposition to IP protection for TK due to certain ideals, policies, and legal instruments which all seek to place the PGRs – *ipso iure* under national sovereignty – into the public domain as global public goods.[52]

All innovations are immediately put into the research commons and are available to other users, enhancing the speed of investment for incremental innovations. Ultimately, the increment of research commons under this regime lowers prices. Investment in research is, indeed, shared by all the actors that participate in the follow-on innovations.

[49] *Ibid.* 219. [50] *Ibid.*

[51] Reichman, "A Compensatory Liability Regime". Reichman and Lewis, "Using Liability Rules", 3.

[52] The introduction of IPRs will inevitably change the very nature of TK in most of its community character. From the TK stakeholders point of view, TK is community heritage that cannot be sold or bought, i.e. it cannot be turned into a commodity or into property.

The TIPRs' model is embedded in the utilitarian theories justifying time-limited legal monopolies of IPRs. Protection through exclusive rights is more likely to create an incentive to innovate among the community of economic actors.[53] Accordingly, in a competitive market conception, the effect flowing from the ownership of an exclusive right is more likely to be achieved by TIPRs than by the CLR. However, competition for follow-on innovation might definitely not be the goal of the TK holders in local and indigenous communities. Therefore, a CLR would spur on economic activity in a more cooperative atmosphere than TIPRs would.

A preference for one model over the other will depend on the economic actor, the TK holder, or on the user. TK holders may be divided in two groups: (i) TK holders who target an economic return must be protected through TIPRs, which implies an exclusive right and a consequently higher royalty; (ii) TK holders who desire to preserve their values and customary rules express their strongest reservations about protecting their TK through distinct and exclusive property rights.[54] In contrast, the user of TK should always prefer the CLR because he would not have to negotiate with the TK right-holder but would simply pay the compensatory royalty.

7.3.2.3 Scope of right and duration

7.3.2.3.1 Compensatory liability regime

Reichman has divided the scope of the right granted by the CLR into three major elements:

(i) The first element characterizes the proximity of the CLR to any regime of unfair competition. This is the right to "prevent second comers from competing on the same market segment for a specified period of years with a product that constitutes a wholesale duplication of the innovator's initial product" unless this is done for an improvement or for a follow-on innovation.[55] This right is qualified as quasi-IP since it has a limited scope of exclusion. From Reichman's description, it might be unclear whether the scope of the exclusive right provided by this regime is limited to excluding

[53] C. Primo Braga and C. Fink, "The Economic Justification for the Grant of Intellectual Property Rights: Patents for Convergence and Conflict" in F. Abbott, T. Cottier, and F. Gurry (eds.), *The Intellectual Property System: Commentary and Materials* (Kluwer, The Hague, 1999) 266–67; K. Arrow, "Economic Welfare and the Allocation of Resources for Invention", in R. Nelson (ed.), *The Rate and Direction of Inventive Activity: Economic and Social Factors* (Princeton University Press, 1962) 609. C. May, *A Global Political Economic of Intellectual Property Rights: The New Enclosures?* (Routledge Publishers, London, 2000).

[54] WIPO/GRTKF/IC/7/5.

[55] Reichman and Lewis, "Using Liability Rules", 21. Article 10 of the *Paris Convention.*

from wholesale duplication only in cases of lack of innovation. I assume that the exclusive right is limited to cases of duplications with a view to further innovation.

(ii) Construed upon the "take and pay" principle, the second right of the TK holder under this regime is the right to be compensated for any access and use of the protected subject-matter. However, the right-holder cannot exclude the second-comer from accessing and using the protected TK for the purpose of adding value to the original know-how through improvement. This second-comer must compensate the first-comer for these uses in one manner or another.

Limiting the scope of the right to compensation would "enable innovators to benefit from their contributions [...] without disrupting the sharing ethos (and the public domain) from which incremental improvements of know-how typically emerge over time."[56]

Under CLR there is no limit of duration of the right while the payment of the royalty is extended to fix a number of years.

(iii) The third right enables the original holder of the first-used know-how to use follow-on innovation in case of improvements to the products that initially qualified them for protection. The first-comer is prevented, as much as the second-comer, from wholesale duplication. Reichman is evidently inspired by the usual mandatory dependent licenses available for patentable improvements to patented products (as also provided by Article 31(1) of the TRIPS Agreement).

7.3.2.3.2 Traditional intellectual property rights

Cottier proposes that this protection be limited to commercially viable information that can serve the potential market. The scope of these TIPRs includes granting exclusive rights to control the information and derived products (subject to rules of exhaustion) for selling, manufacturing, importing, etc. Such right could be limited to the right to compensation for use. The minimum would therefore entail a legal licensing, allowing all persons to use the information, but also to be liable for adequate compensation.[57]

The duration of these rights is strictly related to the commercial value of the information. Hence, the TIPR should be provided for an unlimited duration as long as it is being used in a particular community.[58] In sum, protection begins with commercial exploitation and ends when the protected subject-matter is no longer of any commercial interest to the community. This period is strictly connected with the right to compensation.

[56] Reichman, "A Compensatory Liability Regime", 4, 8.

[57] Cottier and Panizzon, "A New Generation of IPRs", 225ff. [58] *Ibid.* 227.

Regarding royalty payments in connection with the period of protection, Cottier indicates that the compensation should not be paid during the first years of marketing the product based on the licensed TK.[59] He proposes that a double fee solution should be adopted: the first for using the TK for R&D and the second on the sales of the product for a "period of time sufficient to generate adequate benefit sharing," e.g. after 10 years.[60] A similarity with a CLR is provided by institutions of private societies that collect royalties under automatic or template license contracts to then be distributed to the right-holders.

7.3.2.3.3 Comparative analysis

One of the major differences between these two models is the object and content of the right: the TIPRs consist in an exclusive right (*ius excludendi*) while the CLR is based on open access. The scope of the *ius excludendi* controls the use of TK on follow-on R&D of innovation that enables the right-holder to prevent any development of products that may be detrimental to the TK itself and its future exploitation. This stronger right of exclusion provides and entails higher monetary returns from royalties.[61]

As regards the duration, neither TIPRs nor CLR limits the period of protection; this suits the needs of TK holders who view their right as inalienable because their TK stems from time immemorial. Both systems envision an unlimited period of protection, but they seemingly suggest a different period for paying royalties. The two systems will have to clarify these particular elements that are still vague at the present stage. The concept of the commercial value of the TK (i.e. its "information") should be better defined. The user of TK would wish that the rights related to TK would be limited, as envisioned by TIPRs.

It is difficult to determine which model suits the interests of the TK right-holder because of differing preferences between a more participatory quasi-IP or a more IP-like one.

7.3.2.4 Registration of rights and technical support

7.3.2.4.1 Compensatory liability regime

For the CLR to function properly, it requires a collecting agency to collect, distribute, negotiate, and regulate various aspects of the payment schedule.

Reichman and Lewis acknowledge that:

[59] *Ibid.*; J. H. Reichman and C. Hasenzahl, *Non-voluntary Licensing of Patented Inventions, Historical Perspective, Legal Framework under TRIPS, and an Overview of the Practice in Canada and the United States* (Issue Paper No.5, UNCTAD/ICTSD, Geneva, June 2003), www.ictsd.org/pubs/ictsd_series/iprs/CS_reichman_hasenzahl.pdf.

[60] Cottier and Panizzon, "A New Generation of IPRs", 227. [61] *Ibid.*, 216.

crafting modalities for the distribution of royalties among deserving indigenous providers poses well-known difficulties for which I offer no new solutions. I stress, nonetheless, that the collection of royalties under the automatic licenses of a compensatory liability regime raises separate and distinctly different issues from those of distribution. Until and unless such royalties are collected under enabling local legislation, there is nothing to distribute. Any problems of distribution thereafter should not impede early collection of royalties, which can be held in trust for the appropriate beneficiaries however these are to be determined. Care must be taken to keep transaction costs low lest administrators siphon off the benefits at the expense of indigenous providers.[62]

As regards the registration of TK, Reichman does not provide any particular system. It can be suggested that registration of TK be combined with methods already in place like clearing house mechanisms.[63]

7.3.2.4.2 Traditional intellectual property rights

The creation of TIPRs is related to a registration mechanism. The registration can have either a declaratory or constitutive value.[64] Cottier submits that, unlike copyrights that may exist from the moment of their creation, the registration of TIPRs is necessary since they constitute rights from the past.[65] This registration mechanism can be realized in the form of databases and can most effectively be done "on a national level by governmental agencies, NGOs or universities."[66] However, Cottier prefers that the registration be made at the international level through information technology that would connect national agencies, NGOs, and States.

This pattern would be similar to the patent system whereby the information is publicly disclosed. As better explained in the clearing house mechanism, this public disclosure can be a mechanism for marketing such knowledge. Registration also serves to clear opposition procedures and facilitates judicial review.

7.3.2.4.3 Comparative analysis

Both systems leave open the question of whether the registration should have constitutive or declaratory value. I maintain that registration of

[62] Reichman and Lewis, "Using Liability Rules", 26.

[63] Biber-Klemm and Curci, "Clearing House Mechanisms", in Cottier and Biber-Klemm (eds.), *Rights to Plant Genetics Resources*, 269 ff.

[64] *Elements of* Sui Generis *System for the Protection of Traditional Knowledge*, WIPO/GRTKF/IC/4/8, 30 (December 9–17, 2002).

[65] Cottier, "The Protection of Genetic Resources", 555–84.

[66] Cottier and Panizzon, "A New Generation of IPRs", 228. (They point to Alaska TK and Native Foods database, which is a joint effort of the US Environmental Protection Agency, the University of Alaska, and the Alaska Native Science Commission, www.nativeknowledge.com, or the Honey-Bee Network Innovation Database, www.sristi.org/honeybee.html.)

TIPRs as well as of know-how under the CLR should have a declaratory value since this would strengthen claims of traditional communities against infringement even before formal registration, i.e. before acquiring a legal title.[67] The constitutive value of the registration would mean that the exclusive legal title depends on the registration.

Ancillary to registering TK is disclosing the origin of the GR and the related TK. Accordingly, the fact that this registration leads the right-holder to disclose the origin of the GRs can also be crucial for establishing TK as prior art to be taken into account by patent examiners during the application examination (see section 6.2 above).[68]

This registration is crucial for the preservation of TK for future generations.

7.3.2.5 Relations with other intellectual property protected follow-on innovations

Here follows a necessarily brief synopsis of the relationship between such models and the exercise of other IPRs by follow-on innovators, whether local or international, whether industrial parties or small-scale innovators.

7.3.2.5.1 Compensatory liability regime

To illustrate how the CLR functions in relation to the exercise of other IPRs, I shall paraphrase the hypothetical provided by Reichman:[69]

(i) A certain tribe in Ruritania has developed a traditional medicine from the bark and leaves of the "kew tree," that has been used to soothe and cure skin burns. After encouragement from the government, the tribe decides to make this TK available under a compensatory liability regime.

(ii) Registration of this method has been provided under the local CLR whereby no one else can duplicate this remedy for commercial purposes for, say, twenty years. However, the CLR includes a research exemption for using this registered remedy for non-profit public research.

(iii) A local firm is willing to invest "technical knowledge and skills to combine ingredients derived from kew tree bark and leaves with other ingredients known to its researchers, with a view to producing an improved treatment for burns." This ability to do follow-on research is a crucial point of the advantages of the CLR because Reichman observes that, if the tribe possessed instead a patent on its

[67] WIPO/GRTKF/IC/4/8, 30.
[68] Cottier and Panizzon, "A New Generation of IPRs", 228.
[69] Reichman and Lewis, "Using Liability Rules", 10.

TK, a case of anti-commons would occur: on the one hand, the tribe could block the firm's follow-on application; on the other hand, the firm would fear disclosing its own business plan or its formula.

(iv) Instead, by paying low-cost compensatory royalties falling within a specified statutory range for a specified period of time under the applicable CLR, the firm can "borrow" the TK in order to develop an improved derived product.

(v) A multinational pharmaceutical firm, after having observed the local firm successfully marketing its kew-tree-derived burn unguent, starts to invest considerable R&D to identify the specific ingredients in the kew bark and leaves that produce healing effects. The firm ends up developing a new product for the healing of surgical wounds based on the synthesized active ingredients and combines these with other ingredients it has already been using in existing products.

(vi) Since the group of farmers and the follow-on researchers of the local firm did not hold patents, they cannot prevent the multinational pharmaceutical firm from developing its product. The advantage of this system is that the multinational firm can immediately disclose its invention of synthetic processing technology. It can simply borrow the refined burn cure know-how without obtaining permission from the group of farmers or the local firm. Yet it must pay compensatory liability royalties to both farmers and the local firm.

(vii) This system provides an economic benefit because the multinational firm can more easily develop additional products than would be the case if exclusive rights blocked these improvements. In addition, the local firm could also borrow knowledge from the multinational to develop another type of product. This reverse borrowing would also require payment of royalties.

(vii) Another benefit dynamizes the whole system of coexistent applications or follow-on ones. Reichman notes that, for instance, after a specified period, say, ten years, another firm might be allowed to produce competing versions of the surgical wound treatment initially based both on the TK of the traditional farmers and the derived product of the local firm even for head-to-head competition in the same market segment. This competition is possible through negotiating up-front licenses between the new firm and the three former innovators.

7.3.2.5.2 Traditional intellectual property rights

Cottier has sketched in the form of questions the possible relations between TIPRs and other potential IPRs over the same subject-matter. A well-funded research and development industry (plant breeder) will be easily able to locate the PGRFA and capture it for its own uses. The interrelationships between TIPRs and PBRs and other IPRs must be understood in order to determine who will receive the main benefits arising from the exploitation of PGRs. The TIPR, which is a new right, grants a proprietary right to the TK that led to the development of the traditional material proprietary protection before any innovative step by the plant breeder has modified the PGR.

If defined at the international level, a plant breeder in the future will be obligated to file for a license to use the PGR "linked to TK and to compensate the right-holder." The plant breeder must either pay the right-holder or share benefits, as the CBD Guidelines promote, whichever best protects the right-holder. There are various relevant questions to be settled in any international negotiation to shape a TIPR regime:

> [I]f the plant breeding industry isolates a genetic sequence from traditional material, does the genetic sequence also count as information emanating from TK or simply as genetic material barred from IP protection? Does the process of extracting genetic information – absent any modification of the genetic material – amount to an innovation that would confer an IPR to the plant breeder? Should the international system reward a process that requires intellectual creativity, even if the material itself, the genetic sequence laid open is not (yet) modified? As an impartial observer, one could argue that if the international system is preparing to reward the processes of trial and error that lead to the formation of TK, why should it not afford genetic research – the modern type of experimentation and systematization processes – the same status as TK?[70]

Moreover, what are the differences that would lead to treating genetic research on PGR and on TK on PGR differently? Both are more process- than product-oriented; both aim at increasing the efficiency of the crop in general; both might chose to modify genetic information. However, critics of a TIP regime may very well use the argument that opening the door to protecting TK with proprietary rights will instantly invite genetic research to be similarly protected, or at least exempted from applying for licenses. A solution, which would offer science some leeway for research on PGRFA, would be to have the TIP-right protect the TK and the seed at its origin, but not the genetic information itself.[71]

7.3.2.6 Final observations on the comparison of the two models

Besides the comparison of each elements of the two models there are further considerations to be made.

First of all, it seems that Reichman's CLR requires less rules than Cottier's TIPR. Hence, the first is more flexible and thus more likely to

[70] Cottier and Panizzon, "A New Generation of IPRs", 232. [71] *Ibid.*

foster follow-on innovation. There is certainly further work to be done to clarify both models, especially how they operate in an international context. The first element that needs to be clarified is whether an international organization needs to be involved in the administration of such rights or whether the whole administration is going to be left to national entities and how to connect those national entities. Cottier and Panizzon have indicated that international negotiation will determine the exact scope of the rights. They envisage the possibility of creating additional bureaucratic structure even at the international level for this purpose. The internationalization of the TIPRs cannot be avoided if certainty and common understanding are to be reached. Reichman is more determined to reduce the width of the administrative structure of the CLR to the minimum.

The international organization in charge of the administration should be WTO or WIPO. The first is preferable because of its dispute settlement mechanism, but the latter is endowed with its arbitration and mediation center that can represent the most appropriate way to solve disputes in this field.

Cottier's model requires a registration mechanism, an international scrutiny and assignability procedure of the PGRs and the related TK to a particular applicant. It should function similarly to any other IP office granting the exclusive right upon a given subject-matter that the applicant needs to explain. All this involves the work of lawyers assisted by anthropologists, historians, and probably ethics experts. In this sense the CHM can be useful if the CHM function in a multitask purpose can include the administration of the new legal IP or quasi-IP regime.

Certainly the CLR could help defuse the negative effects of the tragedy of anti-commons that is bound to continue and its effects for the TK holder are less detrimental than the total absence of achieving his own right over the subject-matter of his work or ingenuity. The fact that local or indigenous communities might hold a right over the information contained in their PGRs and related TK gives them a voice to be heard and a tool to act, as well as representation in a world of highly concentrated markets, in particular in the sector of PGRFA and, increasingly, biotechnology. The TIPR model can be convincing because holding an IPR creates the tool to enter and participate in global trade. In his pragmatical approach to the international market, Cottier states that "TK associated to PGRFA, its holders and the seed at the basis, will not become players unless their identity becomes a tradable asset."[72]

[72] Cottier and Biber-Klemm (eds.), *Rights to Plant Genetic Resources*, 234.

TIPR regime fosters the objective of conservation and preservation of information concerning GRs. It initiates a process of relation between a geographical area population and a certain type of GR and related TK, with clearer rights over them. It encourages the transparency of the information. The transaction to access it creates human relations, contact, and knowledge sharing. The transaction cost may be outweighed by the benefits that it creates.

Clear rules should be also negotiated regarding the sequential innovators in order to foster competition for innovation. As soon as a new type of IPR is introduced in the legal order new forms of competition laws should also be contemplated so as to modulate the adverse effects of monopolistic rights. A special attention should be devoted to the relation between TIPRs and PBRs and patent rights.[73]

In a nutshell, the main difference between the models is that the compensatory liability regime entails a certain level of assignability to a certain rightholder, while TIPRs provide a clearer utilization of the right than CLR because we are dealing with exclusive rights. This is of particular relevance in light of the evidence from economic analysis that the loss of biodiversity is also caused by poorly defined property rights along with uncertainty and information failure.

In sum, the two models of TIPRs and CLR are both valid but need to be discussed further to find ways of implementation. One model can be more suitable than the other in certain given circumstances.

7.4 The protection of traditional knowledge through unfair competition

Fujichaku observes that "misappropriation performs an interstitial role in protecting the investment in developing intangible goods which are otherwise ineligible for traditional intellectual property protection."[74] Therefore, while the international community tries to define the scope of a *sui generis* protection of TK it is useful to explore the concept of misappropriation in all its potential. This section explores the possibilities of protection of TK through the already available *lex lata* of Article 10*bis* of the Paris Convention. It also observes the option to create *de lege ferenda* an international misappropriation regime at the WIPO IGC on IPGRTKF

[73] *Ibid.*, 144.

[74] Y. R. Fujichaku, "The Misappropriation Doctrine in Cyberspace: Protecting the Commercial Value of 'Hot News Information'" (1998) *University of Hawaii Law Review* 439, quoted in *The Protection of Traditional Knowledge: Revised Outline of Policy Options and Legal Mechanisms*, WIPO/GRTKF/IC/9/INF/5, 14 (April 24–26, 2006).

which includes a *sui generis* IP protection of TK, and thirdly concludes with a few considerations on the existing and possible protection of TK through the broader field of tort law that is already available in every domestic jurisdiction.[75]

7.4.1 *The protection under Article 10bis of the Paris Convention*

To some extent, the international protection of TK is possible through Article 10*bis* of the Paris Convention which is part of the TRIPs Agreement and may be invoked in international disputes before a WTO Panel in case of violation. The concept of unfair competition contained in this provision certainly requires further attention by individual States,[76] regional systems, and the international community. The WIPO IGC on IPGRTKF is basing its attention on this provision without prejudice to any further effort to create *sui generis* IPR for TK. It is evident that the concept of unfair competition is the foundation of all forms of IPR protection, and it is therefore also the intellectual foundation of future specific rights related to TK. Article 10*bis* of the Paris Convention provides that:

(1) The countries of the Union are bound to assure to nationals of such countries effective protection against unfair competition.
(2) Any act of competition contrary to honest practices in industrial or commercial matters constitutes an act of unfair competition.
(3) The following in particular shall be prohibited:
  (i) all acts of such a nature as to create confusion by any means whatever with the establishment, the goods, or the industrial or commercial activities, of a competitor;
  (ii) **false allegations** in the course of trade of such a nature as to discredit the establishment, the goods, or the industrial or commercial activities, of a competitor;
  (iii) indications or allegations the use of which in the course of trade is liable to mislead the public as to **the nature, the manufacturing process**, the characteristics, the suitability for their purpose, or the quantity, of the goods.

[75] T. Taubman and M. Leistner, "Analysis of Different Areas of Indigenous Resources", in S. Von Lewinski (ed.), *Indigenous Heritage and Intellectual Property: Genetic Resources, Traditional Knowledge and Folklore* (Kluwer, The Hague, 2008) 109–111.

[76] DCs, as in the case of any other IPRs, are required to develop a competition law so to limit the harmful effects of too wide monopolistic rights. F. Abbott, "Public Policy and Global Technological Integration: An Introduction", in F. Abbott and D. J. Gerber (eds.), *Public Policy and Global Technological Integration* (Kluwer, Deventer, 1997) 1; J. H. Reichman and P. Samuelson, "Intellectual Property Rights in Data?" (1997) 50 *Vanderbilt Law Review*, 51.

The acquisition of TK in violation of PIC and benefit-sharing principles can be considered as an act of misappropriation that squarely falls within the scope of Article 10*bis*. The precondition to the triggering of the violation is that the two actors are competitors. In a blatant case of biopiracy in which an industrial party has patented a GR and TK from a provider country without benefit sharing or PIC, this provision can be useful for damages against the TK and GR holding community or provider State. However, it will not be easy to prove that the industrial party and the indigenous community are competitors as they must offer similar products to similar customers. It is clear that in the *Basmati* and *Texmati* rice case, the Texan company and the traditional Pakistani rice farmers are direct competitors and this provision could be the legal basis on which Pakistan could initiate a dispute against the US before a WTO Panel.

Paragraph (ii) of Article 10*bis* could be useful in the *Neem* case in which the contract of MTA did not mention anything about the participation of the local communities in the industrial development of the new products, thus not granting any rights to the community over the sales of the final product. As a result, the biodiversity-related innovation based on the traditional germplasm, be it patented or not, will not be profitable to the community whose TEK has been crucial in the preservation of the successful compound. This omission could amount to "create confusion" (Article 10*bis*(i)) in the course of trade because it leads the customer to believe that Monsanto has the exclusive merits over the invention at the basis of the product. In this case, the provider community has played, on the contrary, an essential role in identifying the special relevant characteristics of the Neem GR, i.e. in breeding and preserving this essential element of that invention.

In an international context, the definitions of the "competitor" or of "competition" become more complex than in the domestic legal system. In general, unfair competition practices reflect habits that are within the spirit of a particular community. It is difficult to capture the spirit of the international community on how a competitor in this field is perceived. Moreover, creating a uniform code on this manner of conduct even in this one field can unfold extreme complexities. A useful tool in this regard is the initiative of the WIPO IGC on IPGRTKF to study how TK can be protected more specifically by a misappropriation regime. This endeavor will specify the content of the TK subject-matter.

7.4.2 *The development of a misappropriation regime*

In various jurisdictions, misappropriation can be one particular type of unfair competition. In every jurisdiction, the misappropriation rationale serves as a remedy against methods of imitation that are blatantly unethical or distort the market.

Sui generis IPRs have been developed on the basis of this rationale.[77] In the ongoing effort to protect TK at the international level, a similar misappropriation regime has been proposed by the WIPO IGC on IPGRTKF. Its Secretariat produced a document summarizing the initiatives that have been made in this regard. The main objective of this protection is "to repress the misappropriation of TK and other unfair commercial and non-commercial activities, recognizing the need to adapt approaches for the repression of misappropriation of TK to national and local needs."[78]

The first three articles of the draft provisions set forth the main principles of legal protection of TK. Article 1 reads:

1) Traditional knowledge shall be protected against misappropriation.
2) Any acquisition, appropriation or utilization of traditional knowledge by unfair or illicit means constitutes an act of misappropriation. Misappropriation may also include deriving commercial benefit from the acquisition, appropriation or utilization of traditional knowledge when the person using that knowledge knows, or is negligent in failing to know, that it was acquired or appropriated by unfair means; and other commercial activities contrary to honest practices that gain inequitable benefit from traditional knowledge.
3) In particular, legal means should be provided to prevent:

   acquisition of traditional knowledge by theft, bribery, coercion, fraud, trespass, breach or inducement of breach of contract, breach or inducement of breach of confidence or confidentiality, breach of fiduciary obligations or other relations of trust, deception, misrepresentation, the provision of misleading information when obtaining prior informed consent for access to traditional knowledge, or other unfair or dishonest means;

   acquisition of traditional knowledge or exercising control over it in violation of legal measures that require prior informed consent as a condition of access to the knowledge, and use of traditional knowledge that violates terms that were mutually agreed as a condition of prior informed consent concerning access to that knowledge;

77 "In Europe the misappropriation doctrine and concept has already been applied to develop *sui generis* protection for new types (Recital 39, Directive 96/9/EC of the European Parliament and of the Council of March 11, 1996 on the Legal Protection of Databases)" of IP protectable subject-matter. In the US the "United States Semiconductor Chip Protection Act, which had been based on the misappropriation doctrine, subsequently influenced the Swiss unfair competition law and the Israeli Commercial Torts Law" WIPO/GRTKF/IC/9/INF/5, 14–5.

78 *The Protection of Traditional Knowledge: Revised Outline of Policy Options and Legal Mechanisms*, WIPO/GRTKF/IC/9/INF/5, 12.

false claims or assertions of ownership or control over traditional knowledge, including acquiring, claiming or asserting IPRs over traditional knowledge-related subject-matter when those IPRs are not validly held in the light of that traditional knowledge and any conditions relating to its access;

if traditional knowledge has been accessed, commercial or industrial use of traditional knowledge without just and appropriate compensation to the recognized holders of the knowledge, when such use has gainful intent and confers a technological or commercial advantage on its user, and when compensation would be consistent with fairness and equity in relation to the holders of the knowledge in view of the circumstances in which the user acquired the knowledge; and

willful offensive use of traditional knowledge of particular moral or spiritual value to its holders by third parties outside the customary context, when such use clearly constitutes a mutilation, distortion or derogatory modification of that knowledge and is contrary to *ordre public* or morality.

4) Traditional knowledge holders should also be effectively protected against other acts of unfair competition, including acts specified in Article 10*bis* of the Paris Convention. This includes false or misleading representations that a product or service is produced or provided with the involvement or endorsement of traditional knowledge holders, or that the commercial exploitation of products or services benefits holders of traditional knowledge. It also includes acts of such a nature as to create confusion with a product or service of traditional knowledge holders; and false allegations in the course of trade which discredit the products or services of traditional knowledge holders.
5) The application, interpretation and enforcement of protection against misappropriation of traditional knowledge, including determination of equitable sharing and distribution of benefits, should be guided, as far as possible and appropriate, by respect for the customary practices, norms, laws and understandings of the holder of the knowledge, including the spiritual, sacred or ceremonial characteristics of the traditional origin of the knowledge.[79]

Various provisions within the Article stipulate that legal protection should be granted where acquisition of TK is a result of, *inter alia*, (1) theft or bribery;[80] (2) a violation of legal measures requiring PIC;[81] (3) access to TK without just compensation;[82] or (4) willful offensive use of TK that is contrary to *ordre public* and morality.[83]

[79] *The Protection of Traditional Knowledge: Revised Objectives and Principles*, WIPO/GRTKF/IC/8/5 (June 6–10, 2005).
[80] *Ibid.*, 12; Article 3(i). [81] *Ibid.* [82] *Ibid.* [83] *Ibid.*

Furthermore, it has been recommended that TK should be granted protection against other acts of unfair competition, including acts specified in Article 10*bis* of the Paris Convention, including acts of "misleading information that a product or service is produced or provided with the involvement or endorsement of TK holders" or "acts of such a nature as to create a confusion with a product or service of TK holders."[84]

Commentary on this Article has stressed its emphasis on providing a "common frame of reference for protection"[85] that clearly reflects past "expressions of commitment"[86] in "preventing the misappropriation of TK."[87] It defines the nature of the protection sought by providing "a general non-exclusive description of misappropriation," and supports its claims with Article 10*bis* of the Paris Convention, namely, protection against unfair competition with particular focus on "acquisition by unfair means."[88] In this manner, Article 10*bis*, although primarily for protection against unfair competition, applies directly to protecting TK against misappropriation.

Article 10*bis* allows national systems great latitude in interpreting and implementing it to better protect their TK. In other words, national legislatures may have recourse to various means of protection ranging from breach of PIC to abuse of *ordre public* and morality. The link with Article 27.2 of the TRIPS Agreement is not established in draft Article 1.

The drawbacks of the draft Article reside in its possibilities of application. The commentary to this Article suggests that these standards be applied in accordance with the "customary understanding of the TK holders themselves."[89] This self-referential definition of a TK holder leaves the industrialized countries in the dark about who the TK holders are. The industrialized countries, indeed, do not see how these rights can be exercised in practice mainly because TK is held collectively rather than privately. This is a major difference between some DCs' and the industrialized countries' IP approach.

Given the nebulous and conflicting views over what misappropriation of TK is, however, States have been reluctant to carry on a discussion on this document at the Eighth Session of the WIPO IGC on IPGRTKF. This reluctance demonstrates how far the international community is

[84] *Ibid.* [85] *Ibid.*, 14. [86] *Ibid.*

[87] The summary of the intervention of the US, *Final Report*, WIPO/GRTKF/IC/6/14 paragraph 157 (March 15–19, 2004).

[88] *Ibid.*, 14. [89] *Ibid.*, 15.

from developing positive protection of TK. This is also due to the lack of a legal infrastructure with which to enforce the definitions of misappropriation. National systems are left to grapple with the implementation of *sui generis* systems for the protection of TK (including the implementation of Article 27.3(b)).

By reaching into Article 10*bis* of the Paris Convention as support against misappropriation in the guise of unfair competition, this subsection has demonstrated just how far the international community is from developing a positive protection under a broad, expansive approach – let alone under the structures of classic patent law. Indeed, the existing measures of Article 10*bis* seem more an attempt at accommodating various as-yet incompatible legal systems in an attempt to harness what are primarily economic considerations. Such economically driven activities can hardly afford to wait for a bewildered international legal community to get its bearings on an issue whose solution seems at every WIPO IGC on IPGRTKF to become more and more out of its reach.

Therefore, as mentioned in Article 2,[90] exclusive property rights under *sui generis* systems may not be the best answer to the problem of protection; instead, I will next examine the popular yet controversial notions of equitable benefit sharing and PIC and other such economic mechanisms.

7.4.3 Tort law

The violation of Article 10*bis* is a tort in domestic law. Since the US appears to be prime recipient country of patented GRs and TK it is useful to focus on the tort of misappropriation under US law. The WIPO IGC on IPGRTKF has found out that in US law a person is:

> liable for the taking of publicly disclosed or disseminated intangible objects where that intangible was developed through substantial investment and where such taking caused damage to its original holder. In the US, after the Supreme Court's decision, the courts of a number of states have adopted the misappropriation doctrine to provide a state common law remedy to address unfair commercial practices involving some intangible good.[91]

[90] WIPO/GRTKF/IC/8/5, 17 Article 2.

[91] *The Protection of Traditional Knowledge: Revised Outline of Policy Options and Legal Mechanisms* WIPO/GRTKF/IC/9/INF/5, 13; *International News Service v. Associated Press*, 249 US 215 (1918).

General principles for a regime of "quasi-property right" and principles of tort law can offer methods of protecting TK. Since GRs may not be found solely under the sovereignty of one State it is sometimes impossible or inappropriate to assign them to a particular State or community. In these cases, legal systems may resort to alternative devices, such as "quasi-property rights" (as in the case of trade secrets, where an injunction and restitution are indeed granted, but only to the extent that the third party's behavior may be described as dishonest and only to the extent that the resource possesses special characteristics: see Article 39 of TRIPS) or to tort law. In this regime, the third party tortfeasor is subject neither to injunction nor to restitution but rather must pay a sum determined by courts to the victim of the tort. If a third party has interfered with a GR, the owner may obtain injunctive relief before the interference has taken place or restitution of the GR itself as well as of the assets deriving from it if the interference has already take place. As Ricolfi indicates, to be effective, "the injunctive relief must be complemented by devices which guarantee compliance with the court order, such as contempt of court in common law and in civil law countries."[92]

7.5 Trade secrets

Another possible form of protection for TK is as trade secrets. The element of flexibility that is inherent in the concept of trade secrets can make it a suitable tool for protecting TK. At the international level, there is increasing recognition of and uniformity in the treatment of trade secrets. But because trade secret protection usually depends on the common law or civil law rules of each country, it is relatively difficult to imagine fully harmonized rules in this area.

Trade secrecy, or the protection of undisclosed information, is recognized by the WTO's TRIPS Agreement. However, Article 39 of the TRIPS Agreement resulted in a very limited and loosely worded obligation:

> [n]atural and legal persons shall have the possibility of preventing information lawfully within their control from being disclosed to, acquired by, or used by others without their consent in a manner contrary to honest commercial practices as long as such information [...] is secret, [...] has a commercial value because it is secret; and has been subject to reasonable steps under the circumstances to keep it secret.

[92] M. Ricolfi, "Intellectual Property and Biodiversity: A Review of Legal and Conceptual Issues and of Policy Options", Atti del Seminario, Instituto Agronomico per l'Oltremare Firenze, 40, available at http://brasile.iao.florence.it/documenti/ricolfi.pdf;. G. Calabresi and A. D. Melamed, "Property Rules, Liability Rules and Inalienability: One View of the Cathedral", (1972) 85 *Harvard Law Review* 1089, 1124.

Some basic criteria for this trade secret protection are that (i) the information is not in the public domain; (ii) the level of protection depends on the extent and precision of the information, and how much information has been given to others; (iii) the information can also be a quite simple idea provided that it is sufficiently concrete and original; and (iv) the information received in confidence should have some degree of originality to be protected.

Trade secrets can generally be divided into the following categories: (i) specific product secrets; (ii) technological secrets; (iii) strategic business information; (iv) information about a product.

The duration of the trade secret protection does not imply a fixed term for trade secrecy protection. The length of protection can vary according to the competitive advantage of the right-holders and according to the information itself that may become outdated or be reverse engineered by competitors.

There have already been a number of confidentiality agreements between indigenous groups and institutions on the use of plants and medicines. They are mostly confidential and unpublished.

But trade secrecy law is complex and in many cases uncertain when courts have to deal with the practices governing TK held by aboriginal or other types of communities, in particular because redress focuses on the person who disclosed the confidential information whereas in cases of TK it should focus on an entire community. The problem of territoriality remains the major obstacle to enforcing this protection abroad, since some countries may have weak protection for trade secrets or may not extend this protection to TK at all because it is not industrial knowledge by nature, lacking originality since it comes from time immemorial and is therefore considered to be in the public domain and collective rather than private.

There is certain TK subject-matter, that, by its nature, cannot be protected other than through trade secrets. Traditional medicinal knowledge (TMK) falls into this category, and it poses a serious problem because the steps to keep the information secret may not be sufficient under established standard common law or civil law rules. In fact, secrecy usually follows from the fact that only a few people have access to the information of TMK (the trade secret agreement is based on customary laws and practices). No contract or other "hard" evidence exists.[93]

An example can illustrate the above theoretical considerations on the possibility of protection of "TMK". "Spiritual healing" refers to complex rituals, magic or spiritual beliefs that surround indigenous medicine. When

[93] R. Merges, S. Menell and M. A. Lemley, *Intellectual Property in the New Technological Age* (Aspen Publishers, Gaithesburg and New York, 2000) 120–24. For the matter of the requirement of secrecy and disclosure of the trade secret protection see *ibid.*, 53–59.

added to non-spiritual healing methods, it promotes and diffuses traditional medicinal innovations in local and indigenous communities. Most of these regimes are secret, and the knowledge attached thereto can only be acquired through initiation; the best form of protection would be by customary law or alternatively, in limited cases, trade secrets. For instance, in Cameroon, the *obasinjom* of the Manyu people is a secret society with unimaginable healing powers. It has been impossible so far for an outsider to acquire the knowledge of the healing potentials or secrets of that secret society, unless he has been initiated. And even then, a member of the secret society only possesses the healing powers when he enters into the robe of the *obasinjom*. It is secret information that is well guarded and has been for generations. This regime has so far been given adequate protection by the customs and tradition of the Manyu people, and it would be an aberration against these people to envisage some other form of protection for their knowledge no matter the reasons advanced. In fact, even the people themselves who have not been initiated into the secret society look upon the *obasinjom* in awe. The only reasonable conclusion here is that such regimes can only be adequately protected by the customs of the people concerned, and how they should distribute the proceeds from the practice is their affair.

Alternatively some other forms of spiritual or ritualized healing which are not adequately protected by the customary laws of the people could be given protection under the trade secret paradigm with all that it entails,[94] according to Article 10*bis* of the Paris Convention (1967) and Article 39 of TRIPS.

Finally, rules concerning the protection of trade secrets adapted to TK would have to be reviewed not only in the country of origin, but also in foreign countries.[95]

7.6 Applying patent law to traditional knowledge innovation[96]

Throughout this study, I have described how the patent system has recently come under criticism by TK holders and their advocates as not

[94] M. Dabiri, A. Sadjorno and J. Tambutoh Dashaco, "Traditional Medicine and Intellectual Property Rights – A Move towards Protection in Developing Countries?", Collection of Papers of the Post-Graduate Specialization Course on Intellectual Property 444 (WIPO Worldwide Academy, Torino, Italy 2001). These authors describe the ways in which spiritual and non-spiritual traditional medical knowledge may be protected through *sui generis* IPRs along with domestic customary laws.

[95] Dutfield, *Intellectual Property Rights, Trade and Biodiversity*, 130.

[96] See the possibilities of joint inventorship in the novel uses of TK and its interaction with the existing conditions of patentability, P. Cullet, "Existing Intellectual Property Rights: Avenues for Further Development, in Cottier and Biber-Klemm (eds.), *Rights to Plant Genetic Resources*, 240–42, 244–47.

being "user-friendly." This criticism arises because the patent system procedure facilitates the systematic abuse on the part of private companies acquiring patents on TK-based resources while making the patenting of the TK itself inaccessible to the TK holders. To understand the TK holders' allegation, I examine the nature of the problems TK holders face when applying for a patent.

At present, there are no specific TK patent procedures. Thus, TK holders wishing a patent over their TK-derived innovation must follow the normal national procedures in the countries in which they seek to protect their innovation. These normal procedures pose problems, however. First of all, patent law requires the inventor(s) to be defined. In traditional communities, TK-derived innovations are often community property, and no single inventor or group of inventors is evident. Second, disclosure of the invention is an important step in any patent granting-procedure. But many TK holders are not eager to disclose TK that has been safeguarded within a community for generations and has consequently acquired a sacred significance. In this regard, a detailed description of the innovative process as required for the patent often contravenes the cultural values of the traditional community.

A third obstacle to the patent application of TK is the requirement that an invention must be "novel" in that it possesses some new characteristic unknown in the body of existing knowledge. This fact usually poses a problem for TK holders as TK is generally already well known and thus is prior art to considerable numbers of people in various traditional communities. Finally, because of the complex filing procedures, patent applications often require specialized legal assistance. Keeping abreast of filing and annual renewal fees tends to be beyond the capabilities of most TK holders. Additionally, it may be difficult for TK holders to enforce their rights. Although States may grant patent rights, they do not enforce them. When a patent is infringed, it is up to the patent holder to bring a legal challenge in court. Not only is it difficult for TK holders to know when a patent has been legally breached, it is nearly impossible for them to understand the legal process involved in a suit, much less afford the costs that such suits usually entail. In light of the foregoing, it is clear that the current patent system, while in principle allowing patents for TK-derived innovations, does not take into consideration the special needs and circumstances of TK holders.

National laws can determine the standards of patentability so far as they comply with the basic requirements of TRIPS. A thorough research should comprise the analysis of the main possibilities of application of standards and requirements to protect. Once the standards of patentability are lowered to accommodate grain-sized or merely incremental innovation, the national patent system will inevitably suffer collateral effects. The principles of national treatment and non-discrimination provide that the same advantage of a lax standard of patentability to acquire wide monopolies in all fields of technology will also be available to foreign companies. Hence, it is necessary to balance the benefits of allowing patentability of incremental TK innovation with the costs of allowing wide monopolies on marginal inventions that are immediately applicable industrially.

Specialized features of patent requirements can be accommodated. From this perspective, it stands to reason that the requirement of novelty may be in part redrafted to accommodate a specific case of divulging non-novelty-destroying knowledge, e.g. lowering the standard of novelty thus equating knowledge within a given traditional community to knowledge within the laboratory of a given firm.[97] Accordingly, the patent laws should also take into account the very nature of TK that is not based on individual ownership but rather on collective and public ownership.[98] In this regard, the legislative value of Article 8(j) of the CBD (on the working definition of TK) has to be in line with the realities of joint contribution and conservation of TK and with the mandate of international provisions.

7.7 Overview on the utility of geographical indications and trademarks

What the protection of TK and GIs have in common is not only that both are subject-matters of much painstaking negotiations at the WIPO and WTO. GIs are a subset of TK that are indicative of the geographic origin of a product based on TK. TRIPS mandates GIs protection only for wines and spirits. The extension to other products is one of the most hotly debated issues within WTO. At the regional and national level, the question of relevance here is whether and to what extent this protection should be expanded. The interpretation of Articles 22 and 23[99] of the TRIPS Agreement has even been the object of a recent lengthy WTO Appellate Body decision.[100] In approaching this issue, I will first lay out the parameters and definitions to identify the problem at hand. It will be recalled

97 I. Mgbeoji, "Patents and Traditional Knowledge of the Uses of Plants: Is a Communal Patent Regime Part of the Solution to the Scourge of Biopiracy?" (2001) 9 *Indiana Journal of Global Legal Studies* 163.

98 *Ibid.*, 173.

99 A. Kur, "Use of Collective Marks and Geographical Indications", in Von Lewinski (ed.), *Indigenous Heritage and Intellectual Property*, 127–29 (2008).

100 *European Communities Protection of Trademarks and Geographical Indications for Agricultural Products and Foodstuff*, WT/DS174R (March 15, 2005).

that the working definition of TK within the CBD is "[t]he knowledge, innovations and practices of indigenous and local communities embodying traditional lifestyles" as well as indigenous and local technologies.[101]

This definition essentially seeks to expand the notion of IP to include GIs as a means of protecting the property rights of TK holders who employ certain methods of production in specific geographic regions that distinguish their products as unique and not "generic." Registration of TK related to certain geographical areas may be used as protection, as was the case with "Basmati rice," for instance. Without association to "Basmati" aromatic rice produced by other rice producers, such as Rice-Tec for example, would be less appealing to the public and would not be a competitive product for the real "Basmati."[102] According to a document on France's experience with the IPR of local ecological knowledge, published by Iddri, IPR tools such as GIs, which are based on geospatial considerations, "[allow] those who have decided to play the market game to enhance the cultural and economic value of a certain traditional ecological knowledge."[103]

To elaborate and more fully understand this definition, I will compare GIs with the currently available tools of international IPR. Compared to patents, GIs do not reward innovation, but rather are aimed at rewarding the reputation built up by a group of producers over many years or even centuries without conferring monopoly rights over the use of certain information. The social benefit is to provide consumers with reliable information and assurances of authenticity. Local and indigenous communities, who maintain in perpetuity trademarks and GIs on products based upon sustainable traditional production practices, may be enabled to limit the class of people who can use a certain symbol.

In the WTO, members must prohibit registration of trademarks that are misleading regarding geographical origin and must provide legal procedures for interested parties to prevent competitors from placing designations on their products that mislead the public about their geographical origin (Article 22). The TRIPS Agreement also provides for additional protection of GIs for wine and spirits (Article 23). Obligations regarding GIs are subject to a number of exceptions (Article 24). For instance, if the

[101] CBD Articles 8(j) and 18.4. Moreover, the UN has made significant contributions in this domain.

[102] D. Downes and S. Laird, *Case Study in Innovative Mechanisms for Sharing Benefits of Biodiversity and Related Knowledge, Geographical Indications and Trademarks* (Center For International Environmental Law, Geneva, 1999) 37–38, www.ciel.org/Publications/InnovativeMechanisms.pdf.

[103] L. Berard *et al.*, "Local Ecological Knowledge and Practice: An Original Approach in France" (2005) 8 *Les notes de l'Iddri*.

name of a geographical region has become "generic" – that is, associated with a broader category of products – then it can be used outside the region, even if it originally denominated products from that region. The importance of a registration system in DCs is stressed in paragraph 9 of Article 24, which provides that WTO Members are obligated to provide legal protection of GIs only if they are protected in their country of origin.

7.7.1 *A justification for the use and analysis of the impact of geographical indications*[104]

The justifications for implementing IPR protection of TK through GIs is normative, economic, and political in nature. From the normative perspective, firms in industrialized countries currently enjoy a large market advantage in producing and marketing products in markets where GIs would have impact, such as food and crafts. Producers of TK that would potentially have distinct geographically driven marketing value lack the market power to challenge larger market powers through the traditional market mechanisms of distinguishing products, such as advertising. The very use of the name "Parmesan" is exclusive and cannot be used by firms who do not produce the same product in the geographical area of Parma, Italy. There can be great market power in the exclusive exercise of the appellation of origin of a product.

Even in alleged cases of *biopiracy* and *bioprospecting* (section 1.1.1 above) the ownership of GIs on the TK associated with a specific GR would help to fairly compensate those whose lands and TK produce products which are differentiated by geography through distinguishing their products as unique in the marketplace. Also, GIs can increase the price of such products, thereby increasing profits acquired by local and indigenous communities of these GI regions by differentiating products by their area of origin, restricting supply, and creating barriers to entry into production. Thus, many of those who stand to acquire GIs to differentiate their products would benefit greatly in economic terms, which would, in many cases, aid DCs by expanding and increasing the profits to individuals within their developing markets. GIs serve as a means of leveling the playing field in the market to allow smaller firms holding GIs, cooperatives, or individuals to compete. These economic actors

[104] I owe special thanks to David Newell, for his basic relevant research and writing on this matter while he was performing a short internship in Geneva under my direction, in the summer of 2005, during his Master in Public Policy studies at Brigham Young University.

would also allow for compensation to be given to such GI holders for the use of their geographic resources or TK.

Economic considerations are closely interwoven with political ones. Turning to the economic side of the issue, the Group of Countries of Latin America and the Caribbean (GRULAC) submitted the following to the WIPO Committee on the Relationship between IP, GRs and TK:

> It is the role of the State to protect both intellectual property and the public domain. The interest of the public is in maximizing the material belonging to the public domain in order that there may be an environment of free market competition and so that the community may derive maximum benefit for minimum cost. However, in the interest of progress in art, technology and trade, which also works to the benefit of the public, intellectual property allows certain subject matter, which is precisely defined by the law, to be kept out of the public domain.
>
> As the protection claims, needs and expectations expressed by the possessors of genetic resources and traditional knowledge (including folklore) call for the broadening of the present scope of intellectual property, subject matter that has hitherto been considered public property will cease to be so considered. That subject matter, which once was appropriated, used or exploited without any recognition of ownership, authorization or remuneration, would remain protected in such a way that access to it and its exploitation would be under the control of the person or entity holding the rights.[105]

This statement casts the issue of GIs in the light of private property and the public domain. Under these strict definitions of what can be protected by IP, much of TK has been left out because it is not necessarily innovative upon previously existing technologies and it is not necessarily attributable to a single owner. However, as the statement points out, one of the *purposes* of the creation of IPRs is to ensure maximum public benefit at minimum cost. It also calls for the expansion of the scope of IPR to meet the needs of possessors of GRs. These two ends do not necessarily need to be at odds. While the benefit of GIs to TK holders through increasing demands for "authentic" products is beyond much dispute, the issue of how GIs economically benefit consumers is a little less clear. By implementing GIs, consumers can derive greater benefit through knowing which products are of a particular origin, created with a certain TK, etc. Currently, consumers do not enjoy this benefit and are thus deprived of maximum utility.

In relation to other IP tools, GIs should be viewed in conjunction with other rights like appellations of origin and certification trademarks. Understanding the differences between GIs and trademarks is fundamental

[105] *Traditional Knowledge and the Need to Give It Adequate Intellectual Property Protection*, WO/GA/26/9 (September 14, 2000), www.wipo.int/meetings/fr/doc_details.jsp?doc_id=1482.

to encourage local communities to use these types of IPRs. Discussion of these differences mainly focuses on the characteristics that can create the most adequate existing IPRs to protect the core of TK because it is acknowledged as being collective in nature, i.e. an entire community or region can be the subject of that right. Indeed, these kinds of IPRs may be based upon collective traditions and a collective decision-making process. I consider these IPRs as instruments in the hands of local and indigenous communities particularly suitable to protect their know-how because (i) they protect and reward traditions while allowing their evolution, (ii) they emphasize the relationships between local cultures and their local land and environment, (iii) they cannot be freely transferred from one owner to another, and (iv) they can be maintained as long as a certain TK is maintained.

Thus, protecting and rewarding traditions, while allowing for their evolution through granting IPR is a means of incentivizing TK production and distribution, much as TIPRs and traditional patents incentivize inventors to disseminate their knowledge to the public. This arises because of the profit incentive that may be realized by producers and distributors of TK. When incentivized to distribute their TK goods, services, etc., TK holders will be benefiting society through the increased availability of goods and services currently unique to TK. Since GIs emphasize the relationship between local cultures, their land, and environment they serve as an indirect means of ensuring, promoting, and encouraging the sustainability of TK-affiliated environmental and social structures that have sustainably existed for typically long periods of time. This is because of the usually holistic approach of TK culture towards land, people, etc. Thus, the environmental and social externalities obtained from promoting TK vis-à-vis GIs are potentially positive. Restricting the free transfer of TK from one owner to another also serves the purpose of preserving the very TK nature of the given good or service. When deprived of its environment and context, TK, in essence, ceases to be traditional and becomes mere knowledge. Invariably, valuable elements of TK will be lost in such attempts to communicate TK between cultures or people. Preserving TK integrity ensures its traditionality, and therefore its quality and authenticity. Maintaining TK rights as long as certain types of TK are maintained also serves this end.

Another economic benefit of GIs is that it serves the purpose of creating better free market conditions than currently exist. As has already been mentioned, GIs serve as a means of balancing the otherwise lopsided market power of larger firms in relation to most potential GI holders. GIs would correct this market distortion. In doing so, this balancing would create a more globalized marketplace where more goods would potentially have market access and would overcome barriers to trade under the auspices of GIs, which would provide the increased value and differentiation from other similar products necessary to incentivize consumers to buy over other, more readily available products produced by firms that dominate the current marketplace. By diversifying the global market, consumers would also benefit from greater and superior product and service selection that would be inherent with exposure to TK products and services, which in most cases, due to their typically non-western, non-Northern origins, would offer products and services of an entirely new nature to the market.

In addition, GIs would reduce the information asymmetries inherent in the current system where large firms can claim what would otherwise be GI names and methods as their own when, in fact, they are distinct from products genuinely originating from specific geographic regions employing specific production methods (TK).

Instead of going to the store and not knowing the difference between the Parmesan cheese produced by Kraft and the genuine product, because the Kraft product would not be able to use the term Parmesan, consumers would be able to distinguish between the products, providing them with more information to counter-balance an asymmetry which currently favors larger firms. Based on local and regional products, distinctiveness that provides the consumer with reliable information about authenticity, GIs enhance the power of local producers to sell their products in a global marketplace. Because of this, consumers would benefit, as would the GI holders.

However, it should be noted that larger firms as well as small firms/individuals/local communities would also potentially benefit by designating products that had previously lacked designation. Thus, the advantages afforded are not as one-sided as they might first appear. Appealing to this notion might help to assuage, if not resolve, some of the concerns that are sure to arise in the process of implementing GIs. In other words, GIs such as "made in the USA" could potentially grant the same kind of market power to larger firms from industrialized countries that most would-be GI holders would gain through the implementation of GIs. This aspect could help sway firms that would be opposed to the imposition of GIs as well as the politicians who represent those firms. Politically speaking, GIs need to be sold as a potential benefit to all firms that could differentiate their products to consumers domestically and globally through having better market information and more product/service diversity and quality to choose from, and to TK holders who could potentially be compensated to some degree for losing benefits due to previous and even ongoing exploitation of their TK and associated resources. In conclusion, better use and promotion of traditional GIs would offer better protection of community economic interests, both those of potential GI holders and global consumers, and of traditional products from certain regions of origin.

7.7.2 *An example of the use of geographical indications in protecting TK-based products*

Certain countries have already experienced the benefits of protecting their PGR and related TK through GI. A famous example is Kava in Brazil.[106] I now return to the previous illustration on the usefulness of GIs and trademarks for a traditional product like Basmati. As it has been said, Basmati would qualify for protection as a GI under the TRIPS Agreement if its quality, reputation, or other characteristics were *"essentially attributable to its geographical origin."* Basmati rice is a long-grained, aromatic variety of rice that is cultivated in areas of Northern India and Pakistan, mainly in the Punjab area. Basmati is widely recognized as having specific desirable qualities. It also has a distinctive, rich flavor that is highly prized in the cuisine of the Indian subcontinent and around the world.[107] Protecting the term "Basmati" as a GI in the national legal system requires assembling evidence that Basmati rice – from the Indian subcontinent – has unique characteristics and a reputation based on its geographic origin. Moreover, it should enable the IP holder to counter the arguments from competing producers in other countries of the world.

Basmati can simultaneously qualify for trademark protection, which may offer useful measures for Indian or Pakistani producers or their buyers in importing countries, if they have registered trademarks using the Basmati name. Article 16.1 of TRIPS[108] provides that WTO Members must protect a trademark owner's right to prevent competitors from using similar trademarks on similar goods in a way that is likely to cause confusion among buyers. While names like the "Texmati" term

[106] A. Nascimiento Mueller, "Case Study: Potential Benefits of Geographical Indiciations: The Kava Case in the South Pacific", in Cottier and Biber-Klemm (eds.), *Rights to Plant Genetic Resources*, 253, and 137.

[107] Downes and Laird, *Case Study in Innovative Mechanisms*, 34.

[108] TRIPS, "The owner of a registered trademark shall have the exclusive right to prevent all third parties not having the owner's consent from using in the course of trade identical or similar signs for goods or services which are identical or similar to those in respect of which the trademark is registered where such use would result in a likelihood of confusion. In case of the use of an identical sign for identical goods or services, a likelihood of confusion shall be presumed. The rights described above shall not prejudice any existing prior rights, nor shall they affect the possibility of Members making rights available on the basis of use."

used by the RiceTec Company connote Texas more than they evoke the Indian subcontinent, the current case of the use of the Basmati name among French trademarks can be more successfully challenged, being utterly misleading for consumers. Although TRIPS allows an exception to a trademark owner's right for the fair use of descriptive terms, competing producers can demonstrate that the Basmati term only indicates rice having a certain flavor regardless of where it is produced. However, indigenous producers of Basmati rice can also claim unfair competition. Generally speaking, since in a market economy each trader strives to gain an edge over his competitors by means of innovation, research, or reputation, unfair competition is generally considered to refer to the act of one trader misappropriating the intangible fruits of another trader's skill, time, and labor. This concept can be widely used to protect TK. The legal basis can be Article 10*bis* of the Paris Convention,[109] which obliges Members to ensure that people are protected from unfair competition resulting from (for example) acts that cause "*confusion by any means whatever with the establishment, the goods, or the industrial or commercial activities of a competitor.*" Such a provision could also be relevant to indigenous groups seeking to control the imitation or unauthorized commercial sale of indigenous products. Not providing such protection for indigenous peoples could arguably be a breach of this Convention's Article, which obliges members to provide nationals with "appropriate legal remedies to repress effectively all the acts referred to in Article 10*bis*."[110] Of course, unfair competition can be the grounds for legal actions in US or French courts; such an argument could be used to prevent companies from marketing their competing rice in a way that misleadingly implies that it has its geographic origin in the Indian subcontinent.[111] But in order to be successful, such an action should only follow the creation of a logo and trademark duly registered in a strong trademark national system. Only then will indigenous producers of Basmati rice be better situated to take action to protect themselves against unfair competition in their export markets.[112]

[109] Paris Convention, Article 10(*bis*)

[110] F. Yamin and D. A. Posey, "Indigenous Peoples, Biotechnology and Intellectual Property Rights" (1993) 2(2) *Review of European Community and International Environmental Law*, 141–48.

[111] Downes and Laird, *Case Study in Innovative Mechanisms*, 36.

[112] D. Downes, *Global Trade, Local Economies and the Biodiversity Convention*, in J. Snape (ed.), *Biodiversity and the Law* (Island Press, Washington, 1996); D. Downes, *Integrating Implementation of the Convention on Biological Diversity and the Rules of the World Trade Organization: Law and Policy Discussion Paper: Discussion Draft* (World Conservation Union, Center for International Environmental Law, Gland, Switzerland, 1998).

7.7.3 *Implementation strategies in international law*

Beyond all the efforts at the domestic law level,[113] there is an essential international dimension that GIs on TK need to ensure. At the WTO level, the TRIPS Agreement, already mandates protection for wines and spirits as seen in Article 23. There is the need to extend GI protection to all products that would qualify under the TRIPS definition of GIs. This extended protection would give IPR protection in a method different from that of patents and more in conformity with the cultures and TK that typically surround potential GI products. This protection would also better level the market playing field both for GI-product producers (through the profits realized from selling a uniquely demarcated product) as well as for consumers (who would have better and fuller information concerning purchases made in a global marketplace). In addition, the WIPO Secretariat concluded:

> At some point in the future the Intergovernmental Committee may also wish to undertake additional work with the aim of deepening the understanding of how existing intellectual property mechanisms, with their current standards concerning availability, acquisition, scope, maintenance and enforcement of rights, may be used as effective mechanisms for the protection of traditional knowledge.[114]

It has been suggested that certification marks, which already exist, are a sufficient solution to the problems of IP protection of GIs as a subset of TK. While certification marks are collectively owned, the difficulty in employing this specific tool of IPR is that it does not necessarily indicate the quality, the methods of production, or the geographic area/origin of the product. Thus, it provides less information than GIs and thereby provides fewer economic benefits to producers and consumers, while at the same time it lacks the ability to fully address the normative issues surrounding the lack of TK IPR protection raised earlier.

For sustainable use of GIs or trademarks on TK subject-matter, a national legal registration pattern and system must be developed. Although such a system is expensive to set up, many DCs are already bound under TRIPS to provide for registering and enforcing trademarks, and several of them have been considering registers for GIs, at least for wines and spirits. In addition, the TRIPS Council is discussing the extension of GIs to products other than wines and spirits.

[113] See the important considerations and conclusions on the collective right of certification trademarks at the national level, Kur, "Use of Collective Marks", in Von Lewinski (ed.), *Indigenous Heritage and Intellectual Property*, 131–32 (2008).

[114] *Survey on Existing Forms of Intellectual Property Protection for Traditional Knowledge – Preliminary Analysis and Conclusions*, WIPO/GRTKF/IC/2/9 (December 3, 2001).

There are, however, some difficulties in providing such protection. For GIs, the main difficulty lies in finding the appropriate rights holder, a problem arising in part from the absence of "communal" rights grants under current IP legislation.[115] However, treaties already provide for creative law-making in this field. For example, Article 22.2 of the TRIPS Agreement states:

> [i]n respect of GIs, Members shall provide the legal means for interested parties to prevent the use of any means in the designation or presentation of a good that indicates or suggests that the good in question originates in a geographical area other than the true place of origin in a manner which misleads the public as to the geographical origin of the good.

The use of the term "interested parties" seems broad enough to allow countries to designate who their proper rights holder(s) should be. However, current TRIPS Agreement obligations only apply to "goods," and this would not cover several forms of TK, in particular medicinal knowledge and certain forms of artistic creation. Below are some possible implementation strategies for overcoming these difficulties and protecting TK and GI.

In a proposal submitted by the EC, they suggest that:

> Traditional knowledge is of intangible nature and the obligation to disclose cannot be based on physical access. It could therefore be proposed that the applicant [for geographical indication IPR protection, for example] should declare the specific source of traditional knowledge that is associated with genetic resources, if he is aware that the invention is directly based on such traditional knowledge. In this context, the EC and its Member States refer to Article 8 (j) of the CBD where the notion "knowledge, innovations and practices" is used [...].

An indispensable measure that would make the disclosure requirement outlined in the previous sections an effective incentive to comply with ABS rules would be the introduction of a simple notification procedure to be followed by the patent offices. The latter, every time they receive a declaration disclosing the country of origin or source of the genetic resource and/or associated TK, would notify this information to a centralized body. This could be done, for instance, by means of a standard form. That would facilitate the monitoring – by countries of origin and TK holders – of compliance with any benefit-sharing arrangements they entered into. The relevant information must be made available in accordance with the present rules on the confidential nature of applications.[116]

[115] J. Tunney, "Indigenous People and the Digital Age: Intersecting Circles?" (1998) 20 *European Intellectual Property Review* 335.

[116] *Disclosure of Origin or Source of Genetic Resources and Associated Traditional Knowledge in Patent Applications*, WIPO/GRTKF/IC/8/11 (May 17, 2005).

In other words, this proposal suggests the creation of an international form of centralized registry that would work in collaboration with national registries and patent offices to disseminate information regarding various forms of TK, including GIs. The registry would, as the proposal argues, provide protection for the rights of TK holders and compliance with benefit-sharing mechanisms associated with IPR to be monitored at the international and national level. In addition, GIs would not be exclusive rights held by an individual but rather the rights held by a community of TK holders. This communal right would respect the nature of TK and overcome some of the difficulties of applying traditional western tools for IP protection to TK holders who, by definition, cannot be exclusive, individual holders of IPRs based on innovation, but rather represent the cumulative TK, built and learned through generations. Expanding on the concept of benefit sharing, Portugal submitted the following in the previously mentioned WIPO proposal:

> [GIs] will likewise form the basis for the fair allocation of the benefits generated by the use of this material among the parties involved in their differentiation or maintenance or both. Finally, it will also make a positive contribution to the promotion of the secure interchange of plant genetic resources, at the same time ensuring the protection and preservation of the cultural diversity of local populations that is associated with the plant genetic resources of communities that have not had access to proper intellectual property machinery and so have seen innumerable technical contributions fall into the public domain or be appropriated by third parties without deriving any benefit therefrom.[117]

Although the mechanisms mentioned here cannot be fully developed and need to be fully explored, the work done at WIPO through the proposals submitted by Portugal, the EC, and others has made a good start toward making GIs a practical and realizable IP protection tool for TK both at the national and international level.

[117] *Ministry of Agriculture, Rural Development and Fisheries, Decree-Law No. 118/2002*, WIPO/GRTKF/IC/8/13 (June 1, 2005).

8 Final observations

Article 27 of TRIPS Agreement has "globalized" IPRs on biotechnology and consequently spawned various controversies on its relation with relevant multilateral environmental treaties and customary norms. The controversies that have arisen during the negotiation of multilateral treaties establishing legal regimes of utilization of GRs (TRIPS, CBD and FAO Treaty) have opposed, on the one side, gene-hunting countries (technologically rich but poor in biodiversity), and, on the other side, gene-endowed countries (technologically poor but rich in biodiversity). The issues involved range from ownership of GRs and the protection of derivatives to the phenomenon of biopiracy, from the problem of preservation and sustainable use of biodiversity to the equitable benefit sharing thereof, including the ethical and moral issues that have been analyzed in this book.

The present study has demonstrated how international IP law is intertwined with other fields of public international law. IP law does not stand in clinical isolation, it rather needs to be interpreted in light of the relevant treaties. TRIPS, CBD, ITPGRFA, their derivative laws and their relationship with general international law have created a thick network of obligations that their State parties have to attentively analyze and comply with.[1] There are various degrees in the acceptance of these treaties that often are monitored by different international organizations, thus revealing the complexities of the contemporary highly interconnected world.

The international law perspective has stretched beyond the written treaty law and discovered the relation between some articles of TRIPS Agreement and international customary norms crystallized from the CBD and its derivative soft law (sections 6.1.1.2, 6.1.1.2.4 and 6.1.1.2.5). Various tendencies and theories on the sources of international law have been evaluated to provide a clearer panorama on the positive law. This has been done in order to achieve, to the utmost extent, a solution to the

[1] G. Buzzini, *La théorie des sources face au droit international général* (Graduate Institute of International Studies, Geneva, 2001).

potential conflicts among IP treaties and MEAs through the method of interpretation of mutual supportiveness (section 1.2.3).

This international law perspective on such matters has also inspired the particular structure of this work whereby Part I lays out the main conceptual elements to be used in the other two parts. Part II has focused on the interaction of the treaty and customary obligations of States and international organizations and on how to solve the potential conflicts among norms stemming from different origins. Finally, Part III has dealt with the protection of TK through the application of the principles and concepts explained in Parts I and II.

A major emphasis has been put on the international law-making process that, in my view, has rendered some of the relevant provisions of the CBD as customary norms. This is the case for PIC while doubts have been expressed on the benefit-sharing concept (section 6.1.1.2). Accordingly, this analysis is important for the purpose of determining the level of opposability of the PIC customary norm to the US, that is the only country not party to the CBD and at the same time the country that grants the highest number of patents (and other IPRs) on foreign GRs. The very patentability of biotechnological innovations has started in the US and its current position concerning this problem can potentially influence other industrialized countries. Certainly, US legal scholars will raise concerns and doubts over the customary norm of PIC in the field of our analysis given the restrictive manner in which they usually determine the existence of customary norms. This book intends to launch the scholarly debate on these norms.

The emerging of related customary norms depends on the results of the increasing interconnectedness of the States within their bilateral, regional and multilateral relations, thus facilitating the development of norms that are felt and expressed as binding by the international community. The preparation, organization, and negotiation leading to a meeting of minds of States through the adoption of repeated soft-law instruments having a rather specific language leads to the creation of potential customary norms. Much uncertainty can be expressed on the customary normative value of the concept of benefit sharing given the vagueness of the concept expressed in the CBD itself. The current customary norm of PIC is that GRs can be accessed and extracted from a country for the purpose of commercial use only with the PIC of the provider State exercising its sovereignty over the territory on which the GR has been found and to the extent that an exercise of that State sovereignty can be determined. Commercial use of GRs includes the acquisition of IPRs on the invention based on that GR. The scope of the customary norm does not include, however, the modification of the patent system so to include a requirement of the certificate of origin (sections 6.1.1.2.4 and 6.1.1.2.5).

The opposability to a State of a treaty law obligation that has crystallized into a customary norm opens wide the door to doctrinal discussion and welcomes eventual criticism. The study of its unilateral acts and the *opinio juris* expressed in various international fora has helped to determine to what extent the US can qualify as persistent objector or violator of a customary norm such as PIC, depending on whether the status of "persistent objector" is a valid concept in international law (sections 6.1.1.2.5.1 and 6.1.1.2.5.2).

The interpretation of the concept of State sovereignty over GRs is torn between the two regulated areas of TRIPS Agreement and the CBD. No matter how controversial the tensions between TRIPS Agreement and the concept of sovereignty enshrined in the CBD or ITPGRFA, they cannot warrant any suspension of IPRs, i.e. non-compliance with TRIPS provisions. This means that a WTO Member country cannot reject a patent application for a genetic invention on micro-organisms (found on their territory) on the sole basis that this kind of patenting is contrary to the object and purpose of the CBD. The CBD concept of national sovereignty over biological resources is a simple reaffirmation of the right of States to control exports and imports, or to set conditions for access to biological resources within its borders. Hence, the fact that States, by ratifying the CBD, have committed themselves to preserve biodiversity under their sovereignty and that later they came under the obligation of TRIPS to grant some forms of IPRs over biological resources, does not amount *per se* to a legal contradiction. I have therefore concluded that, in accordance with the principles of *pacta sunt servanda* and *ut magis valeat quam pereat* (that the matter may have effect rather than fail) there is a presumption that both conventions are enforceable without contradiction.

Moreover, the method of interpreting these treaties in a mutually supportive way could not solve the most direct conflict between the TRIPS Agreement and the FAO ITPGRFA in case of the refusal of a State Party to the ITPGRFA and TRIPS to grant a PGR acquired from the Multilateral System (section 3.3.3.1). This conflict is likely to create legal disputes between WTO Members, one having ratified both treaties and one having ratified only the TRIPS Agreement. An amendment in one of the treaties is necessary to harmonize the two legal regimes.

In a much more positive approach, it can be said that all the treaties analyzed can even build up a synergy. The controversy on their incompatibility may indeed be quenched, or at least defused, if national IP lawmakers and officers (in developing as well as in industrialized countries) undertake interpretative efforts to seriously comply with all the applicable international legal instruments. Industrialized countries, on their side, should not deliberately interpret in a restrictive manner the safeguards and flexibilities offered by TRIPS to DCs to strike a balance of interests (chapter 3).[2] On the other side, DCs should start considering national IP protection of TK through existing or even new forms of IPRs able to recognize and compensate the creators and possessors of such knowledge (chapters 3 and 4).

The major objective of Part III has been the protection of TK. Considerable conceptual divergences still exist among regions of the world on the objectives, scope and content of possible rights to be recognized. Therefore, it seems unlikely that a binding international legal instrument on this matter will be rapidly negotiated in international fora (chapter 4). The protection of TK and its relation with the IP system is a difficult task at the national level itself. If a protection needs to be sought within the WTO system, the negotiations in this field need to face the blockage of WTO negotiations. Certainly, paragraph 19 of the WTO Doha Ministerial Declaration – that mandates the TRIPS Council to explore its relationship with the CBD – will stand as an important objective to be achieved as long as a satisfactory revision of Article 27 of TRIPS is not achieved, including a clear incorporation of the PIC concept and possibly the benefit sharing one (chapter 5). The option of amending the PCT and PLT to include the PIC requirement in the patent system as a partial solution of the more complex problem has been presented as a transitory or more easily achievable solution, as Switzerland proposes. Meanwhile, the development of the clarification of the underpinning legal concepts and the related common terminology can assist in finding ways to encourage better drafting of national and regional legislation that will take TK into account.

Meanwhile, the quality of patents granted by Patent Offices in industrialized countries can be improved. The attempt to render TK potential novelty-destroying prior art in case of its misappropriation also provides solutions to the problems caused by extensive monopolistic rights held by certain biotechnology sectors and which have seriously blurred the distinction between discovery and invention in patent law. This can improve the current situation in which biotechnological patent holders merely tinker with the natural substance or traditional practice, making only minor changes.[3] Therefore, it seems unjust that the biotech-patent holder

[2] I have argued that exceptions like compulsory licensing and parallel imports are in compliance with TRIPS Agreement. As such WTO Members should be at liberty to include them in their domestic legislation. This is also in line with the spirit of the Doha Ministerial Declaration, stating that the TRIPS Agreement "can and should be interpreted and implemented in a manner supportive of WTO Members' rights to protect public health and in particular to ensure access to medicine for all."

[3] G. S. Nijar, *In Defence of Indigenous Knowledge and Biodiversity: A Conceptual Framework and Essential Elements of a Rights Regime* (Third World Network, Penang, Malaysia, 1996) 4.

gains exclusive rights for making changes that are either minor or obvious. And when such doubtful patents are based on a flagrant misappropriation of GRs, even the most persuasive justifications – that patents reward the additional time and resources necessary to maintain high standards of biotechnological innovation – hardly support the validity of such a patent. I believe that determining whether such types of patents are non-obvious certainly warrants further investigation.

Even more problematic is it to know to what extent the invention is novel, considering the TK on which it is based. The starting point for such assessment can be the integration of prior art searches on TK databases and some form of PIC requirements in the US and European patent systems. Then these patent systems will be more credible and will better counter the sharp criticisms that argue that novelty and invention step/non-obviousness requirements for biotechnological patent applications have been unduly loosened.

The same reasoning is valid for the compliance of patents with minimal standards of environmental compatibility. The creation of a Board of Public Policy within, e.g., the USPTO or any other patent office, for the examination of the patent's compliance with the concepts of *ordre public* and morality can improve the quality of patents. Indeed, all patent examiners should bear in mind that a patent is an exclusive right that covers only the invention described and nothing more. If the invention based on a certain type of TK is indeed minor with regard to the TK on which it is based, then its exclusive rights cannot prevent the reproduction of this TK. On the contrary, if the invention is a major advance, the patent rights should be more extensive.

These matters do not only concern the patent offices in developed countries but also those of DCs (that are currently being created). It would be wrong to think that this kind of superficiality in granting biotech-patents might constitute a problem affecting solely DCs. If the biotechnological patent quality remains so poor, this phenomenon will eventually stifle innovation, even in the industrialized world. The life science corporations are already complaining about the patenting practices of small biotechnology firms that amass sizeable patent portfolios on basic research tools. Even though many of these patents would struggle to survive a legal challenge, they are asserted aggressively because they are the only assets many of these companies have. Here lie important issues that should increasingly concern IP legal doctrine in industrialized countries.

The patent system needs adjustments in light of the problems that have been raised. However, throughout the short history of international IP protection of the life sciences, it is noticeable that most of the regulatory reforms were undertaken only if powerful industrial businesses pushed for them given the underlying commercial interests at stake. The pessimistic mind would wonder: where is the commercial representative to inspire countries to lobby for the inclusion of TK protection in the TRIPS Agreement? What business entity will accept the introduction in the patent system of the requirement of submission of a certificate of origin or improvement mechanisms of international prior art search? Even governments of biodiversity-rich DCs, that should normally defend the interests of their indigenous peoples, are now bending their knee in negotiating TRIPS-plus standards through bilateral treaties with strong and influential countries such as the US, Japan and Australia, or regional organizations like the EU. This process "may significantly shift the balance of economic interests to the more powerful WTO Members, thereby further exacerbating problems in the global distribution of wealth."[4]

If the relationship between TK, IP and environment protection is not seriously taken into account by the international community, the TRIPS Agreement may fail to achieve its objectives set forth in Article 7. Unless the international patent system undergoes proper modifications in accordance with the CBD principles outlined in Part III, the enforcement of IPRs on GRs will not "contribute to the promotion of technological innovation and to the transfer and dissemination of technology, to the mutual advantage of producers and users of technological knowledge and in a manner conducive to social and economic welfare, and to a balance of rights and obligations."

Besides the positive law of treaty and customary norms, the source of international law of equity should be invoked to rebalance the relationships between TK holders and industrial parties.[5] Equity is the underlying concept at the basis of Article 38.2 of the ICJ Statute which enables the Court "to decide a case *ex aequo et bono* if the parties decide thereto." The source of equity has various means of expression.

The decision-maker (e.g., the arbitrator, the judge, policy-maker, the patent examiner) can refer to the concept of equity when the application of the *lex lata* produces results that are unjust. The fact that patent law does not take into account the rights of TK holders may constitute a rare case of unjust law that needs to be corrected by this "*source additionnelle ou additionable*" (of Article 38.2 of the ICJ Statute).[6] This is the application of *aequitas contra legem*.

[4] *Resource Book on TRIPS and Development*, 35.

[5] T. Taubman, "Genetic Resources", in S. Von Lewinski (ed.), *Indigenous Heritage and Intellectual Property: Genetic Resources, Traditional Knowledge and Folklore* (Kluwer, The Hague, 2008), 288–281 (2008).

[6] G. Abi-Saab, "Cours général de droit international public" (1987) 207 *Le Recueil de Cours de L'Académie de droit international* 189–90.

The utilization of the concept of mutual supportiveness and the teleological interpretation of TRIPS Agreement that have permeated this study have been largely inspired by the broad source of law of equity *secundum legem*. This concept has thereafter been applied to the international legal fields of sustainable development and environment to the point that it is a *leitmotif* in connection with North-South trade and economic dialogue. The Rio Declaration[7] is a soft law instrument undoubtedly inspired by this process. The Preamble states that this Declaration establishes "a new and *equitable* global partnership [...] through the creation of new levels of cooperation among states, key sectors of societies and people." Principle 3 goes so far as declaring that "the right to development must be fulfilled so as to *equitably* meet developmental and environmental needs of present and future generations." The necessary reconciliation between the CBD and the TRIPS Agreement requires the application of the principle of equity.

There is an incremental development in precision and sophistication in the incorporation of the principle of equity in multilateral treaties. This is the case for equity in the field of protection of biodiversity through GRs by means of Article 8(j) of the CBD which requires that in access to GRs there be "fair and equitable" sharing of the benefits arising from the exploitation of GRs. But this provision can be interpreted as a general expression of the principle of equity, especially because the implementation of the concept is left to the parties of the bilateral contracts negotiated on "mutually agreed terms." There is a recurrent risk to frustrate the concept of equity given the unbalanced relationship between the stronger party of the industrial biotechnological company and the weaker party of the indigenous community. A step forward has been made in the more specific field of PGRs for food and agriculture in Articles 9.2, 10 and 13 of the ITPGRFA that creates a multilateral system of exploitation of PGRs and benefit sharing applying equitable principles.

These treaties also mark a shift from the use of equity for its traditional purpose as a set of principles meant to make the law fairer in individual cases to a system to promote global economic and distributive justice. Equity currently appears in the context of benefit sharing in the use of common goods, and as a necessary element of international economic relations and cooperation.

The appropriation of knowledge by industrialized countries, firms and scientists without fair compensation or reward to indigenous and local peoples can be seen as contravening fundamental moral, ethical and

[7] *Declaration on Environment and Development, proclaimed on the occasion of the World Summit on Environment and Development*, UN Doc A/CONF.151/26.

certain legal norms that protect people from any form of economic, ecological, political and social abuse.

Ultimately, this sort of misappropriation can contravene the concept of equity. Although the application of this principle to alleged cases of misappropriation or biopiracy is not yet clearly defined at the international level, the international law-maker and even the national judge should not disregard the concept of "equity" when applying IP laws enforcing specific patents that may raise such concerns.[8] The concept of equity has a direct link to the ethical considerations that have been analyzed in section 6.3 and that are relevant to the international patenting of GRs.

Another variant of the concept of equity is equity *infra legem*. There is a wide range of potential laws and policies that may apply this concept to regulate bioprospecting and access to GRs and the TK related to it. It may be desirable to identify some general underlying principles of equity common to multiple legal systems. Sources of law and of principles of equity may be derived, for instance, from common law, civil law, Islamic law, canon law, Talmudic law, etc., as well as from customary law and the practice of indigenous and local communities.

Meanwhile, much uncertainty hovers over the international community gathered in multilateral negotiating fora with respect to the possibilities of the adoption of a binding international legal instrument on the protection of GR-related TK within the IP system. The most effective way to enforce a treaty on the protection of TK would be within the WTO legal framework. Its dispute settlement mechanism could then clarify concepts and create a case law on the basis of the established norms. To pave the way towards the objective of the revision of Article 27 of the TRIPS Agreement, much preparatory work is still needed within the WIPO IGC on IPGRTKF, to which, it is to be hoped, this book will give its contribution.

[8] S. Biber-Klemm and D. Szymura Berglas, "Problems and Goals", in T. Cottier and S. Biber-Klemm (eds.), *Rights to Plant Genetic Resources and Traditional Knowledge* (CABI, 2006) 37–38.

Index

Cambridge Intellectual Property and Information Law

Titles in the series (formerly known as *Cambridge Studies in Intellectual Property Rights*)

Brad Sherman and Lionel Bently *The Making of Modern Intellectual Property Law* 978 0 521 56363 5

Irini A. Stamatoudi *Copyright and Multimedia Products: A Comparative Analysis* 978 0 521 80819 4

Pascal Kamina *Film Copyright in the European Union* 978 0 521 77053 8

Huw Beverly-Smith *The Commercial Appropriation of Personality* 978 0 521 80014 3

Mark J. Davison *The Legal Protection of Databases* 978 0 521 80257 4

Robert Burrell and Allison Coleman *Copyright Exceptions: The Digital Impact* 978 0 521 84726 1

Huw Beverly-Smith, Ansgar Ohly and Agnès Lucas-Schloetter *Privacy, Property and Personality: Civil Law Perspectives on Commercial Appropriation* 978 0 521 82080 6

Philip Leith *Software and Patents in Europe* 978 0 521 86839 6

Lionel Bently, Jennifer Davis and Jane C. Ginsburg *Trade Marks and Brands: An Interdisciplinary Critique* 978 0 521 88965 0

Geertrui Van Overwalle *Gene Patents and Clearing Models* 978 0 521 89673 3

Jonathan Curci *The Protection of Biodiversity and Traditional Knowledge in International Law of Intellectual Property* 978 0 521 19944 5

For EU product safety concerns, contact us at Calle de José Abascal, 56–1°,
28003 Madrid, Spain or eugpsr@cambridge.org.

www.ingramcontent.com/pod-product-compliance
Lightning Source LLC
LaVergne TN
LVHW011528200926
848484LV00001B/195

* 9 7 8 0 5 2 1 1 9 9 4 4 5 *